Coming Home to Math

Become Comfortable with the Numbers that Rule Your Life

Coming Home to Math

Become Comfortable with the Numbers that Rule Your Life

Irving P. Herman
Columbia University, USA

NEW JERSEY • LONDON • SINGAPORE • BEIJING • SHANGHAI • HONG KONG • TAIPEI • CHENNAI • TOKYO

Published by

World Scientific Publishing Co. Pte. Ltd.
5 Toh Tuck Link, Singapore 596224
USA office: 27 Warren Street, Suite 401-402, Hackensack, NJ 07601
UK office: 57 Shelton Street, Covent Garden, London WC2H 9HE

British Library Cataloguing-in-Publication Data
A catalogue record for this book is available from the British Library.

COMING HOME TO MATH
Become Comfortable with the Numbers that Rule Your Life

ISBN 978-981-120-984-0
ISBN 978-981-121-126-3 (pbk)

For any available supplementary material, please visit
https://www.worldscientific.com/worldscibooks/10.1142/11540#t=suppl

Printed in Singapore by Mainland Press Pte Ltd.

Dedication

To Rowan Elliot Herman, who I hope will love math.

Preface

I think math is fun and usually easy. Whether or not you agree with me, you surely know that we live in a world where numbers and mathematics run much of your life and you need to work with them to control the direction of your life and your happiness.

This book is meant to make adults more comfortable with math, in using it and enjoying it, whether or not they learned much math or forgot what they once learned, and particularly if they have been psyched-out by math. It is also meant for those who want to transition (or come) back to a life with more math and those who want to expand their math horizons. This includes college students majoring in the humanities and arts. It is not targeted to the young first learning math, adults wanting technical training for a specific job or new career, or math savvy adults who want to be challenged. We present important math concepts in simple terms, with frequent connections to the real world to help make you become comfortable in applying math to everyday life.

Being at ease with quantitative thinking is central to many occupations and will be even more so in the future. Many agree that math skills of some sort and math literacy in general are necessary in the new marketplace, but the specific skills needed beyond arithmetic and some algebra and geometry are up for debate.[iii] Quite a few think either statistics, calculus, or programming math should be widely taught. Developing new approaches and instruction sets (algorithms) for computing, analysis, solving problems, and control are skills within our main theme of "computational thinking," but specifically "thinking about computing methods" is not.[iii]

The histories of math and the human race are inexorably linked. As existing math is applied in one area, new math is developed to further that and other areas. The development and use of new number systems led to the improved understanding and ability to do arithmetic. This, combined with the use of abstract symbols instead of words to solve problems by the Alexandrian Greek Diophantus over two millennia ago, propelled the development of algebra; this supplemented progress in geometry by the ancient Greeks. The fusion of algebra and classical geometry by Frenchman Rene Descartes in the 17th century to create analytic geometry propelled the development of calculus independently by the Englishman Isaac Newton and German Gottfried Wilhelm Leibniz later that century. Each of these advances in math became the framework for new advances in the physical sciences.[iv] They are now leading to advances in the biological and social sciences.

In writing this book, I have made generous use of many wonderful and inspiring sources. The column "Behind the Numbers" in the *Wall Street Journal*, over the years authored by Carl Bialik and Jo Craven McGinty, connects timely topics and the world of numbers. *The Books of Numbers* by mathematicians John Conway and Richard Guy[v] addresses numbers in arithmetic, algebra, and geometry at a level that can be appreciated by neophytes, as well as quite advanced readers. A fun survey of math concepts at a bit lower level is presented in *The Mathematics Lover's Companion* by Edward Scheinerman.[vi] *One Two Three... Infinity: Facts and Speculations of Science* by physicist George Gamow, first published in 1947, is a classic overview of math and physics, which inspired me years ago, particularly its foray into the world of infinity.[vii] Significant parts of the original work on key advances in mathematics are reproduced in English in *God Created the Integers: The Mathematical Breakthroughs That Changed History*, with insightful commentary by physicist Stephen Hawking.[viii] Mathematician Ian Stewart adopted a different approach in *In Pursuit of the Unknown: 17 Equations That Changed the World*.[ix] We discuss nine of these equations. Seven of the 17 equations could be classified as basic math; we present six of them (but not Euler's formula for polyhedral, solid figures with flat faces). We do not describe the seven equations that relate directly to physics, but touch on the three equations in the applications of math in information theory, chaos theory, and options trading. Forays into math at different levels can also be found on many web sites, including the less technical https://plus.maths.org and the more technical http://mathworld.wolfram.com/. "Recreational" math has been a source of fun for many, including Martin Gardner's series on Mathematical Games that ran in *Scientific American* for over two decades.[x] His work is still celebrated in a website devoted to his work.[xi] The collection of essays and thoughts from mathematician Gian-Carlo Rota, who was my math instructor in my first term in college, is a treasure trove of insights into math and related topics, and is just plain fun to read.[xii]

We will cover elements of the math taught in grade school through high school and undergraduate college, though never with the depth needed to achieve mastery. We touch on many topics in arithmetic, some core concepts in algebra and in probability and statistics, and only elements of logic, geometry, set theory, ... We will use symbols and equations only as needed to simplify discussions. The approach is to be explanatory with the assistance of examples, and so it differs from that in a textbook or a trade book narrative. Topics are arranged so you can read from start-to-finish or instead skip around. One constant theme is how to address and optimize real-life situations with math.

We begin with **Part I: Our World of Math and Numbers**, which is an overview. In Chapter 1, we highlight the overall goals and motivations and introduce the central themes of how math rules much of our lives and how we can control this math and be comfortable with it. Chapters 2 and 3 illustrate how you are immersed in the world of numbers in serious and less serious ways.

We then journey through three interlocking worlds of math that affect our lives. In **Part II: The First World of Math: The Math of What Is, Always Was, and Will Always Be**, we introduce the world of numbers, the world of the eternal truths of math. It includes basic operations on numbers (Chapter 4), using words with numbers (Chapter 5), ways of expressing large and small numbers (Chapters 6 and 7), approximating numbers (Chapter 8), and different types of numbers (Chapter 9), including really large and small numbers (Chapter 10) and whole numbers (Chapter 11). Chapter 4 reviews some of what you may know (and may want to skip) and overviews some areas that may be new to you, some of which will be used in later chapters.

In **Part III: The Second World of Math: The Math of Doing**, we show ways this math has been developed further to shape our lives. This includes the "math of the digital world" (Chapter 12), the subworlds of "math of what will be" (Chapter 13, progressions), "math of what might be," or probability, and "math of what was," or statistics (Chapters 14-16), and big data (Chapter 17).

The third and final world of math, the "math of making decisions and winning," enables you to take control of your life and the direction of society. This **Part IV: The Third World of Math: The Math of Making Decisions and Winning**, often uses the results of the "math of doing," and includes subworlds where the outcomes may depend on optimization (Chapter 18), interdependent decisions (Chapter 19), or how risk is evaluated relative to reward (Chapter 20).

I warmly thank the many who helped me with some phase of preparing this book, including Daniel, Jonathan and Janet Herman, Paul Blaer, Andrew Cole, Eric Frankfort, Oleg Gang, Paul Glasserman, Pat and Mike Huston, Eddie Lapa, Kyle Mandli, Michael Mauel, Cev Noyan, Sid Redner, Christina Rohm, Jay Sethuraman, Adam Sobel, Michael Tippett, Ward Whitt, David Yao, Drew Youngren, and Datong Zhang.

July 1, 2019
New York, New York

Table of Contents

Part I

Our World of Math and Numbers

Chapter 1

Introduction

We live in a world of numbers and mathematics and need to work with them to control our lives and happiness. While some love to work and play with numbers and math (I do!), others are less comfortable with them.

Some claim **they are not good at math**. Perhaps so, such as the Justin Timberlake character Dylan in the 2011 movie *Friends with Benefits* who claimed he cannot "do math"—even simple arithmetic. But, I bet most can "do" more than they think.

Some maintain that **they do not like math or are afraid of it**. But, I bet that many of them have been psyched-out by society, peer groups, friends, family or bad teachers. For some, math anxiety may be an acquired disorder passed on by frustrated parents trying to help their children with homework.[xiii] Parents should encourage the love of math by talking with their children about numbers.[xiv] Laura Feiveson, an M.I.T.-trained economist now at the Federal Reserve Board has noted, "I owe my love of math to my dad who brought me up on logic and math problems since before I can remember."[xv] Unlike Justin Timberlake's Dylan, there are positive math role models on the silver screen, such as Lindsay Lohan's character Cady in the 2004 film *Mean Girls*, who loved math and was quite good at it. Still, such negative thoughts on math are pervasive. In what may have been a peak of the fear and ignorance of math (and of other things as well), the departure of an American Airlines flight from Philadelphia was delayed in May 2016 when a passenger mistook the math equations being written by a fellow passenger, an economics professor, as potential terrorist activity.[xvi]

The wide range of comfort levels with math is not new. At the U.S. Military Academy at West Point, future World War II general George S. Patton failed math in his first year and was held back the next year. He passed it the second time around by memorizing solutions to common math problems.[xvii] In contrast, though many academic subjects were challenging for to-be U.S. president and Civil War general Ulysses S. Grant at West Point, math was not: "The subject was so easy to me as to come almost by intuition."[xviii]

Jennifer Ouellette notes in *The Calculus Diaries: How Math Can Help You Lose Weight, Win Vegas, and Survive a Zombie Apocalypse*: "I think scientists have a valid point when they bemoan the fact that it's socially acceptable in our culture to be utterly ignorant of math, whereas it is a shameful thing to be illiterate."[xix] This big cultural and professional divide is shameful. It is widely promoted in our society,[1,xx] including through inaccurate statements made in the popular media. In the 1986 film *Peggy Sue Got Married*, 40ish Peggy played by Kathleen Turner faints and is "transported" as an adult back to her high school days 25 years earlier. After taking an algebra test for which she was totally unprepared, she declared to the teacher in front of her classmates: "I happen to know that in the future I will not have the slightest use for algebra, and I speak from experience."

Peggy Sue was wrong, because in the real world she would have needed to at least think numerically and algebraically. This was the experience of entrepreneur Alexandra Samuel, who developed a phobia of math in high school and was later convinced she needed to overcome it to succeed in business. She set out to learn how to think quantitatively, in part by passionately trying to answer numerical questions she needed to address.[xxi] School performance also affected Maryam Mirzakhani, but thankfully only for a while. After doing well but not earning top marks in

---

[1] This includes the seemingly benign yet ultimately arrogant proclamations made by some well-versed in math, such as (1) "This was NOT a proper math talk. Many in the audience could almost understand it.", by Kimmo Eriksson. (2) "Every human activity, EXCEPT Mathematics, must come to an end.", by Paul Erdos. (3) "If you understand something, you understand that it is obvious," by Israel Gelfand. (The references are in the text.)

math in sixth grade, she told a friend that she was not going to even try to do better in math. But, she changed her mind and at the age of 37 won the 2014 Fields Medal, the equivalent of the Nobel Prize in math, the first woman to win this award.[xxii]

Peggy Sue was also wrong because in addition to being the language of physical sciences and engineering for ages, math has now also become the language of biomedical sciences, economics, finance, and social sciences, and is evermore present in many parts of your everyday life. Moreover, being at ease with math and quantitative thinking is central to many occupations and to everyday life.

Math is pervasive even at play, and has been a central theme of several excellent movies. *Good Will Hunting* followed a fictitious, self-taught math genius. *A Beautiful Mind*, adapted from Sylvia Nasar's biography of the same name,[xxiii] followed real math genius John Nash who had problems adapting to life, but nonetheless developed the groundbreaking math now known as non-cooperative game theory, which is the math of controlling decisions (and part of the subject of Chapter 19). The film *Moneyball*, based on the book *Moneyball: The Art of Winning an Unfair Game* by Michael Lewis, follows a different math theme, how advanced baseball statistics are used to field a cost-efficient winning major league baseball team.[xxiv]

Without a modest level of computational ability, you are numerically illiterate, even with access to calculators and computers.[xxv] You need to reason quantitatively or "think math" at work and in everyday life for context and for speed in the simple recall of facts and analysis—and for evaluation. The use of the Internet to assist you in "math thinking" is better only for high volume information recall and very complex analysis.[xxvi]

Interpreting information in math form is a part of the broader area of communicating in math in an organized, logical, and convincing manner. Analyzing math information by making the necessary assumptions and approximations, and arriving at well-thought-out conclusions are equally important parts of "thinking math" both critically and creatively.[xxvii,xxviii] Interpreting math information, as from equations, graphs, tables, and words, and in turn representing it in such forms, is a part of the broader area of communicating in math in a logical and convincing manner. For example, such critical and "math thinking" is necessary when you are presented with statistical data presented in the forms of carefully-crafted graphs and carefully-chosen "averages." It is also necessary when you receive medical results, such as with positive results for a disease that may well be "false" positives, and when you learn of supposedly unlikely coincidences, as noted by neuroscientist Daniel J. Levitin in his *A Field Guide to Lies*.[xxix,xxx]

To "think math" you need to become equally at ease with several quite natural "contrasting" math situations and contexts. As we will see, we live in a world of ***continuous*** numbers, so we may think we weigh 165.5738… pounds and that the distance between two cities may be 1,629.628… miles, but we also live in a world of ***discrete*** numbers, as when we count our 10 fingers and 2 eyes and when we use the logic systems of digital electronics and computers. There are also ***algebraic*** (and so expression-based) and ***geometric*** (and so object-based) modes of thinking; we will focus more on the former. Much of math involves situations that are totally well defined and characterized and so are ***deterministic*** in nature, as with how much money I owe this month to pay rent or the mortgage, while others involve a degree of uncertainty or randomness and so are ***stochastic***, as in gambling. You frequently encounter both in everyday life. Some math is ***exact***, as in "I owe $32.69," while some is ***approximate***, as in "I weigh approximately 164.6 pounds at this very moment" and some entail. ***estimations***, as in "You seem to weigh roughly 160 pounds."[xxxi]

We will focus on using and obtaining numbers, sometimes using a bit of algebra. One theme is to present underlying math concepts and then outline how to use them in simple examples that demonstrate the concepts. *Appreciating how math operates in real life examples is the best way to learn how to use it <u>and</u> to become comfortable with it.* Encountering terminology that you may not know can be a barrier in using math, so we largely avoid the use of symbols and equations, and complex terminology and notation. Still, avoiding all notation can make some math more complicated than it needs to be. Notes for readers who are more comfortable seeing algebraic expressions and want to delve into matters a bit more are provided in the footnotes. They can be skipped without any loss of continuity. Topics are arranged so you can either read the book from start-to-finish or instead skip from place to place.

We have yet to define math and expound on its power.[xxxii] Physics Nobel laureate Richard Feynman, a folk hero to generations of physicists, once noted "Mathematics is a language plus reasoning; it is like a language plus logic. Mathematics is a tool for reasoning."[xxxiii] and added "To those who do not know mathematics it is difficult to get across a real feeling as to the beauty, the deepest beauty, of nature .... If you want to learn about nature, to appreciate nature, it is necessary to understand the language that she speaks in."[xxxiv] Mathematician Jonathan Rosenberg once wisely noted: "Mathematics is asking questions and solving them using logical deduction. There is no reason to be afraid of mathematics, if one understands that this (is) an efficient way to analyze things in a logical way. In fact, everyone is doing mathematics all the time, whether they know it or not. Every time you think about something, and try to understand how it works—this is mathematics."[xxxv] In his famous 1959 lecture "The Unreasonable Effectiveness of Mathematics in the Natural Sciences," Nobel Prize winner in physics Eugene Wigner noted "… I would say that mathematics is the science of skillful operations with concepts and rules invented just for this purpose."[2,xxxvi]

Among academics, there is a distinction between the development of math free of purpose, for fascination and curiosity, called pure mathematics and the use of existing math and the development of new math for needed applications and new opportunities, called applied mathematics. The latter includes the use of computational and numerical methods and mathematical modeling.[xxxvii] We will encounter both "types" of math in our journey that is divided into four parts. **Part I: Our World of Math and Numbers** is an overview of our world of

---

[2] Wigner added that concepts in elementary math were formulated to describe the actual world, while more advanced concepts in math were developed because of the ingenuity of mathematicians and their seeking of formal beauty, and then these concepts were later used frequently in physics. The "unreasonable effectiveness" description of math in the talk title speaks to the amazement that the math equations developed to describe physics were derived by using data from some specific physical situations, but can be used to describe other types of situations as well.

numbers, and is followed by what we will call the three worlds of math. **Part II: The First World of Math: The Math of What Is, Always Was, and Will Always Be** introduces some of the basics of math. **Part III: The Second World of Math: The Math of Doing** shows some of the ways the math of the first world lead to other math used more directly in our lives. **Part IV: The Third World of Math: The Math of Making Decisions and Winning** builds on this with the math that directly controls decisions: those used by society and those needed by you to control your life and the direction of society.

More generally, please try to be at ease when seeing numbers in everyday life and start thinking "That makes sense." and asking "What does this number really mean?" Become comfortable in inserting numbers to estimate and evaluate situations, and therefore to "Do the Math." Learn when you can generalize numerical results from the case at hand or transfer math concepts from one type of problem to another, and when you cannot. "Attack" math problems in general by asking: "What are the steps?", "What do I know exactly or roughly?", and "Does my answer make "sense"? No longer fear giving the wrong math answer ("experts" do it all the time), fear challenging the math results of others (maybe they are wrong!), or feel that a stream of numbers and math talk makes your "eyes glaze over" (it should instead perk up your eyes!). Our discussions should help you develop these skills and traits by teaching you a bit more math and presenting simplified examples that relate to real life. You may well learn that you know much more math than you currently think and are very comfortable with it!

Remember not to let the math in your life rule you. Control and love the math in your life, and welcome home to your world of math!

Chapter 2

We Use Numbers Here, There and Everywhere

We are surrounded by numbers, numbers that guide and sometimes control us.[3,xxxviii]

We use numbers for money. We use numbers to measure time. We use numbers when we count. We use numbers for addresses. We use numbers in recipes. We use numbers in making inventories and budgets.

We use numbers in the data that describe us. Your medical records are lists of numbers characterizing test results. They may describe us as 120/80, 115, and 20/30 (shorthand for having systolic and diastolic blood pressures of 120 and 80 millimeters of mercury, having 115 milligrams of low density cholesterol per tenth liter of blood, and being able to read letters 20 feet away in one eye that someone with really good (20/20) vision can read 30 feet away). They also include lists of numbers that characterize digitized images and scans of us (MRIs, x-rays, etc.), in which each dot (pixel) is a number denoting the brightness of the image at that point. Law enforcement agencies have digitized photos and fingerprint scans of many on record and data on DNA for an expanding number of us. Your financial profile is a stream of numbers that we, credit card companies, and the U.S. IRS maintain.

We use numbers to identify ourselves and to be identified. In the U.S. we were once identified by our nine-digit Social Security numbers. These have largely have been replaced with other ID numbers, credit card numbers, passwords that often need at least one number, and so on. Others use these numbers and our cellphone numbers to identify us and

---

[3] In this chapter, I pay homage to the distinctive writing style of Donald E. Westlake in *Dancing Aztecs*. (The reference is in the text.)

to keep track of us.

We use numbers for serious professional classification. The U.S. currently uses ICD-10-CM codes (ICD stands for International Statistical Classification of Diseases, Injuries, and Causes of Death) to characterize medical conditions and procedures, and deaths.[xxxix] They are also used for medical billing, so these code numbers largely control medicine and medical costs in the U.S. A medical center could bill at a higher rate for treatments described by the more specific Code 428.11 (of the earlier ICD-9-CM system), which stands for "acute systolic heart failure." than for more general Code 428, for "heart failure."[xl] Law enforcement uses codes with numbers, such as 10-40, where the 10 indicates the numerical code of interest is coming next and the 40 means there should be a silent run, with no lights and siren.

We use numbers in other sorts of labeling and characterization. The names of the singer Adele's earliest albums were her age during production, 19, 21, and 25. Most of the albums of the rock band Chicago are known as Chicago "number," the album order, such as *Chicago III, V, VI, VII, ...16, 17, 18, ..., XXXII, XXXIII, XXXIV, ...,* with most in Roman numerals. The 8 largest accounting firms had been called the Big Eight, but after mergers and the like, it became the Big Six, then the Big Five, and now the Big Four. When there were eight teams in the college football "Pac" (Pacific) conference, it was called the Pac-8. When two teams were added it became the Pac-10, and then when two more were added, the Pac-12.[4]

We use numbers to define or describe athletics. Football (U.S.) and ice hockey games are played for 60 minutes, with football played with 4 15-minute long quarters and hockey with 3 20-minute long periods. Professional U.S. basketball games are 48 minutes long, with 4 12-minute quarters. Baseball teams each usually can make 27 outs before the game ends, with no more than 3 outs in their own half-innings. Batters with 3 strikes are out and those with 4 balls advance to first base. The 9 batters per team follow a precise order of batting. In fact, the first

---

[4] However, the Big-10 college football conference, which once had 10 teams, has grown to 14 teams (in football) plus affiliate members, with no corresponding change in the conference name.

professional baseball team to regularly put numbers on the backs of their uniforms were the 1929 New York Yankees, with numbers then indicating the order of the player in the batting lineup. Babe Ruth wore No. 3 and Lou Gehrig wore No. 4. The positions of players involved in fielding plays in baseball are indicated by the number associated with the position: 1 for the pitcher, 2 for the catcher, 3, 4 and 5 for the first, second and third baseman, 6 for shortstop, and 7, 8 and 9 for the left, center and right fielder.

We use numbers to keep score. A score of 10.0 once denoted a perfect performance in gymnastics, the highest a gymnast could achieve. At the highest levels of competition, a different numerical scale is now used which uses 10.0 as its base, with points added for performance difficulty and deducted for execution that is not perfect.[xli]

We use numbers to evaluate our success. People are judged by how many dollars they earn. Students are judged by their numerical course grades and standardized test scores. Scientists are judged by how many times their publications are cited or referenced by other publications. Scientific publications are judged by how many times the papers they publish are cited. Those on social media are judged by how many follow them and like their comments. Universities are judged by how they are ranked in magazines and websites. Restaurants, movies, books, and products are judged by how many stars they receive in reviews. Some employers use numerical metrics to judge their employees, including how much money they bring to the organization and the costs of their salaries and benefits. Some rate and rank employees on numerical scales, though Goldman Sachs, which once rated staff on a scale of 1 to 9, has joined the Adobe, Microsoft, and others in abolishing numerical ratings because that practice can hurt employee morale.[xlii]

We use numbers to evaluate situations. The need to rapidly evaluate the health of newborn babies are made on the basis of their Apgar scores one and five minutes after birth; this is based on a method introduced by Virginia Apgar in 1952. The scores are the sum of 0, 1 or 2 for complexion (2 for pink bodies), pulse rate (2 for >100 beats per minute), reflexes and responses (2 for crying, grimacing, coughing or sneezing on stimulation), muscle activity (2 for active motion), and breathing effort (2 for a strong cry), and so they can range from 0-10. Experience has shown that babies with scores below 7 may need immediate medical attention.

We use numbers in rules of thumb and metrics. The 80/20 (or Pareto) rule of thumb is that 80% of the effects, such as sales in business, comes from 20% of the causes, such as clients.[xliii] The 4/5ths or 80% fairness rule or metric says that the selection rates of any group, such as success in loan or employment applications, should not be less than 80% of that for the most successful group.[xliv]

We use numbers in building and upgrading our homes. Home insulation is characterized in terms of its "R-value" number, which characterizes the resistance to heat flow. For a given type of insulation, it is essentially proportional to insulation thickness. Adding R-30 insulation to existing R-13 insulation gives a total of R-43 insulation.

We use numbers to set (sometimes dubious) goals. We are sometimes overly concerned about reaching specific number milestones, as the world was in 1954 when someone, Roger Bannister, finally ran a mile in under 4 minutes. Was the career of a baseball pitcher with 300 victories, really so much better than one with 299 or even "only" 285 victories? Is a baseball batting average of "300" (0.300, or 30.0% of official times at bat resulting in hits) so much better than one of 299 or 285? Has a hitter with 3,000 lifetime hits had a much better career than one with "only" 2,900 hits? When the University of Connecticut women's basketball team extended its already-record winning streak to 100 on February 13, 2017, was it worth the extra attention it received, compared to its equally remarkable previous 99th straight win and subsequent 101st straight win? (The streak received attention again when it ended at 111.) Has the Dow Jones Industrial Average reached a value of 10,000, 20,000, 30,000, 40,000, 50,000, or …? Have gas prices reached the $3 a gallon barrier (or the $4 barrier or the $5 barrier or the $6 barrier, …)? How about specific society targets of unemployment rates, inflation rates, poverty rates, and so on?[xlv]

We use numbers in much of our casual language. The slang to "deep 6" people means to kill them, because people are commonly buried 6 feet underground.[xlvi] "To the Nines" has long denoted something that is near perfection. Being "Dressed to the Nines" means you are properly attired for a high-level event, like a society party or formal concert, while being

"Dressed to the Eights" means you are improperly attired for a formal event, though you may think otherwise.[xlvii] Sometimes numbers are not used in a literal sense. "Catching forty winks" means a short, restful nap, not in a bed, and one that usually lasts longer than the half a minute or so needed to wink 40 times. Phonograph records represent an outdated technology that appears to be making a comeback as a curiosity. Calling a record a 45 meant that it needed to be played with it rotating 45 times a minute. There were also "33 1/3" and "78" records.

We use numbers to make political pronouncements, which may have an extraordinary impact, intentionally, as in the rallying cry slogan "Fifty-four Forty or Fight!" in the 1844 U.S. presidential election for the 54°40' latitude goal for the Pacific Northwest border or, non-intentionally, as in Mitt Romney's statement of "There are 47 percent of the people who will vote for the (then current) president no matter what …," which is thought to have hurt his bid for the U.S. presidency in 2012.

We also use numbers to tell jokes, both good ones and bad ones. "Seven out of ten people suffer from hemorrhoids. Do the other three enjoy them?."[xlviii]

We encounter numbers all of the time. They affect and sometimes control our actions, but sometimes they play a whimsical role in our lives.

Chapter 3

Numbers Are Some of My Favorite Things

Sometimes numbers are not meant to rule us, but to amuse us, either as part of a recreational activity or by the nature of the number itself. In *Cutting for Stone* by Abraham Verghese, it is asked: " 'Pray, tell us, what's your favorite number?' ... Shiva jumped up to the board, uninvited, and wrote 10,213,223 .... 'And pray, why would this number interest us?' 'This number is the only number that describes itself when you read it, One zero, two ones, three twos, and two threes'."[xlixl] (This is not strictly true: consider 15,333,110.)[5,li]

Numbers seamlessly appear in the titles, plots and descriptions of many books and documents to provide content or set a mood, including *The Three Musketeers* by Alexandre Dumas, *Richard III* by William Shakespeare, *Catch-22* by Joseph Heller, *Fahrenheit 451* (the temperature at which book paper is supposed to catch fire and burn) by Ray Bradbury, *Twenty Thousand Leagues Under the Sea* by Jules Verne, *Fifty Shades of Grey* by E.L. James, *A Thousand Splendid Suns* by Khaled Hosseini, and *1984* by George Orwell. *Six Degrees of Separation*, a 1990 play by John Guare and 1993 movie, is based on the 1929 theory by Frigyes Karinthy that everyone is connected to everyone else by six steps or fewer. There are variations in this theme, such as the game *Six Degrees of Kevin Bacon*.[lii] There are Five Books of Moses and the Ten Commandments, and the first 10 amendments to the U.S.

---

[5] The only 10-digit number that describes the numbers of times the digits 0, 1, 2, 3, 4, 5, 6, 7, 8, and 9 are used in it, in sequence, is 6,210,001,000, which stands for 6 zeroes, 2 ones, 1 two, 0 threes, and so on. (The reference is in the text.)

Constitution, which we call The Bill of Rights. In the 1950 movie *Harvey*, based on the 1944 play by Mary Chase, Elwood P. Dowd, played by Jimmy Stewart, is urged to take Formula 977 so he no longer will see his 6'3½" rabbit companion, named Harvey. (If you do not resonate with these cultural references or those below, please find your own!)

Many songs have numbers in their title and lyrics, with varying degrees of meaning and consistency, including *50 Ways to Leave Your Lover* by Paul Simon (who actually proposes *only* three distinct ways to do this, with each one repeated four times), *When I'm Sixty-Four* by The Beatles (written at 16 by lead vocalist Paul McCartney, who released it when his father turned 64 and who himself had passed that age at the time of this writing), *100 Years* (used as the JAG TV show finale concluded[liii]) written by Five for Fighting (where five denotes the number of minutes a penalty for fighting in ice hockey lasts and not the number of people in the group, which is one: Vladimir John Ondrasik III), and *867-5309/Jenny* sung by Tommy Tutone (whose popularity led to decades of using and acquiring that local phone number). (We will revisit this very interesting number several times, and call it, 8,675,309, Jenny's number.)

Some may have learned to count from songs, such as from the "1, 2, 3, 4" at the beginning of the Beatles' *I Saw Her Standing There* and in the last part of Bruce Springsteen's *Born to Run*. Others may have learned from "The Count" on *Sesame Street*, who loved to count. Some may have learned to add from songs, such as from *(I'm Gonna Be) 500 Miles* by the Proclaimers, who sang that if you walk 500 miles and then 500 miles more, you will have walked a thousand miles.

Some people find significance in numbers for reasons that are clear only to them. This is "numerology." In an episode of the late 1970s police sitcom *Barney Miller*, a man, previously known as Ira Grubb, was arrested after a bank refused to let him open a checking account under what he deemed to be his real name, 1223, and he responded with a degree of passion that was deemed to be excessive.[liv] Sometimes the apparent significance of numbers is one of fear. 13 is considered by some in Western cultures to be so unlucky that many buildings do not have a 13th floor and Friday the 13th is met with dread. In Chinese culture, the number 4 is thought to be unlucky because it sounds like the word for "death" and so floor numbering skips 40-49 in some tall Eastern buildings. This is also true for floor 17 in some Italian hotels, because 17 is XVII in Roman numerals and (the re-ordered) VIXI is common on gravestones. (It means "I lived" in Latin.)[lv] The number 666 evokes fear in some because *The Book of Revelation* calls it the number of the beast. (My laboratory phone number ends with 6666, so ....) Some numbers evoke positive feelings. In Chinese culture, the numbers 6 and 8 are thought to be lucky, with 8 sounding like the Cantonese and Mandarin word for prosperity, "faat." Hebrew letters are also used for numbers and so the number corresponding to the word for life "chai" also stands for 18. So, 18 and its multiples are associated with happy thoughts (and fundraising efforts).[lvi]

Within the rubric of numerology, there is also fascination with numerical versions of various dates, such as Square Root Days, 1/1/01, 2/2/04, 3/3/09, 4/4/16, 5/5/25, 6/6/36, 7/7/49, 8/8/64 and 9/9/81 (as with 9 being the square root of 81 because $9 \times 9 = 81$).[6] There are Palindrome Days, which are the same read forward or backward, such as 5/1/15 (5115) and 8/10/18 (81018).[7] March 14 (3/14) is annually celebrated as Pi Day because the "irrational" number pi (or π, the Greek letter, with value 3.141592653... continuing forever) begins with 3.14.[8] Some prefer celebrating Tau Day, June 28 (6/28), after the Greek letter tau, which denotes twice pi.[lvii] Of course, much of calendar numerology hinges on placing the numeric value of the month before the day (as we usually do in the U.S., unlike in other English-speaking countries), and other

---

[6] There are also Addition Days, such as 5/10/15 ($5+10=15$), Multiplication Days, 3/5/15 ($3\times5=15$), and Power Dates, such as 2/11/2048 (2048 equals the multiplication of 11 2s).

[7] There are longer versions of Palindrome Days with four-digit years that have either seven or eight digits. In the 21st century there are 26 seven-digit Palindrome Dates, such as 1/10/2011 (1102011) and 8/10/2018 (8102018), and 12 eight-digit ones, such as 10/02/2001 (10022001) and 01/02/2010 (01022010).

[8] Pi is the distance around a circle, its circumference, divided by its diameter. Pi Day took on special significance in the years 2015 and 2016 because with four places after the decimal point pi is 3.1415 (3/14/15) and it rounds off after four places to 3.1416 (3/14/16).

calendar details.[lviii,9]

Numbers and math are used as part of the plot throughout the world of entertainment. In the 2005-2010 American TV show *Numbers* (*NUMB3RS*), Prof. Charlie Eppes of the fictitious California Institute of Science used numbers and math to help solve crimes. In the 2012-3 show *Touch*, Jake Bohm, a boy with autism-like symptoms, saw patterns and meanings in a new number each week, thereby helping his father connect seemingly uncorrelated events across the planet.

However, some characters in this fictional world of entertainment bridle at being identified as a number. In the classic late 1960s British TV show *The Prisoner*, Patrick McGoohan played a secret agent who quits and is then brought to and confined on an island, where everybody is called by his or her number.[10] When frequently reminded that his name is now Number 6, the recalcitrant McGoohan bristled "I am not a number. I am a free man." In 1966 Johnny River sang in the song *Secret Agent Man* about secret agents that "They've given you a number, and taken away your name"; however, James Bond readily accepted being called 007.

Yes, numbers flood both the serious and whimsical parts of our lives and we need to be comfortable with them and the math that rules our lives. The first step in doing this is with the following overview.

---

[9] 3/5/15 is a valid Multiplication Day in the U.S. (being March 5, 2015 and $3\times5=15$) and also outside the U.S. (May 3, 2015 and $5\times3=15$).

[10] The island is led by Number 1, who is never seen, and a hands-on administrator for that episode, Number 2, who tries to prevent McGoohan, from escaping.

Part II

The First World of Math: The Eternal Truths of Math

To understand the math that rules your life, you need to know arithmetic and a few slightly advanced concepts, and to be at ease with a bit of math notation. Being comfortable with some eternal truths of math will empower you with the "Math of What Is, Always Was, and Will Always Be."

Chapter 4

Linking Numbers: Operations on Numbers

We first review the core of arithmetic (skip this if you like) and then overview other areas you may encounter, by largely remaining within the framework of arithmetic and by avoiding extensive formalism. In exploring some algebraic concepts, using a little bit of notation can make understanding problems and performing calculations far simpler. This is also true for understanding the methods that underpin many calculations, known as calculus. (*Calculus* is not meant to be a scary word—just one that indicates it helps you *calculate*.) Still, we will usually stay within the notation of arithmetic. Sometimes being uncomfortable with a given type of math may mean you are really uncomfortable only with notation strange to you and not with the math concepts or operations involved.

In arithmetic, we do things with and to numbers called operations. The symbols +, –, ×, and ÷ denote addition, subtraction, multiplication, and division, respectively, but these are represented in other ways as well. 3 multiplied by 4, or the product of 3 and 4, is represented by 3×4, $3 \cdot 4$, 3(4), or (3)(4). "3 times" can be written as 3×. Looking at $3 \times 4 = 12$ in reverse, 12 can be *factored* into 3×4, with *factors* 3 and 4. 3 divided by 4, or the quotient of 3 and 4, is $3 \div 4 = 3/4$ (where the part on top of the fraction, 3, is the numerator and the part on the bottom, 4, is the denominator); it is also 3:4, which is also called a ratio. This same fraction can also be expressed by the decimal 0.75 or 75% (where the per cent symbol % merely means hundredths). 0.75 means 7 tenths plus 5 hundredths, just as 75 means 7 tens plus 5 ones. Furthermore, 3/4 = 1/(4/3), so you can put a fraction in the denominator, with 1 in the numerator, by interchanging its numerator and denominator. The *reciprocal* of a number is 1/(number), so 3/4 and 4/3 are reciprocals of each other, as are 8 (= 8/1) and 1/8.

The *golden rule for fractions* says you can multiply or divide the numerator and denominator by the same number (other than 0) without changing its value; sometimes you can use this rule to simplify the math.[lix] You add fractions by adding their numerators when their denominators are the same, such as 3/7 + 2/7 = 5/7, and subtract them by subtracting their numerators, again when their denominators are the same, such as 3/7 – 2/7 = 1/7. When they are not, you use the golden rule to make the denominators equal, so 1/2 + 5/6 = 3/6 + 5/6 = 8/6 because 1/2 = (1 × 3)/(2 × 3) = 3/6. Moreover, 8/6 = 4/3 (by dividing the "top" and "bottom" by 2), and this can be simplified to 1 1/3, when needed, because 1 1/3 = 1 + 1/3 = 3/3 + 1/3 = 4/3. You multiply fractions by multiplying their numerators and multiplying their denominators, such as 3/7 × 2/11 = (3 × 2)/(7 × 11) = 6/77. You divide one fraction by another, by multiplying the first one by the reciprocal of the second one, such as 3/7 ÷ 2/11 = 3/7 × 11/2 = (3 × 11)/(7 × 2) = 33/14.

A fraction is in its lowest terms if the numerator and denominator have no common factor, such as for 2/3 and 3/5, but not for 14/21 (which have a common factor of 7, because 14 can be factored into 2 × 7 and 21 into 3 × 7) or 30/50 (common factor of 10). It is usually best to convert a fraction to its lowest terms, unless, as we just saw, a given form is needed to help add or subtract the fractions. In any case, 2/5, 4/10, 6/15, …, 0.4, 40%, 1 – 3/5, and (as we will see when we introduce negative numbers) (-2)/(-5), (-4)/(-10), and so on are just different ways of expressing the same number.

The number "1" can be expressed in several ways, such as 4/4, 6/6, and 100%, and they all mean the whole. Baseball Hall of Famer Yogi Berra is renowned for his great baseball play and 10 World Series rings as a player (which is more than anyone else), his shrewd business decisions, and the bewildering quotes attributed to him. Some concerned math, such as, "You better cut the pizza in four pieces because I'm not hungry enough to eat six." The entire pizza is the entire pizza whether expressed as 1 or 4/4 or 6/6. Perhaps he meant something else. He also said, "Baseball is ninety percent mental. The other half is physical." This is usually thought to be nonsense because if it is mental it is not physical and vice versa, and 90% + 1/2 is 90% + 50% = 140%, which is more than the whole, 100%. However, Yogi may have been suggesting that some mental aspects of baseball are also physical and vice versa, which would be insightful.

As noted, the term % or percent means one hundredth (1/100th of). If 4/5 of the people voted for a candidate, that person received 80% of the vote. However, the % or percentage of the vote the candidate received was 80 (and not 80%, which would mean that the person received only one hundredth of 80% or 0.8% of the vote). In their initial report of United Airlines dragging a passenger from his seat on April 10, 2017, the *New York Times* noted that United involuntarily bumped 3,765 of its more than 86 million passengers in 2016, and this was 0.00004% of the total. This was wrong by a factor of 100 (and was later corrected), because 3,765/86,000,000 ~ 0.00004, which is 0.004%.[lx]

For decimals between 0 and 1, it is good form to place a 0 before the decimal point. You don't want 0.52 to be written as .52 and then be misconstrued to be 52 if the decimal point is not readable. Sometimes common parlance trumps good form. A baseball player hitting safely 32.6% or 0.326 of the time (the number of hits divided by the official number of times at bat) is said to have a batting average of .326 (with the zero missing) or just as 326 (which is implicitly in thousandths, but at times it is mistakenly called a percentage).

Negative numbers arise very naturally, as in temperature readings (-10 °F) and bank accounts (an overdraft of $34). When you subtract 2 from 5 you get 3, so 5 – 2 = 3. When you subtract 5 from 2 you get -3 ("minus 3"), a negative number, so 2 – 5 = -3.[11] (The positive number +3 is usually just called 3.) This is the same as adding the negative number -5 to 2, so 2 + (-5) = -3 also (Figure 4.1). Adding a negative number to or subtracting a positive number from a negative number makes it more negative, as with -4 – 6 = -10. Subtracting a negative number from a positive number makes it more positive: 4 – (-6) = 4 + 6 = 10. This is like a double minus (a minus minus or - -), with the first

[11] Though the same "minus sign" denotes subtraction and that a number is negative, we will use "–" to denote the former and "-" the latter. We will also use "-" to denote "to," as across a range of numbers from 2-5.

minus reversing the sign or "direction" of the second one, so it becomes a plus or positive, +. Adding a positive number to a negative number makes it less negative, and it may remain negative (-4 + 3 = -1), become zero (-4 + 4 = 0), or become positive (-4 + 6 = 2). Multiplying two positive numbers or two negative numbers gives a positive number, while multiplying a positive number and a negative number gives a negative number, so 4 × 3 = 12, (-4) × (-3) = 12, (-4) × (3) = -12, and (4) × (-3) = -12, and similarly for division. (Sometimes adding parentheses makes things clearer, even when not absolutely necessary.) The magnitude of a number is its (positive value) with no sign, so that magnitudes of 4 and -4 are both 4. Using the common symbols for "the magnitude of," |4| = 4 and |-4| = 4, and also -4 = -|-4|.

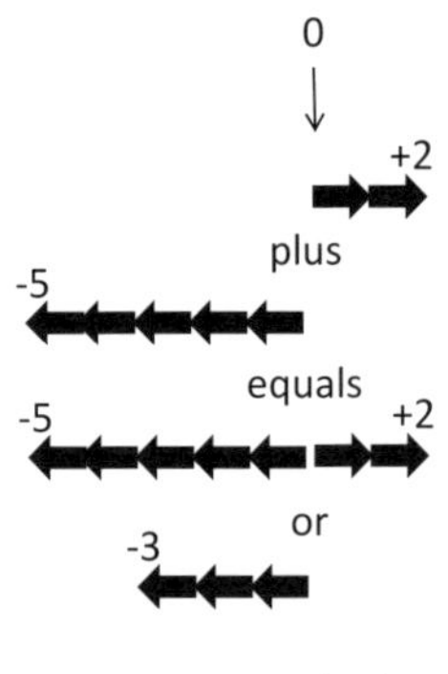

2 - 5 = -3 and 2 + (- 5) = -3

Figure 4.1: Adding negative numbers.

Algebraic notation makes "math thinking" clearer and more concise, especially when addressing more complex problems. When you use Excel spreadsheets to keep track of expenses or income, you are explicitly or implicitly using algebraic relations. Numbers with unknown values are called *variables* and are sometimes denoted by "x" or "y." When the values or variables correspond to locations, they are also called *coordinates*.

When you multiply any number (say, called x) by itself, you get x "to the power of 2" or "to the 2nd power," also more colloquially known as its square: $x \times x = x^2$, where 2 is called the *exponent*. (x^1 is merely x.) x^2 is pronounced "x" squared. The squares of whole numbers or integers, 1, 2, 3, 4, 5, 6, 7, 8, … are 1, 4, 9, 16, 25, 36, 49, 64, …, the perfect squares. Reversing the process, the square root of the squared number is the original number. This can be written as $(x^2)^{1/2} = x$. So, using the exponent 1/2 is one way of denoting the square root of a number, say $y^{1/2}$, as is $\sqrt{y}$. Technically, the square root of 4 is +2 (or merely 2), even though there are two solutions to the equation $x^2 = 4$, which are +2 and -2 (or ±2, which denotes "plus or minus" 2). When you multiply the squared number by the original number ($x^2 \times x$), you get x to the power 3, or its cube (x^3), pronounced "x" cubed. The cube root of the cubed number is the original number, and so on. It is a curious observation that the number of people in the most significant chambers of a national legislature is very roughly equal to the cube root of its population.[lxi]

For $2^3 = 8$, the exponent (or power) is 3 and the 2 is the base. Negative exponents mean that you apply the corresponding positive exponent and then find the reciprocal of the number, so $2^{-3} = 1/2^3 = 1/8$. When you multiply two numbers expressed with the same bases you add the exponents and when you divide two such numbers you subtract the exponents.[12] So, $4 \times 8 = 2^2 \times 2^3 = 2^5 = 32$ and $4/8 = 2^2/2^3 = 2^{-1} = 1/2$. When a base is raised to an exponent and the resulting number is raised to another exponent, the result is the base raised to a power that is the product of the exponents.[13] So, $(4^2)^3 = 4^{2\times3} = 4^6 = 4{,}096$. Any number (other than 0) raised to the power 0 equals 1.[14]

4.1 Order Makes a Difference

The order of arithmetic operations is important. In doing arithmetic, you generally work from left to right, but you perform operations in parentheses first and then multiplication and division before addition and subtraction. When there are parentheses within parentheses the operations in the innermost ones are done first and then you work outward, to avoid ambiguity. So, $6 + (3 \times (4 + 5)) = 6 + (3 \times 9) = 6 + 27 =$

---

[12] such as $a^b \times a^c = a^{b+c}$ and $a^b/a^c = a^{b-c}$

[13] or $(a^b)^c = a^{b\times c} = a^{bc}$

[14] This is clear from expressing $1 = 2/2 = 2^1/2^1 = 2^{1-1} = 2^0 = 1$.

33. Even without added parentheses the meaning of $3 \times 9 + 2 \times 6$ is clear and not ambiguous, $27 + 12 = 39$. Addition and multiplication are commutative, meaning that $3 + 5 = 5 + 3$ and $3 \times 5 = 5 \times 3$ (and so sometimes (!) order is not important); associative, meaning that $3 + (5 + 7) = (3 + 5) + 7$ and $3 \times (5 \times 7) = (3 \times 5) \times 7$; and distributive, meaning that $3 \times (5 + 7) = 3 \times 5 + 3 \times 7$ (and is also $3 \times 12 = 36 = 15 + 21$).

4.2 The Tower of Babel of Numbers

Before you can use and link numbers, you need to know what they represent. And given the extent of international interactions, you should be aware of the "Tower of Babel" of math notation: how different countries express the same big and small numbers. Writing two million one hundred thousand eight and seven hundred and fifty-two thousandths as 2,100,008.752 is the standard U.S.-U.K. way of writing numbers and decimals (and the way we will adopt). It uses a dot or point to separate the decimals, separates every three sets of whole digits—groups of thousands such as thousands, millions, and so on—with a marker, and uses a comma as that marker. Other systems are standard in many countries, which include using a comma to separate decimals, not grouping digits in thousands, and/or using dots or thin spaces to separate the groups of digits.[lxii] For example, the above number is written as 2.100.008,752 in Italy, Norway, and Spain. You will introduce large errors if you enter numerical data using the wrong system, so beware.

4.3 More Symbols for Linking Numbers

You may see numbers and other expressions linked by the equal sign, =, or by several other symbols, such as ≠, which means "is not equal to." 5 "is greater than" 3 can be written as $5 > 3$ or as $3 < 5$, which means 3 "is less than" 5. Also, $-8 < 1$, but $|-8| > 1$ because $8 > 1$. The burden of proof is met in many civil trials when >50% of the evidence, or the preponderance of the evidence, points to the side of the plaintiff, and so it could be 80%, 51% or 50.01% of the evidence (whatever that really means). 9 "is much greater than" 3 is expressed as $9 \gg 3$, or as $3 \ll 9$, which means 3 "is much less than" 9. There is also a symbol for "is greater than or equal to" ≥, so $5 \geq 5$ is true as is $5 \geq 3$; the corresponding symbol for "is less than or equal to" is ≤. The expressions $x \leq 10$ and $y \geq 2$

respectively set an upper bound on x of 10 and a lower bound of y of 2. Declaring that x is between but not equal to 2 and 10, means $2 < x < 10$ (or $10 > x > 2$). In statistics, upper and lower bounds are sometimes indicated by using ±, so the numbers from 8 to 12 are within the range 10 ± 2. Several versions of the squiggly symbol ~ also pop up often: as ~ it means "is approximately or roughly equal to," as ≈ it means "is just about equal to," and as ≅ it means "is very nearly equal to." You can also note that a number or expression "is greater (or less) than or approximately equal to" another by using ≳ (or ≲). Another symbol, ∝, means "is proportional to" (or changes with the same ratio as), which looks like the Greek letter alpha α (so don't confuse them).

4.4 Linking Numbers by Relations

We know that 10 people have 20 eyes and the number of eyes is always 2× the number of people, so the number of eyes is proportional to the number of people. 10 and 5 are in the same proportion as are 6 and 3, because 10/5 and 6/3 both equal 2. When one value increases as another increases, as here, the two values are said to be directly proportional to each other. When one value decreases by the same factor as the other one increases, the two values are inversely proportional to each other (and so their product is unchanged). The numerical designation of sandpaper is approximately inversely proportional to the diameter of its grit.

"Linear" terms are ways of expressing directly proportional relationships. They contain expressions with no variables raised to a power other than 1, such as a constant "a" times x (which is ax). Linear equations contain only such linear terms and an added constant (such as $ax + b = 0$ or $y = ax + b$, with constant "b" and additional variable y) (Figure 4.2), as is used to convert temperature between degrees Fahrenheit (called f) and degrees Celsius (c). In words: a temperature expressed in degrees Celsius is obtained by subtracting 32 from the temperature in degree Fahrenheit, and then multiplying it by 5/9. In algebra, this is more simply $c = (f - 32)(5/9)$, or in the more common and pleasing form $c = 5/9(f - 32)$. So, a recipe calling for the oven to be at the surprisingly precise temperature of 392 degrees Fahrenheit (or 392 °F) could also be set at 5/9 × (392 – 32) = 5/9 × 360 = 200 degrees Celsius (or 200 °C). In the reverse process, a temperature expressed in degrees

Fahrenheit is obtained by multiplying that in degrees Celsius by 9/5 and then adding 32 to it, or more simply $f = (9/5)c + 32$. So, a recipe calling for setting the oven at 200 degrees Celsius, needs to set at $(1.8 \times 200) + 32 = 360 + 32 = 392$ degrees Fahrenheit, 392 °F. (Once you understand the concepts of equations and solving them, you may want to use online equation solvers.[lxiii])

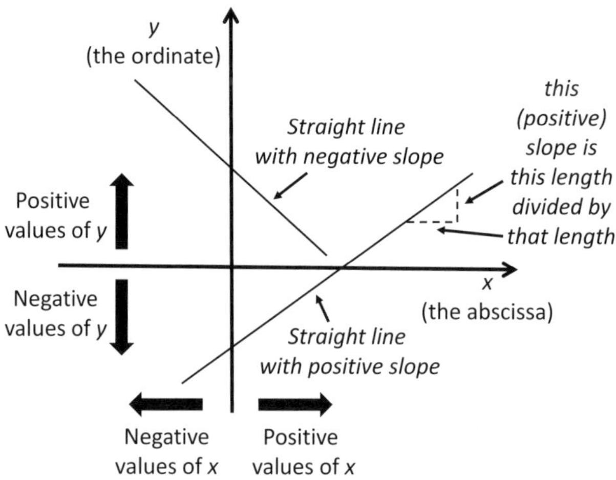

Figure 4.2: Straight lines and slopes.

You may encounter more complex equations explicitly or implicitly. Those that relate squares, linear terms, and constants are called quadratic equations, because of the squared terms, as with $ax^2 + bx + c = 0$. Those with cubes, squares, linear terms and constants are called cubic equations, $ax^3 + bx^2 + cx + d = 0$, and so on. They are collectively called polynomial equations, and as such are *nonlinear* equations; the expressions shown here on the left-hand sides are called polynomials.

Sometimes when there are several parameters, variables and constants are distinguished from another by using different smaller labels below the baseline called *subscripts*. x, y, and z could then be replaced by x_1, x_2, and x_3 (where the 1, 2 and 3 are the subscripts), and a, b, and c could be replaced by a_1, a_2, and a_3. (The smaller "2" above the baseline in x^2 denotes a squared quantity, and so it is a *superscript*. When used only as a label and not to denote a power, the superscript is often written as $x^{(2)}$.)

Equations that convert temperature and currency, such as from dollars to quarters, are linear over the entire range, but others may be approximately linear only in regions. The exam score you would get by studying for a certain amount of time would initially increase with longer studying time, possibly linearly or faster than linearly at first; however, spending much more time studying would not necessarily increase your grade correspondingly. This *saturation* is an illustration of the *Law of Diminishing Returns* (Figure 4.3).

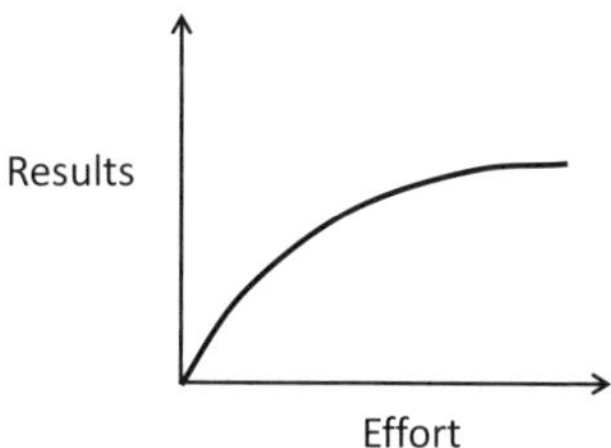

Figure 4.3: Law of diminishing returns.

Words play an important role in describing, relating and using numbers, and so in thinking in a clear quantitative manner. A collection of numbers can be called a *set*, such as the set of all positive whole numbers. A *relation* relates or associates specific numbers in one set to those in another. A *function* is a special type of relation, one that *maps* one number to *only* one number. A relation that connects a number to its square is also a function, mapping x to x^2 (Figure 4.4a). It relates 2 to 4, 2.5 to 6.25, and 3 to 9, but also -2 to 4, -2.5 to 6.25, -3 to 9. This is fine because a function can map more than one number to the same number. However, a relation that maps a number to more than one number is not a function (and is perhaps ambiguous), such as one that maps x^2 to x, which maps 4 into both 2 and -2 (as in Figure 4.4b). (This may seem strange because the square root of a number, such as 4, is assumed to be only the positive value, or 2.)

Functions and relations are often plotted as curves with points that show the vertical position (or the y coordinate or function $f(x)$ (pronounced "f" of "x"), the *ordinate*) vs. its horizontal position (or the x coordinate or *abscissa*) because this helps us understand the dependence. Figure 4.2 shows two plots of straight lines (which, of course, have *linear* dependences). How much the vertical coordinate on the plot changes divided by the corresponding change in the x coordinate is called the slope. For straight lines, $y = ax + b$, the slope is a constant a and b is another constant parameter, called the y intercept (see Figure 16.2a).[15] Figure 4.2 shows lines with positive and negative slopes. In Figure 4.4a, the function $y = x^2$ is plotted (which could be $y = ax^2$ with constant a). This quadratic form is called a parabola, and it varies *quadratically*. We see it is a function. The curve of $x = y^2$ in Figure 4.4b, is a relation that is not a function.

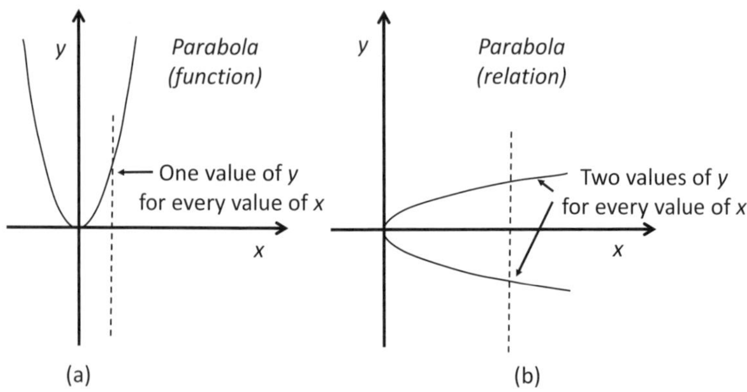

Figure 4.4: Parabola that is (a) a function or (b) a relation.

Some functions become larger, then smaller, then larger, then smaller, and so on. They are said to *oscillate*.[16] The prototypical functions that oscillate in a repeating or *periodic* manner are the sine and cosine waves (Figure 4.5a). In the U.S. electrical power is delivered with

---

[15] It is the value of y when $x = 0$.

[16] Some may have seen that the sine function (often written as "sin", with $\sin x$ being the sine function evaluated at x) and cosine ("cos" or $\cos x$) function oscillate between 1 and -1.

voltages oscillating at 60 cycles per second (also called 60 Hertz or Hz). Outdoor temperatures vary in a rough oscillatory manner, peaking during the day and reaching a minimum at night (Figure 4.5b).

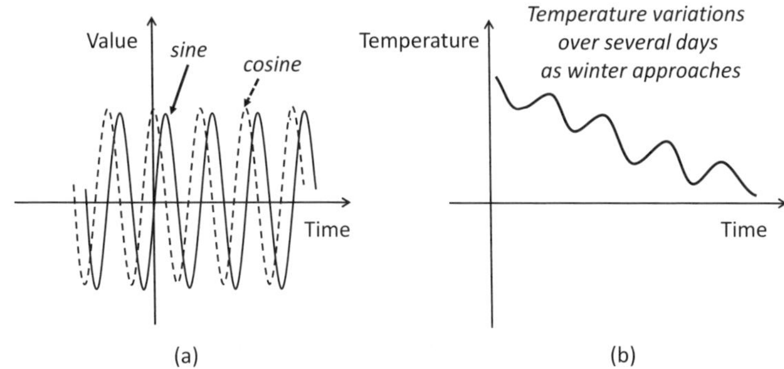

Figure 4.5: Oscillating functions.

These sine and cosine functions play another very important role in math, in *Fourier analysis*. If you add sines and cosines with different frequencies and different weighting amplitudes, you can create any function of time, which we will call *f*. The different weighting amplitudes can be presented as a function of frequency, which we can call *g*. *f* and *g* are equivalent ways of describing the same functional dependence, a version of dual math analysis and thinking, *f* in the so-called *time domain* and *g* in the *frequency domain*. *g* is said to be the *Fourier transform* of *f* and *f* is the Fourier transform of *g* (which is also called the inverse Fourier transform). The equations that perform such math procedures are called *Fourier transformations*, which Ian Stewart has noted are among the equations that have changed the world.[lxiv] Two important Fourier transforms that arise in studying functions in probability math are noted in Chapter 15.

When a sequence of numbers or an expression gets successively closer to a specific value, or its *limit*, it is said to *approach* it. The sequence 0.9, 0.99, 0.999, 0.9999, … clearly approaches 1. The sequence of 0.1, 0.01, 0.001, 0.0001, … approaches 0. A function can approach a given limit that is finite or not (and then it would be infinite) as a variable approaches a given limit. You need not have an infinite number of terms to determine the limit, because you can phrase the goal as coming within the limit by a pre-set very small number, which is often denoted by the lower case Greek letter epsilon (or ε) or by delta (δ). This concept is based on the work of Bohemian mathematician Bernard Bolzano in 1817 and German mathematician Karl Weierstrass ca. 1859. These and other Greek letters are routinely used in math as standard notation, such as the number denoted by the lower-case letter pi (π) and as parameters for a particular problem, such as the lower-case letters alpha (α), beta (β), and gamma (γ). We will encounter others later.

Division by 0 is taboo, usually. Writing 8/0 is a math crime; expressed more properly, it may really indicate infinity, which is denoted by ∞ (such as by $8/\varepsilon$ in the limit that ε approaches 0). However, 0/0, though expressed poorly and improperly, may indicate something meaningful. Consider the fraction 1/4. Dividing the numerator and denominator by 10, gives 0.1/0.4, which has the same value. Repeating this gives 0.01/0.04, and then 0.001/0.004, and so on. The numerator and denominator both approach 0, but in the limit of performing this operation an infinite number of times, the fraction maintains the value 1/4. Following the same procedure with 1/2, the numerator and denominator again both approach 0, which further demonstrates the confusion caused by 0/0.

4.5 What is the Difference?

You have likely been presented numbers or data in ambiguous or outright deceptive forms. Sometimes the goal is to find the best way to convince you that a change or difference is favorable or unfavorable to you. Clarity is essential The difference of 8 and 5 is 3 (= 8 – 5), while the difference of 5 and 8 is -3 (= 5 – 8). The fractional change from 8 to 5 is -3/8 = -0.375 = -37.5% (= (5 – 8)/8), while the fractional change from 5 to 8 is +3/5 = +0.6 = +60% (= (8 – 5)/5); they are not equal in magnitude. So, you can reversibly go from 8 to 5 and back to 8 by subtracting 3 and then adding the same number, 3, but to do the same thing in a fractional sense you need to subtract and then add different percentages; here you would need to first subtract 37.5% and then add 60% (or vice versa). It may seem that the process is not reversible, but it really is because you are multiplying and dividing by the same number, first multiplying by 5/8 and then dividing by 5/8, which is the same as multiplying by its reciprocal, 8/5.

The impact of differences hits home with Mr. Micawber's famous statement in *David Copperfield* by Charles Dickens: "Annual income twenty pounds, annual expenditure nineteen [pounds] nineteen [shillings] and six [pence], result happiness. Annual income twenty pounds, annual expenditure twenty pounds ought [shillings] and six [pence], result misery." In more local and current parlance this could be: "Annual income $50,000, annual expenditure $49,500, result happiness. Annual income $50,000, annual expenditure $50,500, result misery." Such differences control your financial health. Of course, the magnitudes of numbers are as important as such differences: "Annual income $50,000, result some happiness. Annual income $500,000, result much more happiness (perhaps[lxv])."

Sometimes straightforward subtraction does not yield the correct difference. The difference between 2016 and -325 is 2016 – (-325) = 2016 + 325 = 2341; however, this is not the number of years from 325 BC (or BCE) to 2016 AD (or CE), because the year 1 BC immediately preceded the year 1 AD and so there was no year zero. (This is one of many examples of the importance of 0.) So, the year sequence 2 BC, 1 BC, 1 AD, 2 AD, 3 AD does not correspond to the number sequence -2, -1, 0, 1, 2, 3. There are only 4 years from 2 BC to 3 AD (2 – (-3) – 1 = 2 + 3 – 1), and only 2016 + 325 – 1 = 2340 years from 325 BC to 2016 AD.

Do differences, as from subtraction, provide fair comparisons? It depends. Say the national debt was $200 in the last year of President A's term, $1,000 in the first year of the next president, President B, and $300 in President B's last year. During President B's term, did the national debt decrease by $700 (= $300 – $1000)—comparing President B's last and first years, in a way favored by President B's supporters, or did it increase by $100 (= $300 – $200)—comparing the final years of Presidents B and A, as favored by President B's detractors?[lxvi]

Though 0.999999999 is not a small number, it equals 1.0 – 0.000000001 and 0.000000001 is a small difference. Expressing a number such as 0.999999999 by using this difference may, in fact, point to what may really be significant. The probability of losing a lottery may be 0.999999999, but that of winning it is only 0.000000001. To this effect, Princeton University neuroscientist Samuel Wang once suggested to the *Wall Street Journal "The Numbers"* columnist Carl Bialik that "to improve numeracy": "Every policy story should contain an example that is typical, again reflecting true probabilities. 'MAN LOSES LOTTERY' – how's that for a headline? (Run that by your editor as a possible headline!)"[lxvii] We will revisit the world of very small probabilities later.

4.6 Differences, Relative Numbers, Ranking and Comparisons

In golf, your score for a hole is the number of strokes you need to knock the ball into the hole, the score for the round is the total number strokes for say 18 holes, and, for professional tournaments, the total number of strokes for say 4 rounds. These are *absolute* scores. Scoring is unusual in golf. Unlike that in most sports, in golf the higher the score, the worse the performance. Moreover, the score is often presented as the difference *relative* to *par* (differences again!), which is that expected for a very good or "0 handicap" golfer. Par for a hole is usually 3, 4 or 5 strokes. One fewer than par is called a birdie, one more a bogie, and so on. If par for a round is 72, a 70 would be 2 under par and a 75 would be 3 over par for the round and 293 would be 5 over par for the 288 ($= 4 \times 72$) par of the tournament.

Similarly, relating numbers is important in ranking and ordering. If A is greater than B and B is greater than C, then A is greater than C. Sometimes you need more information to rank. Knowing that A is greater than B and that C is greater than B gives you no idea whether or not A is greater than C.

Rankings also depend on how you sort, bin and present the information. Say that songs A, B, and C are top selling (or downloaded) songs in the two-year span of 2016 and 2017. If song A was sold 1,000,000 times in 2016 and never in 2017, song B 600,000 times in 2016 and 900,000 times in 2017, and song C 1,300,000 times in 2017 and never in 2016, history will rank A as the biggest seller in 2016 and C as the biggest seller in 2017, even though song B outsold both of them, with total sales of 1,500,000. (We will revisit ranking several times, including in Chapter 18, and binning several times, including in Chapter 15.)

Still, when the goal is to compare two numbers, sometimes it is better to present their ratio rather than their difference. The comparison had more impact when Hope Jahren stated in her book *Lab Girl* "The ratio of trees to people in America is well over two hundred." than if she had merely provided the numbers of trees and people or their difference (which here would be nearly the same as the number of trees).[lxviii] This is also an example where providing rounded-off numbers improves the impact of a comparison. (See more on this in Section 8.1.1.)

4.7 Differences, Calculating and Modeling

Very small differences are called *differentials*, and they underpin the type of calculus called *differential calculus*. (Finding and using these small differences is all this type of calculus is about.) The difference of 2.000 and 2.001 is technically 2.000 – 2.001 = -0.001, but the convention in calculus is to use the difference between the second number and the first, which is 0.001 here. The squares of 2.000 and 2.001 are 4.000 and approximately 4.004, so the difference of 0.001 in this variable (say x, and so this differential is called by the symbol dx, which is not a product and which is pronounced "d" "x") corresponds to a difference of approximately 0.004 in the function that squares the number (x^2, and so this is the differential of this function). In differential calculus, the ratio of these two differences, here approximately 0.004/0.001 = 4.0, is called the *derivative* of the function and determining the derivative is called *differentiation*. This is a cornerstone of this type of math. Technically, this derivative is this ratio in the limit where the difference in the variable approaches 0.[lxix] The reverse of taking differences of numbers and finding derivatives in this *continuous* mode, is taking sums of them, or determining the *integrals* in the reverse process called *integration*; this is the essence of *integral calculus*. Similarly, the reverse of integration is differentiation. Differentiation of a function follows straightforward rules, while integration sometimes can be more difficult, and so the former has been characterized as the "easy" calculus and the latter as the "difficult" one.[lxx] Sadly, calculus conjures up images in many as being difficult, esoteric, and exotic. It is not. Keep in mind that the purpose of *calculus* is to perform *calculations* by solving problems with well-defined operations, which would be difficult to solve otherwise.[lxxi] (Also, please keep in mind that we will not do any calculus explicitly.)

Equations containing derivatives are called *differential equations*. Though you may not seem them explicitly, they are working behind the scenes in modeling weather, finance, epidemics, airplanes and bridges, and so on. They can contain these derivatives, called first derivatives, and derivatives of derivatives called second derivatives, and so on. They may contain one or more functions, such as a function y or f that varies with one variable, such as a position x, so would be called $y(x)$ or $f(x)$, or time coordinate t, and so $y(t)$ or $f(t)$, so the function could be $y(x)$ or $y(t)$. ($y(x)$ was x^2 above.) These are *ordinary* differential equations, which contain *total* derivatives such as the derivative of the function y "with respect to" x, expressed as dy/dx (pronounced "d" "y" "d" "x"),[17] as well as higher-order derivatives where derivatives are taken more than once, such as twice in d^2y/dx^2 ("d" squared "y" "d" "x" squared).

Some equations contain derivatives of one or more functions of several variables, such as $y(x,t)$. These are called *partial* differential equations, and they contain *partial* derivatives, which are derivatives involving differences of only one of the variables, say x, while the others, say t, are kept constant, to give $\partial y/\partial x$. The symbol ∂ used for partial derivatives, instead of the "d" used for total derivatives, is the analog of the letter d in the Cyrillic alphabet, as used in Slavic languages. Other derivatives of $y(x,t)$ are possible, such as those with respect to t, $\partial y/\partial t$, and second derivatives, including those differentiated with respect to one variable twice, $\partial^2 y/\partial x^2$, and those differentiated with respect to more than one variable, such as with respect to x and t, both once, $\partial^2 y/\partial x\partial t$ ("d" squared "y" "d" "x" "d" "t").

(Pardon this very brief foray into calculus and its notation, notation that makes the math clearer to understand and perform. If you ever happen to encounter it again outside of this book, I hope this demystifies it for you a bit.)

---

[17] The derivative dy/dx (or $\frac{dy}{dx}$) equals $(y(x+h)-y(x))/h$ in the limit that h approaches 0.

4.8 Initial Conditions and Boundary Conditions in Modeling

In modeling, you start with a condition, defined by a number and its units, such as feet or seconds, and use a function or rule to come up with the next number, and in turn use that to get the next one, and so on. Your results will depend on the first number you started with at each position, such as the current weather conditions for a weather forecast. These are the *initial conditions* of the problem.

In other problems there are preset conditions of some sort that apply, often at the outer edges of the object you are analyzing. One possibility is to set value of the function at the boundaries, maybe by setting the temperature there to room temperature or to the appropriate conditions for an airplane wing in contact with air and the fuselage. Such conditions are known as *boundary conditions*.

One goal in math has been to find algebraic or *analytic solutions* to differential equations, which may be exact or approximate, but now *numerical solutions* using computational methods and simulations on high-speed computers often suffice in "solving" them for applications. (See Chapter 13.) Methods of finding analytic solutions and new approaches or algorithms for fast computation are central to basic and applied mathematics. (An algorithm, pronounced "Al-Gore-Rhythm," is a programmed procedure, usually implemented as a computer code.[lxxii])

Calculus involves continuous changes in coordinates. As we will see, you can compound interest every year, or every month, or every day, or every hour, or every minute, or every second, and so on. Such successively shorter time intervals are a transition from a *discrete*-time analysis to segments so short it will use the *continuous*-time system of calculus. However, the discrete approach with sufficiently small segments is sufficient for numerical analysis of differential equations.

The use of intervals of variable size or width means that comparisons may be made more meaningful with smaller "binning" units. For example, in the earlier example of the performance of presidents, it may make more sense to compare the level of the national debt during President B's last and first days in office (rather than his or hers last and first years). However, as we will see, the downside of using very small binning units is bin-to-bin fluctuation (randomness) that may turn out to be too large.

4.9 Making Complex Problems Simpler

Sometimes, problems are best solved by breaking them into separate simpler parts, and then working on them in a piecewise or sequential manner. If the tax rate is 10.0% on the first \$20,000 of your income and 30.0% for amounts in excess of \$20,000, what is your effective (or average) tax rate? For \$15,000 income you pay \$1,500 in taxes; for \$30,000 you pay \$2,000 for the first \$20,000 and \$3,000 on the next \$10,000 for a total of \$5,000; and for \$60,000 you pay \$2,000 for the first \$20,000 and \$12,000 on the next \$40,000, for a total of \$14,000. The effective tax rates in the three examples are \$1,500/\$15,000 = 10.0%, \$5,000/\$30,000 ≈ 16.7%, and \$14,000/\$60,000 ≈ 23.3%, respectively. (Algebra can give you a more general result.) Also, see the discussion of selecting a medical insurance plan in Chapter 20.

4.10 The "Let Me Count the Ways" Numbers

Sometimes problems involve grouping objects or numbers. For example, when you ask: "What are the odds of this or that happening?", you are often really asking: "How many ways can I arrange a set of objects, given specified constraints?". This is why you should care, for example, how many ways can you choose three balls from five of them. This counting problem, called *combinatorics*, is a natural part of the math of coding and passwords and of probability reasoning and analysis, as we will see later.

Grouping and probability analysis have two underlying principles that we use in everyday life with little thought.[lxxiii] The *Addition Principle* means that if you have a set of 3 balls, and so you can choose 1 ball in 3 ways, and then another set with 8 balls, from which you can select 1 in 8 ways, after combining them you have 11 balls, and you can choose any 1 of them in 11 ways. The *Multiplication Principle* means that there are $3 \times 8 = 24$ ways to choose a ball from the first set and then one ball from the second set.

Less intuitive are the different ways you can choose such objects, called permutations when selection order is important and combinations when order does not matter. (This area of math is not always taught in core math classes.[lxxiv]) Order sometimes gives distinct outcomes in real life, as when dialing in the numbers to open a "combination" lock. The sequences 7, 42, 16, 28 may open the lock while the permutation 16, 7, 42, 28 would not. Sometimes order does not matter. In a power-ball-type lottery, the combination 7, 16, 28, 42 could be a winner independent of the order the numbers were chosen.[lxxv] The numbers that describe how many ways objects can be grouped for a given a set of rules have been dubbed the "Let Me Count the Ways" numbers.[lxxvi]

You have 5 different balls and you want to place 1 of them in each of 5 distinct, numbered boxes. You can place any 1 of the 5 balls into the first box, so there are 5 ways to do this, and then any of the remaining 4 balls in the second box, for a total of $5 \times 4 = 20$ different ways of placing the first two balls. Continuing this process, you can choose any of the remaining 3 balls for the third box, for a total of $5 \times 4 \times 3 = 60$ ways of placing the first three balls. Then either of the remaining 2 balls can be put in the fourth box, for a total of $5 \times 4 \times 3 \times 2 = 120$ ways of placing the first four balls. The remaining ball goes in the fifth box, so there are 120 ways of sorting 5 objects with a given order, using each one only once, and this is the number of different *permutations*.[18] This product $5 \times 4 \times 3 \times 2 \times 1$, is written as 5! and pronounced "5 *factorial*." You can find the factorial of any positive integer. It may be surprising that 0! is 1, as is 1!. (If you are allowed to choose a previously chosen ball, and so "repeat" or "replace and choose" it, the number of possibilities is much larger, $5 \times 5 \times 5 \times 5 \times 5 = 5^5 = 3{,}125$.)

Now let's say you have 5 different balls and you want to place 1 each in only 3 boxes. The number of ways you can do this is 5 (for 1 ball in the first box) × 4 (for a second ball the second box) × 3 (for a third ball in the third box) for a total of 60 such different permutations, *arrangements*, or *orders* of selecting them in a given order. This is the same as the first three terms in 5!, excluding the last 2 terms, which equals 2!, and so it is 5!/2!.[19] (If you can repeat a previous choice, the number of possibilities is again much larger, $5 \times 5 \times 5 = 5^3 = 125$.)

---

[18] You can arrange n things, one at a time, in $n!$ ordered ways (permutations), where n is an integer and is 5 here. $n! = n(n-1)(n-2)\ldots(2)(1)$, is pronounced "$n$ factorial."

[19] You can arrange r things, one at a time, out of n different objects in $n!/(n-r)!$ ordered ways (arrangements), where n is 5 and r is 3 here.

You can also place three of these 5 balls in three boxes when the balls are identical, so the order is no longer important. For these 60 arrangements, there are 3 ways of choosing which of the selected 3 is chosen first × 2 ways for choosing the second × 1 for the third = 3! = 6. These 6 ways of arranging the 3 chosen balls cannot be distinguished, one each in three boxes, so the number of such *combinations* or *choices* of 5 identical balls in 3 boxes, without regard to order, is 60/6 = 10, or 5!/(2!3!).[20] Using this approach, you can see that there are also 10 ways to choose 2 items from 5 identical items (= 5!/(3!2!)).[21] Counting the number of combinations underpins much of probability, through the *binomial distribution* in Chapter 15.

Figure 4.6 illustrates permutations, arrangements, and combinations for the simpler case with 4 balls.

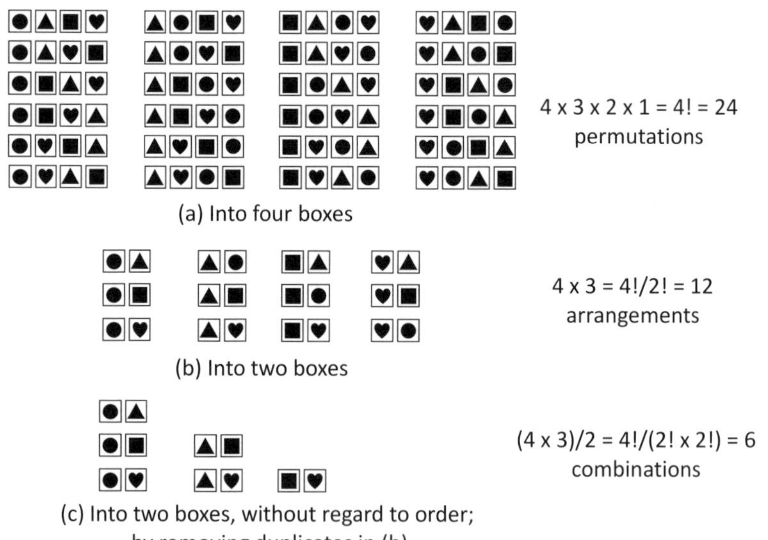

Figure 4.6: Ways of placing four objects, one each into boxes.

[20] You can choose r things, one at a time, out of n objects in $n!/[r!(n-r)!]$ ways without regard to order (combinations). This is also expressed symbolically as $\binom{n}{r}$. n is 5 and r is 3 here.

[21] This is also expressed symbolically as $\binom{n}{n-r}$, which equals $\binom{n}{r}$.

Relaxing the restriction of one item per box changes the sorting possibilities, as with placing 5 identical balls into 3 boxes. If you place the boxes side by side, there are 3 – 1 = 2 identical partitions separating them. This amounts to choosing the combinations of 5 + 2 = 7 objects without concern about the order of the 5 balls and of the 2 partitions. So, there are 7!/(5!2!) = 21 ways to do this, which is more than the 10 ways of placing them, with only 1 per box (Figure 4.7).[22,lxxvii]

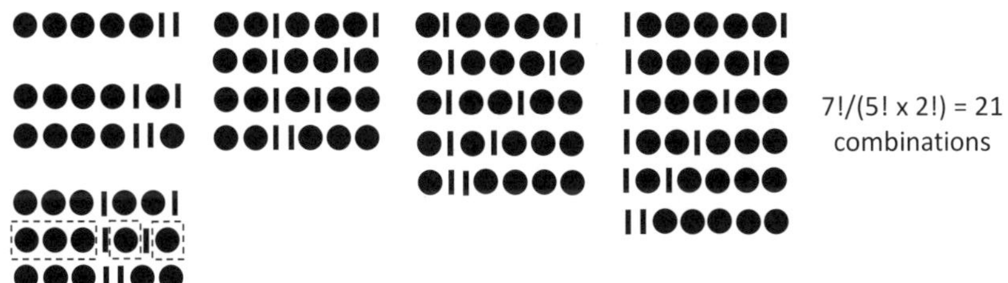

Figure 4.7: Ways of placing five balls, one each into three different boxes (separated by two partitions), as 5 + 2 = 7 objects, arranged by the number of balls in the leftmost box. The three boxes are outlined, along with the two partitions, in this case.

A related problem involves arranging n distinguishable objects into either 1, 2, 3, …, n – 1, or n identical groups (or boxes or bags). More simply, how many ways can you bag 4 different pieces of fruit, say one banana, one apple, one orange, and one pineapple, where all bags used are identical to each other? (So, here $n = 4$.) All can go into the same 1 bag, which is 1 way. They can go into 2 bags, with 3 in one and 1 in the other (there are 4 ways to do this because there are 4 ways to choose the fruit bagged by itself) or 2 in one bag and 2 in the other (there are 3 ways to do this because once you choose the first fruit in the first bag, there are 3 choices for the second fruit), for a total of 4 + 3 = 7 ways. They can go

---

22 You can choose r things out of n objects in $(n + r - 1)!/[(r - 1)!n!]$ or $\binom{n+r-1}{r-1} = \binom{n+r-1}{n}$ ways without regard to order or how many are chosen each time (or put into a box, as for n identical balls into r identical boxes, any number at a time). Here, this is for any number of the 5 identical balls placed in 3 identical boxes, with 2 identical partitions, so there are 7!/(5!2!) combinations.

into 3 bags, with 2 in one bag, 1 in the second, and 1 in the third bag (which can occur in 6 ways because there are 4 ways to put a fruit in the second bag and 3 ways to put another in the third bag and which is in the second or the third bag does not matter), so $(4 \times 3)/2 = 6$ ways. Finally, the 4 pieces of fruit could go into 4 separate bags, which is 1 way. So there are 1, 7, 6, and 1 ways of putting them into 1, 2, 3 or 4 bags respectively, for a total of $1 + 7 + 6 + 1 = 15$ different ways.[23] Figure 4.8 shows that there are 5 different ways to do this for the simpler case of placing 3 distinguishable objects into 1, 2, or 3 boxes.

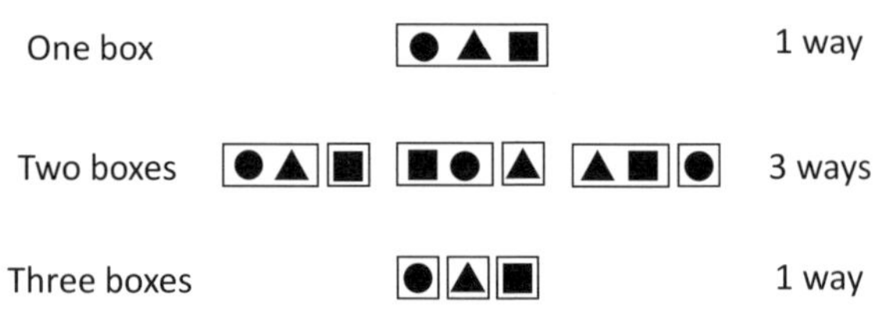

Figure 4.8: Ways of placing three distinguishable objects into one to three boxes.

Some problems seem to involve *combinatorics* or probability, but the math is different and far simpler. If your drawer has 3 blue socks, 30 red socks, and 43 black socks, what is the minimum number of socks you need to take out, while blindfolded, so you can be sure that you have selected a matched pair? This is an example of the *Pigeonhole Principle.* If you have a certain number of pigeonholes, once you place one more pigeon in the set of holes than you have holes, at least one hole will have at least two pigeons. Here the sock colors represent the pigeonholes and the socks are the pigeons. When you choose 3 socks, you will have one of each color or 2 or 3 of one color. With the 4th one, one more than the

---

[23] The 1, 7, 6, and 1 ways of putting them into 1, 2, 3 or 4 bags respectively, are called *Stirling set numbers* (or Stirling numbers of the second kind). Symbolically they are written respectively as $\left[{4 \atop 1}\right]$, $\left[{4 \atop 2}\right]$, $\left[{4 \atop 3}\right]$, and $\left[{4 \atop 4}\right]$. The sum of all of these combinations $1 + 7 + 6 + 1 = 15$, the number of different ways you can bag these objects, is called the *Bell number*, or more specifically the fourth Bell number for this group of 4 distinguishable objects.

number of different colors (the 3 pigeonholes), you will have at least one matched pair of socks.[lxxviii]

Math controls our lives not only by numbers and equations, but by words.

Chapter 5

Words and Numbers: Being Careful

Just as numbers need to be used precisely in the way intended, you need to choose the words you use with the numbers very carefully. Linked to this are two questions: What does the number really represent? What is the real significance of the words being used to describe the number? For example, how governments decide to allocate resources and how you decide to manage your own investments can hinge on how an *average* number is being defined, calculated, and presented.

Aside from such serious matters, there are amusing interplays of words and numbers. As attributed to Margot Gleave in *A Classical Education* by Colin Dexter in the final Inspector Morse detective novel *The Remorseful Day* (though actually written by him) "Different things can add up in different ways whilst reaching an identical solution, just as 'eleven plus two' forms an anagram of 'twelve plus one'."[lxxix]

5.1 Numbers and Math, and Choosing the Correct Words

Mathematician John Conway has noted "… when you're speaking mathematical English, grammar is more important than it is in ordinary English, and people are surprised by that." The two sentences "That's a large set of numbers" and "That's a set of large numbers" have very different meanings.[lxxx] Careful usage of words is equally important, as is clear from the joke: "How many times can you subtract 7 from 83, and what is left afterwards? You can subtract it as many times as you want, and it leaves 76 every time."[lxxxi]

Math inquiries need to be clear and precise. In discussing apples, "How much do you have?" does not make sense, but "How many do you have?" and "What is their weight?" do. If you are asked whether something is big or small, tall or short, or fast or slow, you need to respond: "Relative to what?"

"Fewer than six" means all non-negative whole numbers up to and including five, while "six and fewer" refers to those whole numbers up to and including six. "Less than six" means all numbers, including fractions, up to but not including six (or < 6), while "six and less" means all numbers up to and including six (or ≤ 6). Some statements employ numbers, but have little numerical meaning. The smartphone Google description of my local library notes that people typically spend up to 1.5 hours there. This means very little—only that over half of the visitors spend between 0 and 1.5 hours there. Sales that advertise that everything is up to 30% off, indicate only that no items are more than 30% off. There might be no discounts on any item.

Such ambiguity or what some may call nitpicking is rife in television shows. In one episode of the TV show *Scorpion*,[lxxxii] genius and team leader Walter insisted that the fee offered to his team for a potential job was, as it was verbally relayed, literally: one hundred "and" nine thousand dollars, meaning $9,100, and not what his associate Paige more-reasonably interpreted it to be, one hundred nine thousand dollars or $109,000. In the pilot of the TV show *House*, supposed genius diagnostician and definite misanthrope Dr. Gregory House noted "Tapeworm can produce 20 to 30,000 eggs a day." [lxxxiii] Does he mean from 20 to 30,000, which is what he said, or from 20,000 to 30,000 which could be reasonably implied?

Understanding local conventions is important, particularly in labeling and counting. We use whole numbers to label floors in buildings, but in the U.S. the ground floor is usually called the first floor and in Europe the first floor is usually one floor up from the ground floor. Sometimes this has real impact. In U.S. cooperative and condominium buildings, apartment appraisals increase with the floor level. If you live on the 8th floor in one of those rare U.S. buildings where the entry floor is not the first floor, as I do, your dwelling will likely be appraised as a dwelling on the 8th floor instead of as one on the 9th floor. Being clear is also important in labeling and counting. Floors 2 to 8 encompass a total of 7 floors ($= 8 - 2 + 1$, inclusively from floors 2, 3, 4, 5, 6, 7 to 8), they are separated by 6 floors in terms of height ($= 8 - 2$), and they have 5 floors in between them (= 8 – 2 – 1, the floors 3, 4, 5, 6, and 7). Local conventions are also important in understanding time. Half-past 10 means a half an hour after 10, or 10:30, on both sides of the Atlantic. The British use of "half 10" means the same thing. However, the word-for-word translation of half 10 into German, "halb zehn" or "halb 10," means half an hour before 10, which is our 9:30, so half past 10 translates into halb elf or halb 11 (half before 11). Moreover, some terms may be approximate and exact at the same time, depending on context and convention. In discussing dates of operations during World War II, it was understood in the U.S. Army from the official travel regulations that "about" a date included a period from four days before to four days after the specified date.[lxxxiv]

Context can provide clarity. "I take the fifth" can mean that you are seeking protection under the 5th Amendment to the U.S. Constitution or you are taking a fifth of a gallon of liquor (= 4/5 of a quart). "I want 100% proof" can mean that you want evidence that is absolutely certain or alcohol that is watered down. In the U.S., 100% alcohol is called 200 proof, so 90 proof vodka is 45% alcohol by volume, with the rest being mostly water. Because 100% means 1, 100% proof means 1 proof or only 0.5% or 1/200th alcohol by volume. More generally, be careful with the units and definitely give them when you use numbers in measurements, unless you want to default to the local convention or the context. A temperature of 35 degrees could mean that is cold, as it would if it were in degrees Fahrenheit, or hot, if it were in degrees Celsius. What it means for a glass to be half full or half empty depends on the context and your outlook on life. Ambiguity and context factor in other uses of numbers, as in the joke: "In high school, my girlfriend's dad said, 'I want my daughter back by eight fifteen.' I said, 'The middle of August? Cool!' "[lxxxv]

Words can be used to compare numbers incorrectly. For many years, football announcers commonly characterized a quarterback who threw 36 touchdown passes (which are good events) and 12 interceptions (bad events) as having a touchdown to interception ratio, or turnover ratio, of 24. No! This ratio is 36/12 = 3. What they meant to say was that his turnover difference was 36 – 12 = 24. Some announcers now correctly call this a turnover difference or, less desirably, a turnover margin, though the sign of the margin is usually not given. Others have begun to call it a turnover differential, which is again wrong, because differentials are very small differences.

Still, some seemingly outrageous statements may be mathematically correct or possible, such for "I just ran a loop starting and ending in the same place, and ran uphill most of the distance!" The sums of the elevation changes going uphill and downhill must have the same magnitudes in this round trip, but the magnitude of the average grade or slope could be smaller in the uphill portions, so the statement can be true. Of course, it would not be true if the return path retraced the first half path because the uphill and downhill distances would be equal. The statement "I just ran a loop starting and ending in the same place, and ran uphill most of the time." could also be true even if the uphill portions were shorter than the downhill parts, because you ran them slower.

5.2 Averaging Numbers

Sometimes the most information you receive about a set of data is its average, but an "average" can be defined and presented in many ways to control the message you hear. In a distribution or set of numbers, the *mean* is the sum of all values divided by the number of values, and this is also called the *average*. When listed in ascending or descending order the middle value is called the *median*, so there are as many values higher than the median as there are ones lower than it (for an odd number of values). The value that appears most frequently is called the *mode*. The mean, median and mode are equal when the distribution of numbers is symmetric, as occurs in the most common distribution of numbers and values, the normal or Gaussian distribution, which we examine in Chapter 15. However, many distributions are not symmetric.

For the sequence of numbers 1, 2, 2, 3, 7, 9, 102, the average is 18 (= 126/7), the median is 3 and the mode is 2, so they are unequal, as they are in the housing market prices. From January 1 to March 16 in 2016, the median selling price of apartments in Manhattan was \$1.17 million, but the average of \$2.1 million was even higher due to very steep prices on the high end,[lxxxvi] so the distribution of prices of sold apartments was clearly *skewed* (not symmetric) toward higher prices.

Which of the five New York City boroughs has the highest household income? In 2007, the New York borough with the highest median household income[lxxxvii] was Staten Island, \$66,985, with Manhattan having the second highest median income, \$64,217. The borough with the highest mean household income was Manhattan, \$121,549, with Staten Island second, \$81,498. Percentage-wise, Manhattan had many more rich people, many more very, very rich people, and many more poor people than Staten Island (17.6% considered to be in poverty, relative to 9.8%),[lxxxviii] so the income distribution was skewed and the "average" income needed to be defined well.

So, when salaries or housing costs are reported to be increasing, does it mean that the average value is increasing because the highest ones are skyrocketing, even though most of them, including those near the median, may be decreasing? When prices are reported to be decreasing, does it mean that the average price is decreasing because the highest ones are decreasing rapidly—perhaps for select seating at a sporting event, the opera, or on a plane, while most prices, including those near the median, are increasing?[lxxxix] Statistics can be presented in misleading ways, as we will see again later.

Sometimes it makes sense to count or weigh some of the numbers more heavily than others to obtain a weighted average. Each number is multiplied by the weighting factor, which can be larger or smaller than 1, before summing them. (These values need to average to 1 to properly "normalize" the sum.) The U.S. Department of Labor, quite reasonably, weights the employment data it obtains from a fraction of the U.S. population to correct its reported unemployment rate for any differences in age, sex, race, and so on between the population it samples and the true general population.[xc]

The U.S. government reports and sets average mileage standards for cars to control emissions and energy usage, but the many ways such averages can be weighted affects your perceptions and actions. When a hundred people drive vehicles that can get 20.0 miles per gallon (mpg), for 1,000 miles each, the total mileage is 100 × 1,000 = 100,000 miles and total gasoline consumed is 100,000/20 = 5,000 gallons. If another hundred people drive vehicles that get 40.0 mpg, for 1,000 miles each, their total mileage is 100 × 1,000 = 100,000 miles and the total gasoline consumed is 100,000/40 = 2,500 gallons. The total mileage for the two groups is 200,000 and the gasoline consumption is 7,500, so the average fuel economy is 200,000/7,500 is ~26.7 mpg, the properly weighted average for this example, and not the simple average of 20.0 and 40.0, which is 30.0 mpg. If the fuel economy rates of each vehicle were to improve by 10%, to 22.0 and 44.0 mpg, then the new weighted average would also improve by 10%, to 26.7 mpg × 1.1 = 29.3 mpg. However, if the fuel economy rates instead each improve by 2.0 mpg, respectively to 22.0 mpg (a 10% increase) and 42.0 mpg (a 5% increase), the new weighted average improves to ~28.9 mpg or by ~8.3%, which is not the simple average of 10% and 5%, or 7.5%. (This is also why total gas consumption improves more when the fuel economies (in mpg) of gas guzzlers increase by a certain number of mpg than when those of more fuel-efficient cars increase by the same amount (or even a bit more). It further suggests that it may be more helpful to present a quantity proportional to the reciprocal of the mpg rate, the number of gallons consumed by a car traveling, say 1,000 miles, rather than its mpg rate to car buyers and when assessing overall average fuel efficiency.[xci])

What controls your life more than the decisions you make in managing your finances? What are you actually being told about the average return on your investments? It depends on which type of average you are being given and how it is being weighted, as we will see in Chapter 13.[xcii]

What do the data really mean? This can be confusing, even when no deception is intended. For example, the averages of the parts of something often do not sum to the average of the whole. A basketball team plays 80 games a season, with 5 players on the court at any time during the game. If the team has only 5 players, each playing each entire game and averaging 20.0 points a game, the team averages the sum of the averages for each player, or 5 × 20.0 = 100.0 points a game. The team still averages 100.0 points a game if it has 6 players, 4 who play each game entirely, averaging 20.0 points a game, and 2 who play half of each game, averaging 10.0 points a game. The sum of the averages of points per game by these players is again 100.0 points. Now let's say the last two players each play the entirety of 40 games, and average 20.0 points in those games they play, as is how the player statistics are presented. The team still averages 100.0 points a game because only five players play each game, and each averages 20.0 points a game. But now, the statistics show six players averaging 20.0 points a game, which sums to 120.0 points, and not the team average per game.

Unconventional ways of averaging might have a rational basis or may be intended to distort the statistical analysis results. The sequence of numbers 2, 4 and 6 have a mean of 4. For each of them individually, the reciprocal of their reciprocal is the initial number, so 2 leads to 1/2, and 1/2 leads to 1/(1/2) = 2, and so on. However, the average of the reciprocals of these three numbers is 1/2, 1/ 4 and 1/6 is ~0.3056, which equals ~1/3.2727. So, the reciprocal of this average of reciprocals is 3.2727, which is not equal to the average of the numbers, 4.

Averages help define distributions of numbers and are at times linked to rankings. You may be asked to rank people, such as an instructor assessing students, as being in the top 5% (which we will see means the top five percentile) for a given attribute, top 20%, and so on. Sometimes such *bins* are also described by terms such as exceptional, superior, and so on and others as above average, average, and below average. However, such assessments may be linked and plain wrong, such as "average" being associated with the range 51-70%, the upper 50%, or even the top 10% (as I have seen). Average could mean 40%-60% or 30-70%, but not any of these.

IQ (Intelligent Quotient) scores are scaled so the median, mean, and mode are 100. So, statements such as the average IQ has increased by 3 points per decade over the last century are misleading. What is really meant is that raw data show a 3-point-per-decade increase, before rescaling; this is called the Flynn effect. In the very funny (and often very crude) 2006 movie *Idiocracy*, two average people with IQs of 100 are mistakenly transported from 2005 to 500 years into the future. The human race has become less bright with each generation and it is claimed that the IQ of the average person has dropped precipitously. This is incorrect. The average IQ has remained at 100, but the two transported people are now by far the smartest people in the word and their IQs have zoomed up, off-scale.[xciii]

5.3 Numbers Needed to Win

Who wins? In a vote or poll, over 50% of the votes cast means you have a majority, while you have a plurality if you have more votes than anyone else, be it 5%, 40% or 55%. The headline in a report that likely referenced a 2005 AP-AOL News poll: " 'We Hate Math,' Say 4 in 10—a Majority of Americans" is nonsense.[xciv,xcv] In elections for government officials, a mere plurality is often needed to win, though a threshold percentage is sometimes needed to avoid a run-off election and sometimes a majority vote is needed, as in the Electoral College vote to elect the U.S. President. A supermajority is a specific fraction needed that is greater than a simple majority, such as 60% (3/5), 66.66666…% (2/3) or 75% (3/4) of all votes. Eligible former baseball players need 75% of the votes cast by eligible members of the Baseball Writers' Association of America to be elected to the Baseball Hall of Fame.[xcvi] Counting only votes that are cast, gives an unqualified majority. With qualification, the reference total might mean votes cast plus abstentions or that of the entire membership, present and voting or not, and this may mean including or excluding currently vacant positions in the membership. To conduct business and conduct votes—but not government elections—a quorum of the membership is sometimes needed, and this can be a majority of the membership, perhaps qualified in some manner. A majority vote with a majority quorum means requires only 25% (= 50% × 50%) of the entire membership needs to support the issue or candidate. (For odd numbers of voters and members, this 50% is really half of each, rounded up, and for even numbers it is half plus 1.) We will return to the math of winning elections in Chapters 16 and 18 on election polling and different modes of voting, respectively.

5.4 Words Describing Words with Numbers

Words that describe large and small numbers and those in between often include prefixes that specify the number. Many of them were derived using the Latin and Greek words for numbers modified to be prefixes, and you likely recognize them. In their "combining form" the Latin/Greek forms from 1 to 9 are commonly: 1: uni/mono, 2: du, bi/dis, dy, di, 3: tri/tri, 4: quadr/tetra, 5: quint/penta, 6: sext/hexa, 7: sept/hepta, 8: octo/octo (okto), 9: nov, non/ennea, 10: decim/deca (deka). Use of these forms is very common, as in: Let us do this as one, in *uni*son. The House and Senate are the two chambers in the *bi*cameral U.S. Congress.

Let us mount the camera on the three-legged stand, the *tri*pod, and so on.[24] Our usual *decim*al number system has 10 digits, while the *bi*nary system uses 2, as we will see in the next chapter. The *pent*athlon and *dec*athlon in the Olympic Games consist of five and ten different events. Other prefixes include those for 0 (Latin: nihil (nothing), nullus (none)), 100 (centum/hecaton, hecto), and 1000 (mille/kilo). Myriad means 10,000, but is commonly used to less precisely to mean "many, many."[xcvii]

Prefixes are commonly used to describe objects in geometry. Two-dimensional figures, or polygons, with 3, 4, 5, 6, 7, and 8 sides and internal angles are called *tri*angles, *quadr*ilaterals, *pent*agons, *hex*agons, *hept*agons, and *oct*agons, respectively. Three-dimensional figures, or polyhedrons, are called Platonic solids when they have the same polygon for each face, with the same number of faces joining to make each vertex (point) and with each polygon being "regular" (all sides and vertices being equal).[xcviii] There are five of them, each denoted by its number of faces, such as *tetra*hedrons (4 equilateral triangles (regular triangles)).[25]

Also, prefixes are commonly used to describe time periods. *Dec*ade, *cent*ury, and *mill*ennium can correspond to any span of 10, 100, and 1,000 years, or to a specified span. The decade of the 1970s was the 198th decade and it spanned from 1971 to 1980, while the 1970s started in 1970 and ended in 1979.[xcix] *Cent*ennials and *bicent*ennials celebrate the 100th and the 200th anniversaries of an event, while *semicent*ennials and *sesquicent*ennials celebrate the 50th and 150th anniversaries. *Millen*nial means the 1000th year anniversary, and as well someone born from the early 1980s to the early 2000s. Our current millennium started with the

---

[24] *Du*els are pistol or other fights between two people, while three-person "duels" are called *tru*els. A general figure with four sides is called a *quadr*ilateral or, because it also has four internal angles, a *quadr*angle (and so it has one more angle than a *tri*angle). *Sept*ember, *Oct*ober, *Nov*ember, and *Dec*ember were the 7th, 8th, 9th and 10th months in the ancient Roman calendar. That calendar started the year with March (until 153 B.C.E). The names of the 5th and 6th months in the earlier calendar, *Quint*ilis, and *Sext*ilis, were later changed to July and August, in honor of Julius Caesar and Augustus Caesar.

[25] The other four Platonic solids are *hex*ahedrons (also known as cubes) (with 6 squares (regular quadrilaterals)), *oct*ahedrons (8 equilateral triangles), *dodec*ahedrons (12 regular pentagons), and *icos*ahedrons (20 equilateral triangles).

year 2001, and not with 2000.

*Bi*ennium (a span of 2 years), *tri*ennium (3 years), *quadr*ennium (4 years) use these Latin prefixes.[c] *Bi*ennial is derived from biennium and denotes something occurring once every two years. Because *semi* means half, it is not surprising that *semi*annual means twice a year (or once every half year). Confusion is common in using *bi*annual, which also means twice a year.[ci]

Words without these prefixes are also used to indicate specific spans of years. *Lustrum*, meaning 5 years, denoted the time between successive Roman censuses and is used rarely now. *Score*, meaning 20 years, though seldom used now is etched in our memories. In the Bible "three score years and ten" means "seventy years old" (Psalm 90).[cii] Abraham Lincoln began the Gettysburg Address with "Four score and seven years ago …". This sounds better than "Seventeen lustrums and two years ago …" or "Seventeen lustra and two years ago …".[ciii]

Words are also used to denote whole numbers and fractions. A *do*zen means 12 of something. It ultimately comes from the Latin word for the number 12, duoděcim. A dozen dozen ($12 \times 12 = 12^2$) is 144 or a gross. A dozen dozen dozen or a dozen gross ($12 \times 12 \times 12 = 12^3$) is 1,728 or a great gross. Ten dozen (10×12) is 120 or a great hundred or a small gross. A baker's dozen is 13 of something.[civ] Semi is but one prefix used to denote fractions. The liquid measure *quart* is short for a quarter of a gallon, as is *quart*er for quarter of a dollar. A quarter is also called two bits, with one bit once meaning an eighth of a Spanish dollar. *Di*mes, *cent*s and the less-commonly used *mil*ls, are tenths, hundredths and thousandths of a dollar.[cv] Cents are also used to characterize relative frequency changes of hundredths between notes in music (as in Chapter 6). Per*cent* means hundredths. Percentile, *dec*ile, and *quart*ile describe rankings in a distribution in terms of hundredths, tenths, and quarters. The top percentile, decile, and quartile are the top of hundredth, tenth and quarter. A score in the 75th percentile means that 75% of the scores are below and 25% are above it. The median is the 50th percentile. The 1st to 10th percentiles span the 1st decile, the 11th to 20th percentiles span the 2nd decile, and so on. Upper and lower half obviously refer to those in the upper and lower halves of a distribution, which brings to mind: "It is a mathematical fact that fifty percent of all doctors graduate in the bottom half of their class."[26,cvi,cvii]

5.4.1 Words describing numbers, large and small (and in-between)

In technical matters, you commonly see these and other forms of numbers to denote very small and large numbers, either as words or in abbreviated form and often as prefixes. Deci (or d) means tenths (10^{-1}), centi (c) means hundredths (10^{-2}), milli (m) means thousandths (10^{-3}), micro (the Greek letter mu, μ) means millionths (10^{-6}), nano (n) means billionths (10^{-9}), pico (p) means trillionths (10^{-12}), femto (f) means quadrillionths (10^{-15}), atto means quintillionths (10^{-18}), and so on. These numbers are also expressed appending hundredths, thousandths, millionths, billionths, trillionths, and so on. 3 millionths is also 3 ppm (parts per million) and 5 billionths is also 5 ppb (parts per billion), which might be the level of a polluting impurity in air or water. A hundredth of a meter is a centimeter. An object that is 4×10^{-9} meters wide, would also be 4 nanometers wide or 4 billionths of a meter wide. The unit denoting sound intensity (as in the next chapter) is a *dec*ibel or dB, which means a tenth of the less-useful, coarser unit of Bel (named after Alexander Graham Bell). The word "*deci*mate" now means to annihilate, but once described the punishment forced by Roman generals on unruly soldiers, in which 1 in 10 were slated to die.[cviii]

Before you invest in precious metals, such as gold or silver coins or jewelry, you should check their purity to make sure you are getting what you think you are. The purity of precious metals, such as gold, silver, and platinum, are commonly given in thousandths (as are batting averages in baseball) as a guide to their purity. In the *mille*simal fineness system, 999 means 99.9% pure gold by weight fraction, such as for 24 karat gold. This is also called three nines fine. A metal that is 6 nines pure would be 99.9999% pure or, for precious metals, also 999.999 in millesimal

---

[26] A ranking in the 82nd percentile on an exam and receiving an 82% on it mean very different things, the former being relative ranking and the latter an absolute characterization of test performance (except in the rare occasions when they both happen to be true). This distinction was not clear when Benny Colon told the jury in an episode of the TV show *Bull* that a student who retook an exam scored in the 82nd percentile and scored an 82%. Which did he mean? In context, it seemed to be the former. (The reference is in the text.)

fineness or 6 nines fine. The significant feature is the small differences from 100% purity. The more common 18 kt gold is only 75% pure, and corresponds to a millesimal fineness of 750. Because it is only 3/4^{ths} pure, 4 ounces of 18 kt gold has the same amount of gold as 3 ounces of 24 kt gold.

For big numbers, the prefix kilo (or k) means thousands (10^3), mega (M) means millions (10^6), giga (G) means billions (10^9), tera (T) (from tetra) means trillions (10^{12}), peta (from penta) means thousand trillions or quadrillions (10^{15}), exa (from hexa) means million trillions or quintillions (10^{18}), and so on. This is known as the USA and Modern British short scale, or simply the American system. This is used in the U.S. and, since 1975, in Britain. Some in the world may still use the European system or the British system, which, for example, calls 10^{12} (which is our trillion) a billion, so be careful when reading older publications and those outside the U.S.[cix] A 5K race is one that is 5 kilometers (5,000 meters) long, which is ~3.1 miles. (Technically, a 5K race is run on a road, while a 5,000-meter race is run on a track.) A *kilo*gram is a metric unit of mass that corresponds to a weight of ~2.2 pounds on Earth.

You can use either upper or lower cases in typing e-mail addresses but not when you use these prefixes! A very reputable pharmaceutical vendor once sent me a bottle labeled for ninety 50 Mg tablets. This meant that each tablet was supposed to contain 50 million grams of active ingredient, and so each tablet was supposed to have a mass of 50,000 kilograms, or a weight of 110,000 pounds or 55 tons of active ingredient. I don’t think so. These tablets were not that heavy. The label should have noted 50 mg (50 milligram) tablets, which weigh ~0.0018 ounces each. Sometimes a factor of a billion, 1,000,000,000, is a big deal.

Even in everyday life, the numbers that control us can be very large or very small.

Chapter 6

Writing Really Big and Really Small Numbers, and Those In-between

We frequently encounter very large and very small numbers, which are those near zero (and not large and negative). As we just saw, words can help express such numbers. When more help is needed, the method of scientific notation is very handy.

6.1 Scientific Notation

On Valentine's Day in 2017, the U.S. national debt was approximately $19,980,000,000,000 (and was steadily increasing) or, in words, nineteen trillion nine hundred eighty billion dollars. Clearly, this number is very big and unwieldy expressed either way. *Scientific notation* makes it easy to present very big and small numbers and to use them. A large or small number is expressed as what we will call the *prefactor* from 1.0 to 9.99999… times 10 to the needed integral power or exponent; we use 10 as the *base* because in everyday life we use a "base 10" number system with digits from 0 to 9. So, the national debt on that particular day was 1.998×10^{13} (in dollars). (The prefactor is technically the *coefficient, significand*, or *mantissa*; you will see a quite different way this last term is also used later this chapter.) You should be relieved when you see scientific notation used for even bigger numbers, such as the number of grains of sand on Earth, 7.5×10^{18}, the 2003 estimate of the number of stars in the observable universe, 7×10^{22}, and the number of molecules in a balloon, $\sim 10^{24}$, and in an 8 ounce glass of water, 1.4×10^{26}.[cx] Extraordinarily large numbers also pop up in games, such as chess. The Shannon number, named after "the father of information theory" Claude Shannon, 10^{120}, is the lower bound of legal chess positions he estimated to be reachable from the initial position.[27,cxi] Scientific notation is equally adept in expressing small numbers. The radius of the smallest atom, the hydrogen atom, is approximately 5.3×10^{-11} meters, which is superior to saying it is 0.000000000053 meters.

Scientific notation is usually presented with a certain number of *significant figures* (Chapter 8). If the IRS says you owe them \$423,008.72, you should not think you can round it off using scientific notation to 4.23×10^5 dollars and pay this amount, because it wants that last 8 dollars and 72 cents as well.

Numbers in scientific notation are multiplied by multiplying the prefactors and adding the exponents, and then simplifying as needed, so $(8 \times 10^5) \times (4 \times 10^{12}) = 32 \times 10^{17} = 3.2 \times 10^{18}$ and $(8 \times 10^5) \times (4 \times 10^{-12}) = 32 \times 10^{-7} = 3.2 \times 10^{-6}$. In division, the prefactors are divided and the second exponent is subtracted from the first, as in $(8 \times 10^5)/(4 \times 10^{12}) = (8/4) \times 10^{-7} = 2 \times 10^{-7}$. In addition and subtraction, the prefactors are added or subtracted after the exponents are first made to be the same, so $3.1 \times 10^9 - 170 \times 10^7 = 3.1 \times 10^9 - 1.7 \times 10^9 = 1.4 \times 10^9$.

A googol is the product of a hundred 10s, which is 1 followed by 100 zeros or 10000000000000000000000000000000000000000000000000 00000000000000000000000000000000000000000000000000, or conveniently expressed in scientific notation as 1×10^{100} or simply as 10^{100}. This number was coined in 1938 by the 9-year-old Milton Sirotta, nephew of Edward Kasner, who then coined the term googolplex, which is 10^{googol}.[cxii] The name of the Internet search engine and company Google came from a misspelling of googol.[cxiii] Googol and even a googolplex are no way "near infinity." Chapter 10 describes how to present numbers that are larger or smaller than we can imagine.

6.2 Logging In

You hear about an earthquake of magnitude 8 on the Richter scale and wonder why it is not twice as "strong" as one with magnitude 4, but is 10,000 times stronger and why a sound with a strength of 80 decibels is not four times as strong as a 20 decibel sound, but 1,000,000 times stronger. The reason is they are being presented through the world of

---

[27] It is thought there are 10^{40} "sensible" chess games.

logarithms, an area of math that is closely related to scientific notation.

When you "move" the prefactor in scientific notation to become part of the exponent, it is usually no longer integral, and this new exponent is called the logarithm of that number n, or $\log n$ (pronounced "log" "n"). This is technically the common logarithm or decadic logarithm (because the base is 10), or, colloquially, log. In words, a positive number equals 10 raised to a power equal to the logarithm of that number.[28] As we will see, $2 \times 10^8 \sim 10^{8.3}$, so its log is ~8.3, and the integer part, here 8, is the *characteristic and the fractional part, here 0.3, is the mantissa.* Logs of numbers that are > 1 are positive, $\log 1 = 0$, logs of numbers between 0 and 1 are negative, and logs of negative numbers are not defined. As a number approaches infinity so does its log, though it does more slowly. As a positive number approaches zero, its log approaches minus infinity.[cxiv]

Analogous to handling exponents in scientific notation, the log of the product of two numbers is the sum of their respective logs and the log of their quotient is the difference of their logs. Numbers can be converted to their logs, such operations can be performed, and the answers in log form can be converted back to numbers. This is an easy way of doing approximate multiplication and division if your calculator and the internet are not handy, and as was done for centuries by hand and by using the now obsolete "slide rule." Large ranges of numbers are more conveniently plotted in terms of logs of the numbers (at times as scientific notation) as "log scales," as we will see, rather than the usual "linear" scales." Logarithms are one type of math that Ian Stewart has said has changed the world.[cxv]

Because $1{,}000 = 10 \times 10 \times 10$, $\log 1{,}000 = \log 10 + \log 10 + \log 10$. Since $10 = 10^1$, $\log 10 = 1$ and so $\log 1{,}000 = 3$. Also, because $1{,}000 = 10^3$, it is clear that $\log 1{,}000 = 3$. More generally, this shows that the log of a number that is raised to a power, say x, equals that power times the log of that number.[29] The log of the square root of a number is $0.5 \times$ the log of that number.[30] Also, $10^{1/2} = 3.162\ldots$ and, as expected, $\log 10^{1/2} = 0.5$.

---

[28] So $n = 10^{\log n}$. Taking the logarithm of both sides gives $\log n = \log(10^{\log n}) = \log n$.

[29] Because $\log xy = \log x + \log y$, $\log x^n = n \log x$. Also, $\log x/y = \log x - \log y$.

[30] Or, $\log n^{1/2} = 0.5 \log n$ and more generally, $\log n^x = x \log n$.

Knowing that $\log 2 = 0.3010\ldots$ and $\log 3 = 0.4771\ldots$, both *irrational* numbers with decimals that continue forever (Chapter 9), can be useful for determining other logs (by hand). For example, $\log 4 = \log 2^2 = 2\log 2 \approx 0.6020$, $\log 5 = \log(10/2) = \log 10 - \log 2 \approx 1.0 - 0.3010 = 0.6990$, and $\log 6 = \log(2 \times 3) = \log 2 + \log 3 \approx 0.3010 + 0.4771 = 0.7781$.[31]

A number raised to the power of a second number usually does not equal that second number raised to the power of that first number. For the numbers 2 and 3, $2^3 = 8$ does not equal $3^2 = 9$.

Curiously, a number raised to the power of the log of a second number exactly equals that second number raised to the power of the log of that first number.[32] Again for 2 and 3, $2^{\log 3} = 2^{0.4771\ldots} \approx 1.392\ldots$ and $3^{\log 2} = 3^{0.3010\ldots} \approx 1.392\ldots$.

6.2.1 Other uses of logs, including for star gazing and common cents

Logarithms are widely used to denote values spanning many orders of magnitude. The intensities of sound and electrical signals are sometimes expressed in units of decibels (where "bel" is named for Alexander Graham Bell) or dBs, which is the logarithm of intensity times 10 (from the "deci" in decibel). It is used in two ways, in a relative sense and in an absolute sense, for the sound relative to a barely audible sound. So sound that has a strength of 80 decibels (as does an operating garbage disposal unit) is 60 dB stronger than a 20 decibel sound (rustling leaves), or $10^{60/10} = 10^6 = 1{,}000{,}000$ stronger.[33] Logs are used as well in many other fields.

The Richter scale is the log of the amplitude displacement of an earthquake, relative to a standard, so an increase in 1 in the Richter scale corresponds to an increase in displacement by a factor a 10, and an earthquake of magnitude 8 on the Richter scale is $10^4 = 10{,}000$ times

---

[31] Also, $\log 1.5 = \log(3/2) = \log 3 - \log 2 \approx 0.4771 - 0.3010 = 0.1761$ and $\log 2/3 = \log 2 - \log 3 \approx -0.1761$.

[32] This means $a^{\log b} = b^{\log a}$. The logarithms of both sides are both equal to $(\log a)(\log b)$. Because the logs are equal, the numbers are also equal. In contrast, as noted a^b usually does not equal b^a, and so also $b \log a$ and $b \log a$ are usually not equal.

[33] The 10 in the 60/10 exponent is needed to convert from decibels back to bels, and so back to the log scale. Also, the absolute intensity for, say 80 dB, is $10^{80/10} = 10^8$ times 10^{-12} watts per square meter.

stronger than one of magnitude 4.[34,cxvi]

The measure of the acidity or basicity of an aqueous solution is usually denoted by its "pH," with the lower the number, the more acidic the solution. It is the negative of the logarithm of the concentration of hydrogen ions (H^+) expressed in units of the number of "moles" of hydrogen ions per mole of solution.[35] An acid with a pH of 2 is $10^3 = 1{,}000$ times stronger than one with a pH of 5. Solutions with pH between 0 and 7 are acidic and those with pH between 7 and 14 are basic.

Stargazing has implicitly used logs for millennia. Around 129 B.C., the Greek astronomer Hipparchus ranked the brightest observable stars as being "of the first magnitude," the next brightest "of the second magnitude," and so on to the faintest stars "of the sixth magnitude." In 1856 the Oxford astronomer Norman R. Pogson proposed that a difference of five magnitudes be defined as a brightness ratio of 100 to 1, because it was known by then that stars of magnitude 1 were approximately 100 times brighter than those of magnitude 6. So, star magnitude is a log scale, with a factor of ~ -2.5 thrown in. With this scale, the brightness of the brightest planet, Venus, with magnitude -5, can be compared to that of a star of magnitude 2; it is $10^{(-5-(2))\times(1/(-2.5))} = 10^{7\times 0.4} = 10^{2.8} \approx (2.512)^7 \approx 631$ times brighter.[cxvii]

When the musical scale repeats from A, B, C, D, E, F, G, and back to A, the frequency, or oscillation rate of the sound wave of the note increases by a factor of 2, and this is called an octave (Figure 6.1). A note 3 octaves higher than another has a frequency $2^3 = 8$ times that of the original note. Including all flats and sharps, there are 12 half steps in an octave, so called because on a piano spans the eight white keys from C to C (including both).

Logs are also used in the everyday world of music. The most common way of designing and tuning musical instruments in the Western world, such as pianos, is by *equal temperament*, with the frequency of successive half notes increasing by $2^{1/12} = 1.0595\ldots$, so with the 12 half steps of the octave the frequency increases by this factor

---

[34] However, the energy released by the earthquake increases by a factor of ~32 for every increase in scale by 1.

[35] A mole is 6.02×10^{23} molecules.

raised to the 12th power, $2^{12\times(1/12)} = 2^1 = 2$. This factor can also be written as $2^{1/12} = 2^{100/1,200}$, and neighboring notes are also (though less commonly) said to be separated by 100 cents, which is the frequency range between notes subdivided in hundredths, just as a dollar can be subdivided into 100 cents. A cent corresponds to a frequency ratio of $2^{(1/12)/100} = 2^{1/1200} \cong 1.0005778$. So, an octave also spans 1,200 cents. Because $\log 2 = 0.3010\ldots$, $2 \approx 10^{0.3010}$ and this factor $2^{1/12}$ is also $\approx 10^{0.3010/12} \approx 10^{0.02508583}$, again in the log scale. This math of stars intensities and note frequencies is similar to that of geometric progressions, as in Chapter 13.[cxviii]

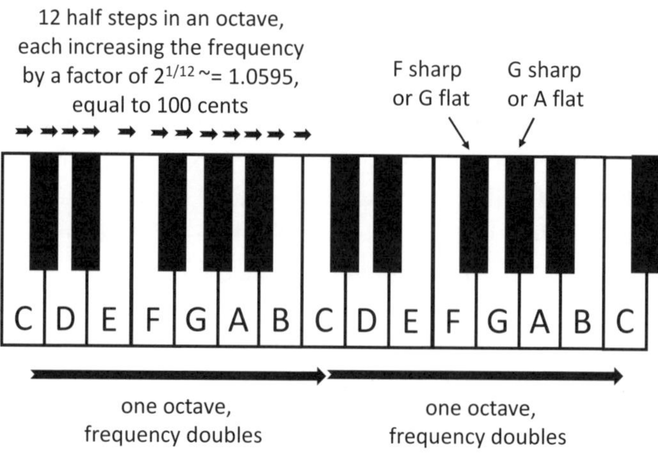

Figure 6.1: Piano keys.

When the ratios of frequencies of notes differ by a simple fraction, such 2/1 (= 2) for the octave, the sound is known to be pleasing and harmonious, and it can be produced from strings or columns of air whose lengths form simple fractions. So, another method of tuning is by *just intonation*, for which the ratio of the frequencies of the scale notes are set to a simple fraction, so the ratio for the perfect fifth (7 half steps) would be set to $3/2 = 1.5$, corresponding to 702 cents or 2 cents higher than the 700 cents for equal temperament. The ratio for the major third (4 half steps) would be set to $5/4 = 1.25$, corresponding to 386 cents or 14 cents lower than the 400 cents for equal temperament.[36,cxix,cxx]

Logarithms also play important roles in other areas of science and technology. Entropy quantifies the amount of disorder in a system and is very important in thermodynamics, the science of heat, work, and energy. Within a numerical factor, entropy equals the logarithm of the number of "states or configurations" a system has. As we will see in the next chapter, logs play a central role in quantifying how much information or data a system has.

6.2.2 Plotting with logs

Many graphs are said to be *linear* or *linear-linear* plots because they show the value of the actual function (say y) in the vertical direction and the actual variable in the horizontal direction (say x) (Figure 6.2). When y varies over many orders of magnitude it is more practical to plot its log, $\log y$, vs. x in a *log-linear* plot. When the function is an *exponential function* such as 10^x (see the next chapter), the log-linear plot is a straight line (linear) because $\log(10^x)$ is x (Figure 6.2a). Using this type of plot is one way to see if a dependence is exponential. You can see if 10^{ax} is exponentially increasing (a is positive) or exponentially decaying (a is negative), and how "fast" these dependences are by the magnitude of the slope of the plotted line, which is $|a|$. When both the function and variable vary by orders of magnitude, it is useful to plot the log of one variable vs. that of the other, $\log y$ vs. $\log x$, in a *log-log* plot. This plot is linear when the function is the variable to a power, $y = x^n$, so $\log y = n \log x$ is plotted and the slope is the power n (Figure 6.2b). Log-log plots are usually used in hearing test audiograms, where barely audible sound intensity (in the decibel or dB log scale) in each ear is plotted the log of

---

[36] Notes separated by such 7 equal half steps (equal temperament) differ in frequency by a factor of $2^{7/12} = 2^{700/1,200} \approx 10^{7(0.02508583)} \approx 1.4983\ldots$, which is very nearly $1.5 = 3/2 \approx 2^{702/1,200}$ or 702 cents. This interval of 7 half steps is known as a perfect fifth, because it spans the five white keys from C to G. It also spans other sets of notes, such as from B to F sharp. Notes separated by 4 equal half steps differ in frequency by $2^{4/12} = 2^{400/1,200} \approx 10^{4(0.02508583)} \approx 1.2599\ldots$, which is nearly $1.25 = 5/4 \approx 2^{386/1,200}$ or 386 cents. The interval of 4 half steps is known as a major third, because it spans the three white keys from C to E, as well as others such as from E to G sharp.

the sound frequency, often from 20 to 20,000 Hertz. To readily identify hearing loss, the log of the sound intensity that is barely audible to a person with perfect hearing is subtracted from that for the patient, and plotted for each frequency.

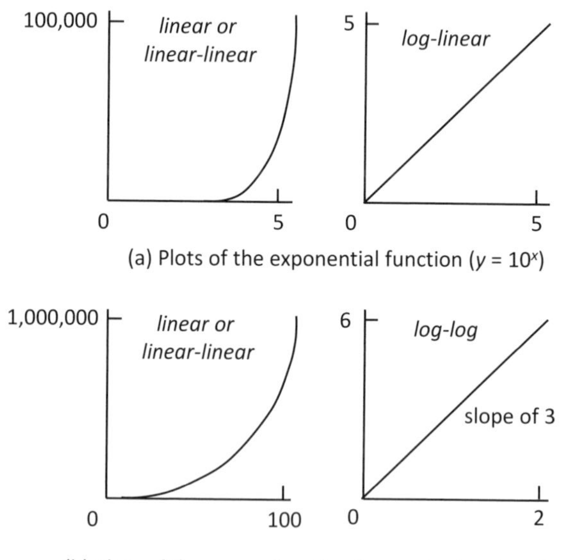

Figure 6.2: Linear and logarithmic (log) plots of y vs. x. The linear or linear-linear graphs plot y vs. x; the log-linear graph plots $\log y$ vs. x, and the log-log graph plots $\log y$ vs. $\log x$.

There are yet other ways to express very large and small numbers that provide us the needed control over them, as we will now see.

Chapter 7

Touching All Bases, At Times with Logs

"There are 10 types of people in this world: those who understand binary and those who don't."[cxxi] This is a joke to those who understand the binary, base 2 system, but it may not be to you right now because in everyday life we work directly with numbers expressed in base 10, the decimal system. This means that we use 10 digits in forming numbers, 0, 1, 2, …, 9, and so the *radix* of this base is 10. The number 4137 means $4\times10^3 + 1\times10^2 + 3\times10^1 + 7\times1^0$ (where 10^0 equals 1) or $4000+100+30+7$. Other base systems are important as well and in operation behind the scenes, though we rarely use them directly. Without information to the contrary, we assume numbers are in base 10. When there is uncertainty, we write the number as, say, $(4137)_{10}$ and the *common or decadic log* as $\log_{10}$.

In the binary or base 2 system, there are 2 digits, 0 and 1, so the radix is 2. The numbers and logic in this binary system form the basis of all current computation, computer systems, smartphones and the like because computer logic depends on a device level being either on or off—as with a voltage being above or below a threshold. In logic and computer operations, 1 can be used to represent on or true and 0 off or false. The number 23 in base 10 is 10111 in base 2, which (using base 10 numbers) means $1\times2^4 + 0\times2^3 + 1\times2^2 + 1\times2^1 + 1\times2^0$ (where 2^0 equals 1) or $1\times16 + 0\times8 + 1\times4 + 1\times2 + 1\times1$. This is an example of how to convert from one base system to another. Because 10,111 is also a perfectly good number in base 10, when bases other than 10 are used they are indicated to avoid confusion, such as in $(10111)_2$ here, where the subscript 2 is in base 10. You should now understand the above joke; 10 in base 2 is 2 expressed in base 10.

Basic computer software commonly uses base 16 or hexadecimal notation because they are convenient strings of four binary digits (or 4 *bits*, which is an abbreviation of "binary digits" [cxxii]). It has 16 digits, 0, 1, 2, …, 9, A, B, C, D, E, F, which correspond to 0, 1, 2, …, 9, 10, 11, 12, 13, 14, 15 expressed in the decimal system and 0000, 0001, 0010, …, 1001, 1010, 1011, 1100, 1101, 1110, 1111 expressed in the binary system. A hexadecimal digit is called a nibble, which is half of an octet or byte (8 bits). Current computers commonly use strings of 32 or 64 bits.

In the 2015 film *The Martian*, overcrowding prevented stranded astronaut Mark Watney from accurately pointing his still-shot camera to one of 26 spots, one for each English language letter, around the 360° circumference of a circle. However, he was able to point to any one of the 16 slots, which he used to represent the hexadecimal digits from 0 to F. Two such digits in succession give $16 \times 16 = 256$ possibilities, which are clearly enough to represent all letters and more by using the standard "ASCII" code (pronounced "ASS-kee"). This code includes each letter in the alphabet (A to Z is from 41 to 5A (hexadecimal) in upper case and from 61 to 7A in lower case), each number digit (from 0 to 9 in the decimal system is from 30 to 39), punctuation and more.[37] ASCII stands for the American Standard Code for Information Interchange, which is the character encoding standard used to represent text in computers, telecommunications equipment, and so on.

Arithmetic is conducted the same in base 10 and in other bases. For example, addition entails the same "carryover" rules. Adding $(30)_{10}$ and $(5)_{10}$ gives $(35)_{10}$. In base 16, this corresponds to adding $(1E)_{16}$ (which is $1 \times 16 + 14 = 30$ in base 10) and $(5)_{16}$ (which is the same in base 10). Adding the E and 5 in the "1s" or digits columns (or $14 + 5$ in base 10, which $= 19 = 16 + 3$) gives a 3 in that column and a carryover of 1 to the "16s" column (which is like the "10s" column in base 10). This adds to the 1 that is there, to give $(23)_{16}$, which is $2 \times 16 + 3 = 35$ in base 10. In

---

[37] In hexadecimal, the numbers 41 to 4F (corresponding to the ASCII letters A to F) in base 16 constitute 16 numbers (in base 10) and those from 50 to 5A (P to Z) to 10 numbers (in base 10), so they provide the needed $16 + 10 = 26$ numbers (in base 10) denoting the 26 letters.

base 2, in any column or pair of digits $0+0=0$, $0+1=1$, and $1+0=1$, as in base 10, but $1+1$ gives a 0 in that column or digit, and a carryover of 1 to be added to the next higher column (as from the "1s" column to the "2s" column, and so on).[38]

Decimal numbers in base 10 have analogs in other base systems, with the "decimal" point now called the radix point. The number 2.25 in base ten means $2\times10^0 + 2\times10^{-1} + 5\times10^{-2} = 2\times1 + 2\times0.1 + 5\times0.01$. The base 2 number $(10.01)_2$ expressed in base 10 is $1\times2^1 + 0\times2^0 + 0\times2^{-1} + 1\times2^{-2} = 1\times2 + 0\times1 + 0\times1/2 + 1\times1/4 = 2 + 0 + 0 + 0.25 = 2.25$ (again).

We will see in Chapters 11 and 12 that such other base systems are central to the math of the digital world and modular arithmetic (even when they expressed in base 10).

7.1 Logic and the Math of Computing

Beginning in 1847, English mathematician George Boole formulated the mathematics of symbolic logic in terms of systems of two states. The two states 0 and 1 in Boolean arithmetic or algebra represent false (or off) and true (or on). Though they are the same digits as used in base 2, the rules of Boolean or logic addition and ordinary binary arithmetic differ. An input variable, A, can be either 0 or 1. Two input variables A and B can be each be either 0 or 1, and so the four potential sets of input are ($A=0$, $B=0$), (1,0), (0,1), and (1,1). For a given *operation* there are specific outcomes for these possible inputs. The basic operations are AND, OR, and NOT. The AND operation, as in A AND B (also written as $A\wedge B$ or $A\times B$), has an output that is true (or 1) only if both A and B inputs are true (or 1); otherwise it is 0. This corresponds exactly to multiplication in binary arithmetic, in which $0\times0=0$, $0\times1=0$, $1\times0=0$ and $1\times1=1$, and to finding the lower of the two input values. The OR operation, as in A

---

[38] Adding $(30)_{10}$ and $(5)_{10}$ in base 2, means adding $(11110)_2$ [or $(16+8+4+2)_{10}$] to $(101)_2$ [or $(4+1)_{10}$]. $0+1$ in the first column gives 1. $1+0$ in the second column gives 1. $1+1$ in the third column sums to 0, carrying over a 1 to the fourth column. This carryover adds to the 1 there to give a 0 and a 1 carryover to the fifth column, which again sums to a 0 in that column and a 1 brought to the 6th column, to give $(100011)_2$ which again equals 35 in base 10. This may seem overly complex, but it is seamlessly compatible with on-off states in the prevalent circuit components that can be driven on or off.

OR B (or $A \vee B$ or $A + B$), has an output that is true (or 1) when either one or both are true (both 1 or one 1), and so it is 0 only when A and B are both 0. This Boolean "addition" examples $0 + 0 = 0$, $0 + 1 = 1$, $1 + 0 = 1$, and $1 + 1 = 1$ correspond to finding the larger value of the two input values. The first three examples correspond to binary addition, but in binary addition the fourth is $1 + 1 = 0$ plus carry the 1 to the next higher column, which is $1 + 1 = 10$; so Boolean and binary addition are different operations. In Boolean arithmetic, there is also a negation or NOT operation, that operates on A (or B) to give NOT A (or $\neg A$). It changes a 0 input to a 1 output and a 1 input to a 0 output. More complicated logic can be built from these three components. For example, the XOR (the exclusive OR, denoted by the symbol $\oplus$) gives a 1 if one input is a 1 and the other a 0, and a 0 if both are either 0 or 1, and so $0 \oplus 0 = 0$, $0 \oplus 1 = 1$, $1 \oplus 0 = 1$, and $1 \oplus 1 = 0$.[39] The commutative, associative, and distributive laws of regular arithmetic also apply to Boolean arithmetic.

Though the elements of Boolean logic are not identical to those of arithmetic, they can be combined to perform ordinary arithmetic. This is what is at the heart of computer arithmetic. In binary addition $0 + 0 = 0$, $1 + 0 = 1$, $0 + 1 = 1$, and $1 + 1 = 10$; this can also be represented as $0 + 0 = 00$, $1 + 0 = 01$, $0 + 1 = 01$, and $1 + 1 = 10$, where the first number in the result can be called the *carry* and the second can be called the *sum*. The *half adder* in Figure 7.1 has 2 binary digit inputs (0 or 1) that enter an XOR gate to give the sum and an AND gate to give the carry. A *full adder* combines two half adders to allow these 2 inputs plus a carry from another circuit.[40,cxxiii,cxxiv]

Boolean algebra is closely related to the widely used Venn diagrams that illustrate how different sets of objects differ, by highlighting their common and differing members. In Figure 7.2 the elements of set A (such as those people with brown eyes) are those inside the circle and are

---

[39] $A \oplus B$ is formally equivalent to (NOT $A \times B$) + ($A \times$ NOT B) or $(\neg A \wedge B) \vee (A \wedge \neg B)$.

[40] In a more realistic system, for example, two two-digit binary numbers (each 0 to 3 in base 10) are multiplied in a "2-bit binary multiplier" to obtain an output with up to four binary digits (0 to 9 in base 10). The 4 inputs are connected to 4 logical AND circuits through a series of interconnections and then their outputs go into 2 other AND circuits and 2 XOR circuits, leading to the 4 outputs. (The reference is in the text.)

a subset of all objects (all space within the rectangle, such as all people) (Figure 7.2a). The corresponding diagram illustrates all objects that are not in A, called or A', which is the logical equivalent of NOT A (Figure 7.2b). The union of sets of A and B, or $A \cup B$, includes elements in either A or B, which includes those only in A, only in B, and those in both A and B (Figure 7.2c). It is equivalent to the logical OR, $A \vee B$. The intersection of sets of A and B, or $A \cap B$, includes only those elements in both A and B (Figure 7.2d). It is equivalent to the logical AND, $A \wedge B$. The analog of the logical XOR, $A \oplus B$, are those in A or B, but not in both (which is not shown).

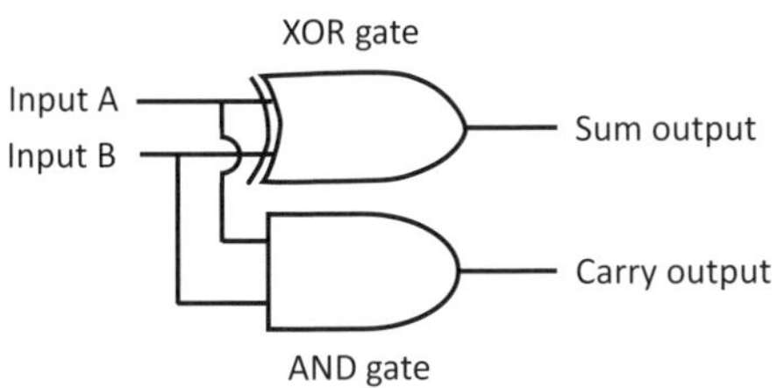

Truth Table

Inputs		Outputs	
A	*B*	*Sum*	*Carry*
0	0	0	0
0	1	1	0
1	0	1	0
1	1	0	1

Figure 7.1: Half adder, with the "Truth Table" that connects the inputs to the outputs.

Nineteenth century English mathematicians Charles Babbage and Ada Lovelace are credited with developing mechanical computers and computer programming. In 1936, English mathematician Alan Turing developed the math of computation and programming (algorithms) to execute instructions by a computer using a stored program written for a specific type of computation. He is generally considered the founder of computer science.[cxxv] During WWII he built a Universal Computing machine now known as a Universal Turing machine, and used it for cryptography. It was a forerunner of modern general-purpose computers. In 1945, Hungarian-American mathematician and physicist John von Neumann built on these concepts to develop the basic architecture of modern computers, with separate central processing units and memories.

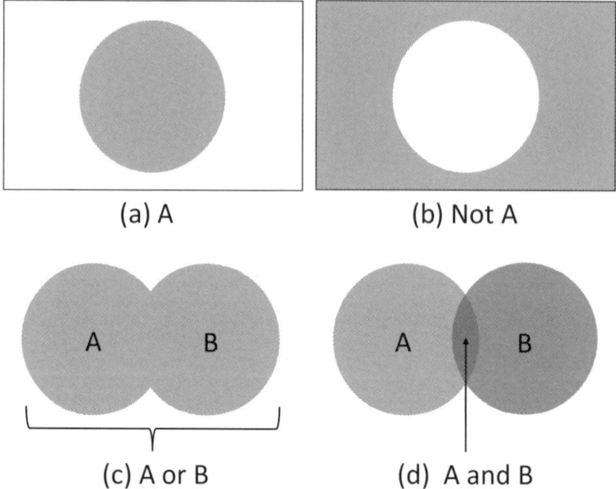

Figure 7.2: Venn diagrams, with the results shaded in (a), (b), and (c), and the noted, darker shaded region in (d).

These logical operations are certain in the sense that entails states that may be 0 or 1, or true or false. In Chapter 14 we will see how the uncertainties that may be inherent in other areas of logic characterization and analysis, are handled by using *fuzzy logic*.

7.2 Other Bases for Logs

Sometimes it is useful to express numbers in base 10, but express the logarithms in other bases, such as base 2 (as $\log_2 x$), 3 (as $\log_3 x$), etc. So $\log_2 8$ is 3 because $2^3 = 8$, with a base of 2 instead of 10. When no base is explicitly given, assume a log to be in base 10.[cxxvi]

For some excellent math reasons, there is also another very important base, which is the irrational and transcendental number *e*. (See Chapter 9.) It is called *Napier's* or *Euler's number*, and equals 2.71828182845904523536028747135266249…, and so is very roughly equal to 19/7. It equals the sum of the infinite series (one that continues forever) $1 + 1/1 + 1/(2 \times 1) + 1/(3 \times 2 \times 1) + 1/(4 \times 3 \times 2 \times 1) + \ldots$, which is $1 + 1/1! + 1/2! + 1/3! + 1/4! + \ldots$. This number comes about quite naturally in calculus. It equals the sum of 1 and the reciprocal of a positive number, which is then raised to the power of that number, as that number is increased more and more (in the limit as it approaches infinity).[41] Starting with the number 2, this expression would be $(1+1/2)^2 = 2.25$; increasing it to 3 gives $(1+1/3)^3 \approx 2.3704$; to 10 gives $(1+1/10)^{10} \approx 2.593$, to 100 gives $(1 + 1/100)^{100} \approx 2.7048$, to 1000 gives $(1 + 1/1000)^{1000} \approx 2.7169$, and it approaches the value of e given for larger and larger numbers. Also quite common is the function e raised to the power x, which is written as e^x (or $\exp(x)$). It is the sum of the infinite series $1 + x/1 + x^2/2 + x^3/6 + x^4/24 + \ldots$, which comes from $1 + x/1 + x^2/(2 \times 1) + x^3/(3 \times 2 \times 1) + x^4/(4 \times 3 \times 2 \times 1) + \ldots$ or $1 + x/1! + x^2/2! + x^3/3! + x^4/4! + \ldots$.[42,cxxvii]

e^x is very special because it (along with its multiples) is the only function that changes at a rate equal to its value, for any value of the exponent.[43] Commonly the function e^{ax} appears in studies, where a is a constant and x is a variable, which may indicate distance or time; this comes from e^x (and the sum it equals) after replacing each x by ax. This e^{ax} variation is called *exponential growth* when a is positive (Figure 7.3a), and is used to model population growth, epidemics, and interest that is compounded frequently (for which x or t represents time), and *exponential decay* when a is negative (Figure 7.3b), which is used to model radioactive decay. Such decay also models the exponential approach to a constant, such as with $1 - e^{-ax}$ vs. x (Figure 7.3c).

The log in base e, $\log_e$, is called the *natural logarithm* and has a special symbol "ln" ("l" "n"). It is colloquially called the "natural log" or just log (and then pronounced "log") when the context is clear. Just as $\log_{10} 10 = 1$ and $\log_{10} 10^x = x$, $\ln e = 1$ and $\ln e^x = x$.[44] It is often necessary to

---

[41] This is $(1 + 1/N)^N$ in the limit of N increasing to infinity.

[42] More formally, $e^x = (1 + x/N)^N$ in the limit of N increasing to infinity.

[43] In other words, e^x and its multiples are the only functions that when plotted versus the exponent x has a slope (which is how fast the vertical coordinate changes relative to the horizontal coordinate) that is equal to its value. In the language of calculus, it is the only function that is equal to its first derivative. In fact, this description specifies this function, with $de^x/dx = e^x$. Also, $d(Ae^x)/dx = Ae^x$ for constant A and $de^{ax}/dx = ae^{ax}$ for constant a.

[44] Also, it is interesting and important that the natural logarithms of numbers near 1 can be expressed in terms of how much the numbers differ from 1, say by x, with $\ln(1 + x) = x - 1/2\, x^2 + 1/3\, x^3 - 1/4\, x^4 + \ldots$ for x between -1 and 1, but not equal to either. Remember that ln 1 is 0.

convert logs in base e to those in base 10 and vice versa. Because $\log e \approx 0.4343$, raising 10 to this power gives $e \approx 10^{0.4313}$.[45]

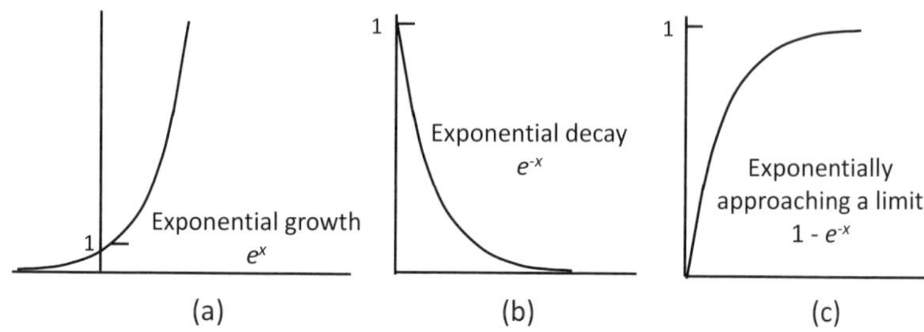

Figure 7.3: Uses of the exponential function, e^x (where x could mean rate x time (Rt)).

Plots with log scales other than base 10 can be very useful as well, in base e (ln-linear and ln-ln) and others. When you plot the ln of the exponential function e^{ax} you see the straight line ax vs. x, with slope a, because $\ln e^{ax} = ax$ (as in the log-linear plot in Figure 6.2a).

There is a curious, yet often very important and useful confluence of this natural log and the factorial (!), Stirling's formula, that for large N, $\ln N! \sim N \ln N - N$, from which you can also get $N!$.

7.3 Coding Information

In the middle of the 20th century, American mathematician and electrical engineer Claude Shannon conceived and developed the field of information and information transfer in communications.[cxxviii] Its key concepts are that communications can be phrased in terms of information that can be measured in bits, which in turn are digital signals, such as on

---

[45] We can estimate $\log_{10} e$ by first approximating e as 2.7. Because $2.7 = 27/10 = 3^3/10$, $\log 2.7 = 3 \log 3 - \log 10 \approx 3 \times 0.4771 - 1 = 0.4313$, and so $\log e \approx 0.4313$ and $10^{0.4313}$ should approximately equal to e. Now raising both sides of $e \approx 10^{0.4313}$ to the power $1/0.434 \approx 2.303$ (with $10^{0.4313 \times 2.303} \approx 10^1 = 10$ and exchanging the left and right sides), gives $10 \approx e^{2.303}$. (This also means that ln 10 is approximately equal to 2.303.) So, $e \approx 10^{0.4313}$ and $10 \approx e^{2.303}$, and $e^x \approx 10^{0.4313x}$ and $10^x \approx e^{2.303x}$. Therefore, that logarithms in base 10 are converted to those in base e by multiplying them by ≈ 2.303 and those in base e are converted to the corresponding ones in base 10 by dividing them by ≈ 2.303.

or off, or heads or tails;[cxxix] this uses binary, base 2 math. The tour-de-force advances of this information theory pioneer have been likened to the feats of Einstein.[cxxx] A related key concept in this field is that to convert analog (non-digital) communications signals into digital signals you need to measure (or sample) the analog signal at a rate fast enough that at least 2 points are sampled in the fastest oscillation cycle of that signal. This is called the Nyquist-Shannon sampling theorem.[cxxxi] (The below Figure 8.1c suggests why this is reasonable.)

The amount of information conveyed in communications, such as in the transmission of a JPEG file, is given by the Shannon information entropy. This is the negative of the product of the probability of a given value of a variable being transmitted times the log of this probability, summed over all possible values of the variable.[46,cxxxii] Because probabilities are always between 0 and 1, the logs are negative and this additional negative sign cancels the negative sign making this entropy a positive quantity (or 0 when the probability is 1, because log 1 = 0). Messages are often compressed digitally to promote faster transmission for a given transmission rate. This Shannon entropy is the lower bound of the information that must remain in a compressed message so it does not lose information content.[cxxxiii] You use this principle frequently, without realizing it, when you use your smartphone.

The units of Shannon entropy depend on the base of the log used to evaluate it, bits or shannons in base 2, nats in base *e*, trits in base 3, and hartleys in base 10. The value of the final entropy is usually given in base 10 for each.[47] For an unbiased coin, the probability of heads and tails are both $1/2 = 2^{-1}$, so in base 2 the log is just the exponent, which is -1 for both possibilities. The Shannon entropy is then $-(1/2)(-1) - (1/2)(-1) = 1/2 + 1/2 = 1$ bit. This is what is commonly known as 1 bit of information, with it being equally probable that the bit is 0 or 1 For a

---

[46] The Shannon information entropy is the negative of the sum over of $p(x_i) \log p(x_i)$, or $-p(x_1) \log p(x_1) - p(x_2) \log p(x_2) - p(x_3) \log p(x_3) + \ldots$, where $p(x)$ is the probability and the log could be in base 2, *e*, 3, or 10.

[47] One bit is approximately equal to 0.693 nats because $e^{0.693} = 2$, 0.631 trits because $3^{0.631} = 2$, and 0.301 hartleys because $10^{0.301} \approx 2$. Just as two equal probabilities of 1/2 gives 1 bit, three equal probabilities of 1/3 gives 1 trit, and ten equal probabilities of 1/10 gives 1 hartley.

heavily weighted coin for which heads has a probability of 0.9 and tails 0.1, using $0.9 \approx 10^{-0.0458} \approx 2^{-0.0458/0.3010} \approx 2^{-0.1520}$ and $0.1 = 10^{-1} \approx 2^{1/0.3010} \approx 2^{-3.3223}$, the entropy is $-(0.9)(-0.1520) - (0.1)(-3.3223) = 0.1368 + 0.3322 = 0.469$ bits, which is smaller than 1. It becomes even smaller for heads with probability 0.99 and tails 0.01, being 0.0808 bits. So, there is less and less information when the coin (or die) is more and more biased, and it approaches 0 bits as the probability of heads (or tails) approaches 1.0. When there is equal probability for each possibility there is maximum information. For an unbiased die, the probability of each face is $1/6 \approx 10^{-0.7782} \approx 2^{-0.7782/0.3010} \approx 2^{-2.5852}$ so the log in base 2 is -2.5852 and summing over the six faces, the information entropy is $-6(1/6)(-2.5852) = 2.5852$ bits. A biased die conveys less information than this.

Your DNA is coded with your genetic information. Single strands of DNA and RNA consist of a series of amino acid bases. There are four types of base pairs, each which can attach to one of four specific bases to make a base pair, so the pair has the same information entropy as the first base itself. The information contained in each base on one strand is 2 bits if they are equally likely, because there are four base pairs, each with probability $1/4 = 2^{-2}$, and $-(4)(1/4)(-2) = 2$. (The information entropy is a bit smaller than this because the sequence of bases is not entirely random.) A series of three bases is a codon, which corresponds to one amino acid (a component of a protein). If all three base sequences were possible and equally likely, there would be 6 bits per codon or $4^3 = 64$ possible triplets, which is enough to encode the 20 different amino acids. [48] Two base sequences would give only $4^2 = 16$ possibilities and so would not be sufficient.

Language also codes information. Most of written language consists of a series of 27 characters, the 26 letters in the alphabet and the spaces separating words. If each character were used with equal probability, 1/27, you would need to know log(1/27) in base 2, which is $\approx$-4.7554

---

[48] 61 of the 64 triplets correspond to the 20 different amino acids (most amino acids can be represented by more than one triplet) and three triplets correspond to a "stop" in reading the code, so there are 21 outcomes in reading a codon. (If all were equally likely, this would correspond to -(1/21)21(-4.3927) or $\approx$4.4 bits per codon, because $1/21 \approx 10^{-1.3222} \approx 2^{-1.3222/0.3010} \approx 2^{-4.3927}$.)

because $1/27 \approx 10^{-1.4314} \approx 2^{-1.4314/0.3010} \approx 2^{-4.7554}$. The information entropy would be -(1/27)(-4.7554) bits summed over all 27 characters, and so there would be ≈4.76 bits per character. But not all characters are used with equal probability in the English language, with spaces being used with probability 0.2, E with 0.105, T with 0.072, and so on to J, Q, and Z with 0.001. After including this "bias" (or tendency), the entropy is only 4.03 bits per character; however, little of this is the *useful* information conveyed by words and sentences. There are specific correlations in letter patterns in words of a given language, such as how often groups of two and three character patterns are used, and this reduces information to ~3 bits per character. When larger groups of characters and real-life writing patterns are examined, this decreases it to ~2 bits per character.[cxxxiv]

7.3.1 Data compression and checking

All of this information can be transmitted or stored, but you may want to purposely lose some due to *compression*. In fact, data compression and checking are at the heart of communications math. Data can be compressed to remove repetitive information, which is lossless compression, and/or to remove what is deemed to be less important information, which is lossy compression. For example, it would be simpler and lossless, to denote a blank page in a pdf file with a specific code rather than coding a page of repeating blank spaces. Also, the letter u that follows q in most words, as in "quiet," is not needed.[cxxxv] Deteriorating the quality of the symbols, letters, diagrams, and the like on the pdf page, would increase transmission speed and decrease storage needs, but at the expense of quality, and so there is a tradeoff and need for optimization (Chapter 18).[cxxxvi]

There are errors in any data transmission and storage, so data checking and correction are essential. Whereas compression lessens the amount of signal data that needs to be communicated, data need to be added to the communicated signal to minimize errors in transmission and analysis. Say a data or communications line can transmit 1,000,000 bits a second, each a binary 1 or 0. Most but not all of this is the Shannon information signal or a compressed signal, perhaps 900,000 bits a second, while perhaps the rest, 100,000 bits per second, is code needed to minimize or eliminate errors. One way to detect and correct errors is by transmitting repetition codes and using the majority detected signals. The four-bit block "1101" can be repeated three times. The received code 110111011101 would be considered correct because the same pattern is repeated three times. 100111011101, with one error, would be considered incorrect because the patterns 1001, 1101 and 1101 are not the same, but 1101 would be treated as being correct because it appears twice. Alternatively, each bit can be repeated three times in succession, so the correct code for 1101 would be 111111000111. The signal 101111010101 would have three errors, but would still be deciphered correctly to be 1101, because the first three bits 101 would suggest a 1 (with one error, the 0), 111 a 1 (with no errors), 010 a 0 (with one error), and 101 a 1 (with one error). These methods are too inefficient because only a third of the transmission speed is effectively used.

This is improved by adding a *parity bit* to the stream. One way to do this is to add after seven bits a 1 if there are an even number of 1s in the first seven and a 0 if there are an odd number. So, 1001110, with an even number of 1s, would be transmitted as 10011101. If there are an odd number of 1s in the first seven, the parity bit is 0, so there still would be an odd number of 1s in the transmitted code in this "odd parity bit" version. (Alternatively, the parity bit could be added so there would be an even number of 1s, in the even parity bit version.) This method can detect 1, 3, 5 or 7 errors but not 2, 4, 6 of them, but the hope would be there would be at most 1 error in this short stream. The more complex, and useful, *Hamming codes* can detect one- and two-bit errors and also *correct* one-bit errors, by adding specific sets of 3 parity bits to 4 data bits to give a 7 bit string that is transmitted, 4 parity bits to 11 data bits for 15 total bits, 5 parity bits to 26 data bits for 31 total bits, and so on.[49] In these three codes, called Hamming(7,4), Hamming(15,11) and Hamming(31,26) codes respectively, the fraction of total bits that are useful data increases the longer the bit pattern, from $4/7 \approx 57.1\%$, to $11/15 \approx 73.3\%$, to $26/31 \approx 83.9\%$.[cxxxvii] There are also "checksum"

[49] With m parity bits, there are $2^m - m - 1$ data bits, for a total of $2^m - 1$ bits that are transmitted or stored for the Hamming $(2^m - 1, 2^m - m - 1)$ code. It has a useful data rate fraction of $(2^m - m - 1)/(2^m - 1)$.

procedures to check for errors in transmission and storage, similar to those described in Chapter 12 for hashing functions, as used for confirming book identification ISBN numbers.[cxxxviii,cxxxix,cxl] The Reed-Solomon correction codes use fits to polynomials to correct errors in mass storage systems, CDs/DVDs, and bar codes, and for data transmission, and were used for transmitting digital pictures from the Voyager space probe.[cxli]

Using the same single processing "code" to minimize error throughout a network could be inefficient, and too poor in some places (giving an error rate that is too high) and too good (and wasteful in a processing sense) in others. Efficient procedures to reduce error have been developed, such as those advanced by the 2016 IEEE (Institute of Electrical and Electronics Engineers) Medal of Honor recipient, David Forney. He developed the sequential concatenation method, in which a moderate error rate in the fairly complex code at the point of reception and transmission that is generated by the "inner code," is reduced by using a less complex "outer code" both before and after that point. (This is an example of optimization, which we will visit later.) He also advanced the graphical Viterbi algorithm, which can be used to eliminate mistakes by using "prior knowledge," such as changing a signal detected incorrectly as "hqnd" to the correct signal "hand." (Prior knowledge in also used probability analysis, as in Chapter 15.) We use such processing algorithms—and take them for granted—in our frequent communications with smartphones, internet activities, and so on.

We have been assuming that the numbers and math that rule much of your life are exact, but they need not be.

Chapter 8

Numbers Need to be Exact, But It Ain't Necessarily So

Knowing when you need to be exact and when it is just fine to obtain and use approximate numbers is key in thinking and using math.

8.1 Measurements and Significant Figures

Say your thermometer gave you a reading of 68.67264739112 °F (degrees Fahrenheit). Of course, thermometers do not give so many figures, but if they did would this reading be useful? Would it be correct? There are different ways to answer this question, and each is important.

Is the value *precise*? This means if you measured it over and over again under identical conditions would you get exactly the same value? Probably not. If successive measurements were 68.65274739112 °F, 68.7245293377 °F, 68.7589273518 °F, 68.6254887294 °F, ..., you would suspect that the best thing you could say that the temperature is about 68.7 °F.

Is the value *accurate*? If so, it would be absolutely correct, independent of the precision. If the calibration were off, it could be that the readings are all too high by 0.2 °F. Sometimes this absolute error is not known and the accuracy needs to be estimated.

With more measurements (called samples in data analysis, as in Chapter 16), the average value of the temperature would be estimated better. Data uncertainties include the statistical standard deviation in the data (Chapter 15) and non-statistical sources of error. If you are told that the average temperature in a given location is colder by 0.3 °F in year 2 than in year 1, you can be confident that it was truly colder only if the carefully-analyzed uncertainty each year was much smaller than this difference. If you just cared about what to wear or whether or not to turn the heat up or the air conditioner on, all you care about is that it is ~69 °F.

In any case, a temperature reading of 68.67264739112 °F is nonsense. If you know that the precision is actually only good to one decimal place, then the temperature should be reported as 68.7 °F after *rounding off*. This has three *significant figures*, because there are three numbers. The temperature 103.6 °F has the same precision, but technically four significant figures, so the numbers of reported significant figures and decimal places and the measurement precision are interrelated.

8.1.1 Rounding off

Any reading can be *rounded* off to a certain number of digits. In rounding off, it is common to look at the next to last digit. If it is followed by a digit from 0 to 4, it is often rounded down and so stays the same. If it is followed by a digit from 5 to 9, it is often rounded up and is increased by one. So, rounding the 6-digit number 68.6527 to one fewer digit gives 68.653 (which would be thousandths here), and then to one fewer digit gives 68.65 (to hundredths), and then to 68.7 (to tenths) (which may be fine for official readings), and then to 69 (to ones) (as in weather reports), and finally to one digit to 70 (to tens) (which is not advised). In calculations involving money, values are usually rounded off to the nearest cent, which could lead to a significant error. This method of rounding actually skews (or tilts) numbers higher on the average. To avoid that, when the last digit is 5, the next to last digit is rounded up when it is odd and down when it is even, which is known as bankers' rounding, because bankers often use it. So, 3.5 rounds up to 4, 4.5 rounds down to 4, and 5.5 rounds up to 6 and so on, which explains why this method is also known as the round-to-even rule. This method moves random numbers higher as much as it does down, so there is no skewing.

Zeros in numbers may or may not be significant. The zero at the beginning of 0308.73000 is not significant, so it should be written as 308.73000. The zero between the 3 and the 8 is definitely significant. The three zeroes at the end should be indicating precision to five decimal places, the five numbers after the decimal point, so the number could not be 308.72999 or 308.73001. If the precision is only to three decimal places, then the number should be presented as 308.730, with six significant figures. Only if the integer 28,000 were not rounded off would it have five significant figures and the same meaning as 28,000.0. When adding or subtracting numbers, the results should be rounded off to the fewest decimal places of the initial numbers, so 6.42081 + 12.37 = 18.79081 should be reported as 18.79. When multiplying or dividing numbers, the results should be rounded to the fewest significant figures of the initial numbers, so 6.3718 × 4.6 = 29.3103 should be reported as 29.3. However, this is not always done.

Sometimes numbers that seem to be rounded-off may just be plain wrong, particularly when integers are involved. Say you ask 28 people to rank an opinion with an integer from 1 (strongly disagree with it) to 7 (strongly agree with it), which is a known as a Likert-type scale. You add the scores to get a sum that is naturally an integer, and then divide it by 28 to get the mean, which you are told was rounded off to 5.19. Could this mean be correct or is it incorrect for sure? Working backward, 28 × 5.19 = 145.32, which means the original total might be 145 or 146. If it were 145, the average would be (to two decimal places), 145/28 ≈ 5.1786, which rounds to 5.18. If it were 146, the mean would be 146/28 ≈ 5.2143, which rounds to 5.21. Since both differ from the reported mean of 5.19, the presented results must be wrong. This relatively new test by Nicholas Brown and James Heathers is called the GRIM test. GRIM stands for Granularity-Related Inconsistency of Means, which just means that using integral or other "granular" number scale imposes constraints on the possible answers.[cxlii] Working backward using a related approach, a stated mean of 5.19 means it could range from 5.185 to 5.195. The total score could then range from 28 × 5.185 = 145.18 to 28 × 5.195 = 145.46, and so the given information is incorrect, because there is no integer in this range. We will learn other ways to detect mistakes in our modular arithmetic discussions of casting out nines and ISBN numbers in Chapter 12.

Sometimes rounding off can be very advantageous. In his biography of Mikhail Gorbachev, William Taubman conveyed that in the late 1980s the Russian Soviet Federative Socialist Republic (Russia) was the largest and dominant of the 15 republics in the Soviet Union "accounting for half of the union's population, two-thirds of its economy and three-quarters of its territory."[cxliii] In this instance, presenting rounded-off fractions confers more meaning and insight than more accurate percentages would.

8.1.2 Significant figures are significant in making calendars

Seasons reoccur every 365.242199 or so days, which defines a year. Because this is very close to 365¼ days, solar calendars with 365 days for three years and 366 days in the fourth, leap, year work pretty well. The period of the phases of the moon is 29.530588 days, which defines a lunar month, and so there are ≅ 365.242199/29.530588 ≅ 12.36826707 lunar months a year. Given the many significant figures in the number of lunar months per year, months need to be constructed cleverly to make calendars work. Our Gregorian calendar is constructed of 12 months of length 28 or 29, 30, and 31 days, and so is not strictly a lunar-based calendar. In 19 years, there are ≅ 12.36826707 × 19 ≅ 234.99707 lunar months, so calendars based on the lunar month, such as the Jewish calendar, have cycles of 235 months that repeat every 19 years. This Metonic cycle, discovered by Athenian astronomer Meton, is also the basis of determining the date of Easter. The Jewish calendar consists of 12 months of 29 or 30 days, with an extra month of 30 days added every leap year. There are 7 leap years (years 3, 6, 8, 11, 14, 17, and 19) in this 19-year cycle.[50,cxliv] No cycle with fewer than 235 months has a smaller error than this error, which is (235 × 29.530588) – (19 × 365.242199) ≅ +0.086399 days per cycle,[51] so this system is pretty good, except that the years are quite different from each other. The next longest cycle with an even smaller error is a cycle is 4,131 months in 334 years, with error ≅ -0.035438 days.[cxlv]

Using lunar calendars is more complex than adding a leap day to (almost) every four years. We will revisit making ordinary (solar) calendars work well when we discuss modular arithmetic.

---

[50] It gets a bit more complicated than this: The length of one month goes from 30 to 29 days in a leap year, and the lengths of two other months (either 29 or 30 days) are affected by other factors, but it all sums up correctly.

[51] This is the difference in the number of days in 235 lunar months and in 19 years.

8.1.3 Significant figures are significant in computers

The number of significant figures affects the accuracy of a calculation, by computers and otherwise. Such numbers are represented as *floating-point numbers*, and the same signed number could be represented in different ways, depending on where the decimal point is placed.[cxlvi] So, -1.234×10^5, which is in scientific notation format and base 10, could be written as that or as -12.34×10^4, -123.4×10^3, and so on. The more the digits in the significand or mantissa digit string, the higher the precision of the computation and the smaller the error. This is important when differences of nearly equal numbers are involved and in other cases as well. On February 25, 1991, during the Gulf War, Patriot missiles failed to track and intercept incoming missiles because timing calculations were performed with too few digits.[cxlvii] Such errors would not occur with higher precision calculations, as with the now frequent *double precision* calculations, in which numbers are represented by 64 (base 2) computer bits.[52,cxlviii]

8.2 Estimating

"Estimating" denotes several modes of "math thinking." For example, "getting an estimate" means to obtain a rough calculation of costs for, say, for a construction job. Estimating the number of socks in New York City is a quite different endeavor. We know there are roughly 8 million people who live in New York. Each person may own 15 pairs of socks—which is a wild guess—so each would have 30 socks, and so maybe there are very roughly 240 million socks in New York City, excluding those in stores and warehouses and those being worn by those who are vacationing or working in the City, but who do not live there. Perhaps the best we could say is there are estimated to be hundreds of millions of socks in New York City. Rough rounding off of numbers can provide an

---

[52] The magnitude of the significand is described by 52 of these bits, its sign by 1 bit, and the exponent by 11 bits (which includes its sign). This means that numbers with magnitudes between 10^{-308} and 10^{308} can be represented with 15 to 17 decimal digits. This makes sense because $2^{10} = 1024$ (which is larger than the noted range of $617 = 308 + 1 + 308$) and $2^{52} \approx 4.5 \times 10^{15}$ (which gives at least the noted 15 digits).

estimate of an answer to assess if the result and arithmetic make sense. 2,874 × 4,166,834 can be crudely estimated as 3,000 × 4,000,000 = 12,000,000,000.

8.2.1 Order of magnitude estimates

Rough estimates, within, say, a factor of ten or so, are order of magnitude estimates.

The order of magnitude sizes of the 50 mg and 50 Mg (*sic*) tablets we spoke of in Section 5.4.1 can be estimated by dividing these masses by the density of the ingredient, which is on the order of 1 gram per cubic centimeter. Their respective volumes would be on the order of 0.05 and 5×10^7 cubic centimeters. If we approximate the tablets as cubes, their volumes equal the cube of their side lengths, which would be the cube roots of these values, or roughly the reasonable 0.37 cm or about 1/7 inch and the unreasonably large 370 cm, 145 inches or 12 feet.

You can determine the crowd size at an event by directly counting people, perhaps by using aerial imaging. If this is not feasible, you can estimate it by multiplying the area of the crowd by the estimated crowd density. As part of this *Jacob's Method*, you can assume 1 person per 10 square feet for a light crowd, 1 per 4.5 per square feet for a dense crowd, and 1 per 2.5 square feet a very dense crowd, as in a mosh pit.[cxlix] So, assuming a hopefully reliable estimate of 100,000 square feet for the crowd area, the estimated crowd size could range from 100,000/10, to 100,000/2.5 or from 10,000 to 40,000 people. The actual estimate you announce might be biased by whether or not you want to announce a larger or smaller crowd. Some think the crowd sizes at the Times Square New Year's Eve celebrations in New York City are overestimated.[cl]

8.2.2 Order of approximation

More rigorous mathematical thinking can also be used to get successively better approximations in estimating. The lowest order of a calculation, a crude estimate, may use rough values as input variables or a crude model. Using more accurate inputs or a more refined model would give the next order approximation, and then the next one would be the next higher-order approximation, and so on. These successive answers can be called the zeroth-order approximation or an answer that is good to 0th order, and then a first-order approximation or an answer that is good to 1st order, and then a second-order approximation or an answer that is good to 2nd order, and so on. However, there is no universality in this nomenclature. Some refer to such respective approximate solutions as being 1st, 2nd, and 3rd order, and so on. This is yet another example about how words and concepts associated with numbers need to be clear, as we discussed in Chapter 5.

Approximate solutions might entail a sum of terms, with successive terms having smaller and smaller magnitudes. Consider the sum of a variable, its square, its cube and so on ($x + x^2 + x^3 + x^4 + \ldots$) for a variable between 0 and 1 (but not 1). For $x = 0.1$, the sum would be 0.1 to 0th order (using the first definition of order), 0.11 (= 0.1 + 0.01) to 1st order, 0.111 (= 0.1 + 0.01 + 0.001) to 2nd order, and so on.[53] The answer up to and including the x^2 term can be said to have an error of order of magnitude of the cubic term (or $O(x^3)$). The successive corrections are successively much smaller, so the result converges close enough to the limiting answer for most needs after just a few terms. In this particular case, the algebraic sum converges to a function (Chapter 13).

8.2.3 Estimating with little prior knowledge

Part of being able to "think mathematically" is having a good handle on the approximate numerical value of things (the number of people in the country, the size of the government budget, and so on), being able to assess whether an estimate you hear makes sense, and being able to estimate a given value even with little information. Many cannot do this.[cli] Over half of the college students who were questioned in class in and before 2012 about the average number of people in a congressional district said it was fewer than 100,000. The U.S. population was roughly 300 million people (really 314 million in 2012). There are 435 members of the House of Representatives, so there were ~300,000,000/435 or roughly 700,000 people per district.[clii]

In Chapter 15 we will see how to estimate probabilities when there is little prior information, particularly when the probabilities are thought to

[53] In Chapter 13, we will this that the sum of this *infinite geometric progression* approaches $x/(1-x)$, which here equals $0.1/(1-0.1) = 0.1/0.9 = 1/9 = 0.11111\ldots$.

be very small. We now address making estimates by using prior, though incomplete, information.

8.2.4 Estimating inside and outside a range: interpolation vs. extrapolation

There is an equation that relates the current world record running speeds for distances from 400 m (400 meters) to 10,000 m that is based on the best times at the standard running distances, 400 m, 800 m, 1500 m, 5,000 m, 10,000 m, 42,195 m (the marathon), and 100,000 m.[54,cliii] Would it be reasonable to use this formula to *interpolate* what the world record times would be for other distances within this range, say for 3,000 m? Yes. Should you use this same formula to *extrapolate* outside this range, say to 100 m? Maybe not. Extrapolation is generally a much shakier endeavor. In this example, it would be definitely wrong to do so for distances shorter than 400 m. Within the given range, the average speed decreases with increasing distance because of metabolic and respiratory factors. For shorter distances, the formula would predict faster running speeds, but would not include the acceleration time at the beginning of the race, which becomes very important for these shorter distances. Of course, you can modify this formula so it would work in both regimes.

You may know the value of a function (y) at n specific values of the variable (x). Drawing straight lines from a given point to the next point and finding the interpolated values on those lines is *piecewise linear interpolation* (Figure 8.1a). Drawing the best straight line through all these points to find y for an x within the measured range of x, *linear regression*, also often works well within a region (Figure 8.1b). (See Section 16.2.) It is exact only if the dependence of all points is linear. More generally, a function given at n points can be expressed exactly for those points only, with a polynomial of order $n-1$, which means it includes terms (a constant times) x^{n-1}, (a different constant times) x^{n-2}, down to x^2, x, and a constant term.[55] You can sometimes, but not always, reliably estimate the value of y for x between the smallest and largest values of x by using a fit to polynomial of order $n-1$ or smaller.

---

[54] The speed is $v = k/d^n$ for distance d, and positive parameters k and n.

[55] For $n = 6$ terms, this could be $y = ax^5 + bx^4 + cx^3 + dx^2 + ex + f$.

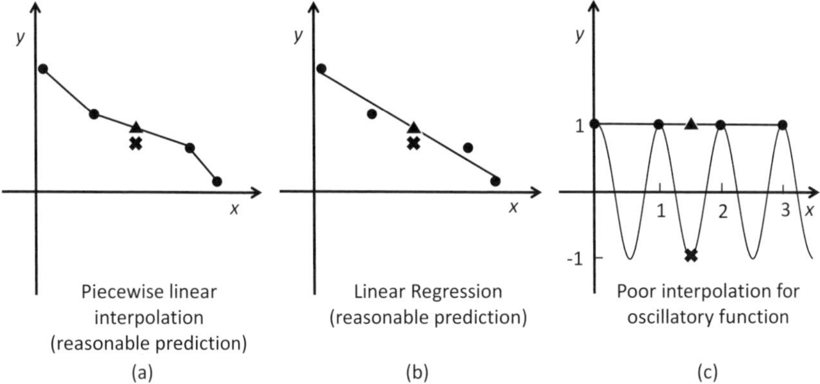

Figure 8.1: Comparisons of a data point (x) with the predictions of interpolations (triangles) using four data points (dots).

However, at times you can make big mistakes using these approaches. Consider a function with value 1.0 at x = 0.0, 1.0, 2.0, and 3.0. This is a fit with either piecewise linear interpolation or linear regression to the function $y = 1.0$ within this region, which is independent of x in this range (Figure 8.1c). Using this linear fit for interpolation would be a disaster if the function were really the oscillating, *cosine* function, which has values -1.0 at x = 0.5, 1.5, 2.5, and so on (Figure 8.1c).

The dangers of interpolation and extrapolation are also illustrated in a puzzle posed in 1968 in a booklet of math puzzles: "A pure mathematician employed in an aircraft plant uses the fairly complex formula $(480 - 196N + 19N^2 - N^3)/(120 - 34N)$ as a mnemonic. For what?". The answer was the number of engines in a Boeing 7N7 jet.[cliv] At the time this formula worked for the then existing Boeing 707, 727, 737, and 747, which had 4, 3, 2, and 4 engines respectively, but it was no longer correct for the subsequent Boeing jets 717, 757, 767 and 777, with 2 engines each.

8.2.5 Scaling

When properties scale with "size," scaling becomes a rational basis for estimating for mathematical, scientific, engineering, business, or societal insight.[clv] Sometimes you may expect a property to depend on an important factor linearly, but it may be slower than linear (*sublinear*) or faster than linear (*superlinear*).

Why are the legs of ants skinnier than those of elephants? Mass is proportional to volume, which varies as the characteristic linear dimensions (such as height or width) raised to the power 3, and so you would expect that the dimensions of animals, such as the lengths of their bones would vary as their mass to the power 1/3. The ability of bones and muscles of an animal to support a weight varies as their cross-sectional areas or widths to the power 2, and so the widths of bones need to vary as the mass to the power 1/2. So the ratio of the widths of bones to their lengths is expected to vary as the animal mass to the power $1/2 - 1/3 = 1/6$. So, smaller animals (such as ants) need to have "relatively" thin legs and big animals (such as elephants) have relatively thick legs. The weights of elephants is 6,000 to 12,000 pounds, corresponding to 3,000 to 6,000 kilograms or 3×10^6 to 6×10^6 grams masses, while that of ants is ~0.003 grams = 3×10^{-6} kilograms, so elephants are roughly $10^{12}\times$ more massive than ants and the width to length ratio of their legs would be expected to be roughly $(10^{12})^{1/6}$, or $10^2 = 100$ times larger.

The energy usage of animals at rest is known as their basal metabolic rate (BMR). More massive animals would be expected to consume energy faster than smaller animals. Kleiber's law says that the BMRs of mammals increase as their mass to the power 3/4, which is sublinear. A mammal with 16 times the mass of another is expected to have a BMR that is ~8 times larger. (Because 16 is 2^4, $16^{3/4}$ is $2^{(4)(3/4)} = 2^3 = 8$.) This "law" appears to hold for mammals from mice to elephants, and perhaps to whales.[clvi]

In Chapter 13 we will explore how computation needs scale with the size of the problem.

8.3 Random Numbers

Use of *random numbers* is entrenched in mathematical modeling, as with Monte Carlo simulations (Chapter 15), modern encryption (Chapter 12), and in lotteries. Such lotteries are used to select jurors from available pools candidates for public housing, in public games of chance such as lotteries and Bingo games, and in the drafting of soldiers in times of war, as during WWI, WWII, and the Vietnam War in the U.S.[clvii]

Random numbers can be generated by physical means, such as coin flipping and dice tossing, and from random noise fluctuations in measurements of physical phenomena. They are more commonly obtained by using computational methods; however, then the sequence of numbers depends on the first number chosen, so the resulting numbers are not truly random and so they are called *pseudorandom numbers*. One early method of generating pseudorandom numbers, the *middle-square method*, was developed by von Neumann in 1946. Though no longer used, it illustrates the thinking behind such number generation. You choose an integer of the length n you want, square it, select the middle n numbers—which is the first pseudorandom number, then square it and select the middle n numbers for the second number, and so on. (When squaring leads to an odd number of digits, a 0 is placed in front.) So if you start with 8675 (the first four digits of Jenny's number 8,675,309), squaring it gives 75255625 and so 2556; squaring 2556 gives 6533136 or 06533136 and 5331; squaring 5331 gives 28419561 and 4195; which gives 17598025 and 5980, which gives 35760400 and 7604; which gives 57820816 and 8208; and so on. The pseudorandom number sequence would be 2556, 5331, 4195, 5980, 7604, 8208, ... If your goal were random number generation between 0 and 1, as is convenient for simulations, this sequence would be 0.2556, 0.5331, 0.4195, 0.5980, 0.7604, 0.8208, ... If you instead started with the last four digits of Jenny's number, 5309, this sequence would be different: 0.1854, 0.4373, 0.1231, 0.5153, 0.5534, 0.6251, ... The problem with this very simple method, aside from it depending on the first number chosen, is that the numbers eventually repeat,[clviii] and so other methods of modular arithmetic (the math of integers) are now used to obtain pseudorandom numbers, as seen in Chapter 12.

Deviations from random behavior can affect you in both trivial and more dire ways. On the trivial side, when you shuffle a playlist of say 10 songs on your smartphone, you may think you are listening to these songs in random order, but you are not because you hear each one exactly once during a cycle of 10 songs; each song is not being selected randomly from the complete set of 10 songs. When it loops over and over again, it is still not random because each song is still played once in each cycle of then, whether or not the same cycle is played each time (the first song played may well also be the 11th, 21st, and so on) or the songs are reshuffled each loop (the first song could be in slots 1, 17, 23, 39, …). It is random only if the next song is always chosen at random from the complete set of 10 songs (as with the first song played in slots 1, 7, 15, 32, 38, 44, …).

On the more impactful end of the spectrum, on December 1, 1969, the U.S. Selective Service System conducted a supposedly "random" lottery by a physical method to choose the order of the men born in 1944 to 1950 slated to be drafted into the military in 1970, according to their birthdays. With January 1 called 1, January 2 called 2 and so on, for a total of 366 days (including the leap years), slips of paper with numbers from 1 to 366 were each placed in plastic capsules, put into a shoebox, mixed up, and selected one by one. The first one denoted the birthday of the first set of draftees to be called, and so on, until the draft needs of 1970 were satisfied by drafting all with numbers up to number 195. Unfortunately, it was later shown that the numbers were biased and not random, because the capsules were not mixed sufficiently well, particularly after the later months were added to the box.[clix] You would expect the average for each month would be near and vary a bit about 183.5 (= (1 + 366)/2), but the averages for each of the first six months were higher than 183.5 and those for each of the last six months were smaller,[56] so on average those born later in the year were drafted earlier than those born earlier. Statistical methods soon showed that it was very unlikely that the selection was random,[clx] and "less random" than that using rotating lottery wheels in bingo games.[57] Draft lotteries were again

---

[56] The average numbers for the months were: January, 201.2; February, 203.0; March, 225.8; April, 203.7; May, 208.0; June, 195.7; July, 181.5; August, 173.5; September, 157.3; October, 182.5, November, 148.7; and December, 121.5, which fit well to a straight line that decreased from ~220 in January to ~140 in December (with added scatter).

[57] Even with this unintended bias, a new and fairer lottery was not conducted.

conducted from 1971 to 1976 for men born respectively from 1951 to 1956 (and used for draft induction in only a few of those years), but with superior ways of ensuring randomness, including the use of tables of random numbers and the selection of pairs of numbers from two drums each with numbers from 1 to 365, one in the pair denoting the date and the other the order of draft selection (and so it was a better physical method). (The draft number for the author's birthday October 18 in the lottery for his birth year 1951 was 340, but would have been 5 if he had been born on that day in 1950.)

These numbers, random or not, impact the lives of many people. There is a wide range of other types of numbers that do also.

Chapter 9

The Different Types of Numbers Have Not Evolved, But Our Understanding of Them Has

We have a pretty good idea of what whole, fractional, and negative numbers are. But, there are other types of numbers that pop up in many problems, though you may not deal with them directly at home and work, with names such as *irrational*, *transcendental*, *imaginary*, and *complex*. They may conjure up obscure or upsetting images. But, "irrational" numbers are not irrational in the common use of the term. "Transcendental" numbers are irrational numbers and not related to meditation. "Imaginary" numbers are numbers that are not "real" numbers, but they exist and are real and not illusory, and they have an impact in much analysis. "Complex" numbers are simple and not complex, being only combinations of real and imaginary numbers.

"Are mathematical ideas invented or discovered?" This question has been posed by philosophers for many years, as noted by mathematician Gian-Carlo Rota, who was the author's math instructor in his first term in college.[clxi] It also applies to the various types of numbers. One view is that different types of numbers have always been there. It just took people some time to uncover, rather than discover, them and to learn how to use them.[clxii] There are counting positive integers (whole counting numbers 1, 2, 3, …), also called natural numbers which we will sometimes just call numbers. Some include 0 in the definition of the set of natural numbers. It took the introduction of our Hindu-Arabic numerals, along with the introduction of the zero by Indian mathematicians, to spur the development of sophisticated analysis with numbers.[clxiii] Even simple arithmetic was challenging or cumbersome using earlier numeral systems, such as Roman numerals (I for 1, V for 5,

X for 10, and so on).[58] Numbers defined by their values, such as 1, 2, 3, and so on, are called *cardinal* numbers. *Ordinal* numbers, such as first, second, third, and so on, define "order" or rankings.[clxiv]

There are positive, negative and zero integers, which are sometimes collectively called *"natural" numbers. Rational numbers* can be expressed as fractions, and are sometimes called fractions. They have corresponding decimal forms that terminate, such as 5/4 = 1.25, or decimal forms that never end, but that "eventually" repeat in one of several ways. For example, there are 1-digit repeats for denominators 3 and 9, with 1/3 = 0.3333…, 2/3 = 0.6666…, 1/9 = 0.1111…, 2/9 = 0.2222…, and so on.[clxv] The same 6-digit pattern, "142857," repeats for the denominator 7, such as 1/7 = 0.142857142857142857… and 6/7 = 0.857142857142857142…. Several different 2-digit patterns repeat for the denominator 11, such as 1/11 = 0.090909…, which has the same "90" repeating pattern as does 10/11 = 0.909090; 2/11 = 0.181818…, which has the same "18" repeating pattern as does 9/11 = 0.818181818…; and so on.[59]

Because they did not use such decimal systems, the ancient Greeks expressed many rational numbers in terms of *continued fractions*, which are fractions in which the denominator contains another fraction, and that fraction might contain another fraction and so on.[clxvi] For rational numbers, the fractions in the denominator eventually terminate. 61/9 or 6 7/9 are perfectly good forms for this fraction, as is the continuing fraction in Figure 9.1a.[60] You can analyze the continued fraction in this diagram by starting at the lower right, and working upward: 3 + 1/2 is 7/2; inverting it gives 2/7 and 1 + 2/7 is 9/7; and inverting it gives 7/9,

[58] In Roman numerals, I stands for 1, V for 5, X for 10, L for 50, C for 100, D for 500, M for 1000 and so on. II is 2, III is 3, but IV (five minus 1) is four; similarly, VI is 6, VII is 7, VIII is 8 and IX (ten minus 1) is 9; XL is 40 and so on.

[59] There are two different repeating patterns of 6-digit patterns for the denominator 13: "076923" for 1/13 (= 0.076923076923…), 3/13 (= 0.230769230769…), 4/13, 9/13, 10/13, and 12/13; and "153846" for 2/13 (= 0.153846153846…), 5/13 (= 0.384615384615…), 6/13, 7/13, 8/13, and 11/13.

[60] This continued fraction is also written as: 6 + 1/(1 + (1/(3 + 1/2)) or $6 + (1 + (3 + (2)^{-1})^{-1})^{-1}$, or more schematically expressed by the shorthand notation 6 + 1/1 + 1/3 + 1/2; in this context, this is not the sum of the four terms.

which is then added to 6, to give 6 7/9.

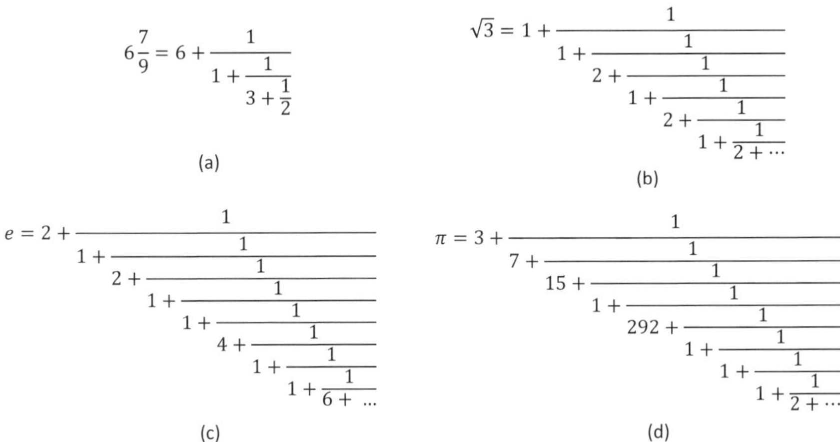

Figure 9.1: Examples of continued fractions.

Irrational numbers cannot be expressed as a fraction; in decimal form they continue forever, but without the recurring pattern seen for rational numbers. They can be positive or negative. The set of real numbers includes all rational and irrational numbers. Irrational numbers pop up in life very frequently and, in that sense, are as "real" as any other types of real numbers. Some pop up in geometrical figures. A triangle has three sides and three angles. The sum of the lengths of any of its two sides exceeds the length of the third. The sum of the three angles is 180 degrees. It may be surprising that the only type of triangle that can have rational numbers for the lengths of each side and for each angle (in degrees) is the equilateral triangle, a triangle with three equal sides and three equal (60 degree) angles. For all others, at least one of these 6 numbers must be irrational.[clxvii]

A right triangle (a triangle with a right or 90-degree angle) with shorter sides of length 1, has a hypotenuse (the longest side, which is opposite the right angle) of length the square root of 2. This comes from the *Pythagorean Theorem*, which states that the sum of the squares of the lengths of the two shorter sides of a right triangle (the sides next to the right angle) equals the square of the length of the hypotenuse.[61,clxviii] This gives $1^2 + 1^2 = 2 = (2^{1/2})^2$. The square root of 2 (expressed as $2^{1/2}$, $\sqrt{2}$ or sometimes sqrt 2) is ≈1.4142, an irrational, number, as is the cube root of 2 (or $2^{1/3}$ or $\sqrt[3]{2}$) ≈1.25992. The square root of 3 is also irrational, ≈1.73205. It is easy to remember the square root of 3 to four digits, 1.732, because George Washington was born in 1732. The square root of 4 is 2, which is rational and an integer, and, because it can be factored as 4×2, the square root of 8 (or $8^{1/2}$ or $(2^3)^{1/2}$ or $2^{3/2}$) equals the product of the square roots of these factors, or 2× the square root of 2 ($2 \times 2^{0.5} = 2^{1.5} = 2^{3/2}$), and it is irrational.

If you jog or run in hilly areas and use GPS to track how far and fast you are moving, you may have misjudged your performance by ignoring the Pythagorean theorem and irrational numbers. When your tracking device notes you have run 1.00 miles, it means that you have moved laterally by a total of 1.00 miles. If the grade were 20%, uphill or downhill, you also moved vertically, up or down, by 0.20 miles.[62] So, you really moved along a hypotenuse of length $(1.00^2 + 0.20^2)^{1/2} = (1.00 + 0.04)^{1/2} = (1.04)^{1/2} \approx 1.0198$ miles. This means you traversed a distance ≈1.98% longer than you had thought and ran at a speed faster by ≈1.98%. If you thought you ran a mile in 10.00 minutes, you actually ran it in ≈ 10.00/1.0198 ≈ 9.8058 minutes or 9 minutes and 48.3 seconds if your route consisted exclusively of uphill and downhill 20% grades. This correction becomes more significant for steeper hills.[63]

Square roots (of numbers that are not perfect squares) often pop up in simple geometry. The length of the diagonal (the line from one vertex angle to the opposite one) of a square with sides of length 1 is $2^{1/2}$, which is the same as the length of the hypotenuse of a right triangle with shorter sides of length 1. The shorter diagonal of a regular hexagon (with all 6 sides equal and 6 angles equal) with sides of length 1 is $3^{1/2}$. The diagonal of a regular pentagon (with all 5 sides equal and 5 angles equal)

---

[61] This is $a^2 + b^2 = c^2$, where c is the length of the hypotenuse, and a and b are the lengths of the other two sides.

[62] Trigonometry shows this corresponds to an angle of ≈11.3 degrees.

[63] This is because when the vertical rise is x miles for a length of 1 mile and $|x| << 1$, the length of the hypotenuse is $(1 + x^2)^{1/2} \approx 1 + x^2/2$ miles.

with sides of length 1 is the famous *golden ratio or number,* $(1 + 5^{1/2})/2$, which is approximately 1.61803398 (Figure 9.2a).[clxix] Consider two line segments, the longer one being x times the shorter one. Now consider a line segment equal in length to the sum of this longer and shorter one. If that line segment is also x times longer than the original longer one, then x equals this golden ratio (Figure 9.2b). (This is clear from algebra.[64]) In Michelangelo's fresco *The Creation of Adam* in the Sistine Chapel, the ratio of the lateral extents of the God and Adam portrayals and the ratio of the overall extents of both of them together to that of God both appear to be the golden ratio.[clxx] As we will see in Chapter 13, this golden ratio also plays a prominent role in the number series called the Fibonacci sequence.

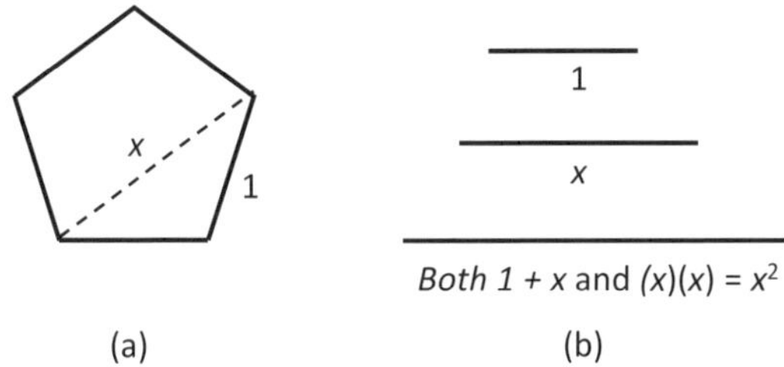

Figure 9.2: The golden ratio, $x = (1 + 5^{1/2})/2 = 1.618\ldots$.

Square roots, cube roots, and so on are types of irrational numbers that are solutions to algebraic equations, such as the square of x equals 2, with solution $x = 2^{1/2}$, or the cube of x equals 2, with solution $x = 2^{1/3}$. Irrational numbers that are not solutions to algebraic equations with rational coefficients are called *transcendental numbers.*[clxxi] Pi is a transcendental number, usually expressed as π, the lower-case Greek letter pi. It is the ratio of the circumference of (the perimeter around) a circle and its diameter, and is approximately 3.141592653. It is sometimes called Ludolph's number, after the German-Dutch mathematician who first calculated it to 35 places circa 1600. π is larger

[64] This is equivalent to saying that $x = (x + 1)/x$, which is the quadratic equation $x^2 - x - 1 = 0$. It can be solved by using standard solution to the quadratic equation in algebra, apparently developed by Indian mathematician Sridhara ca. 900 and others.

than 3 10/71 = 223/71 and smaller than 3 1/7 = 22/7, and is closer to 355/113. Napier's number, $e = 2.7182818284\ldots$ (in Section 7.2), is also transcendental.

Every irrational number can be expressed as sums of an infinite series of fractions. Sixteenth and seventeenth German mathematician Gottfried Leibniz showed that $\pi = 4 \times (1 - 1/3 + 1/5 - 1/7 + 1/9 + \ldots)$, with increasing odd denominators and alternating signs.[clxxii] As we saw in Chapter 7, e can be expressed as the infinite sum of fractions $1 + 1/1! + 1/2! + 1/3! + 1/4! + \ldots$. Seventeenth-century English mathematician John Wallis showed that π can also be expressed as an infinite product $2 \times (2/1 \times 2/3 \times 4/3 \times 4/5 \times 6/5 \times 6/7 \ldots)$, with staggered increasing pairs of even numerators and odd denominators.

The development of ways to determine the values of irrational numbers to more and more decimal places has occupied mathematicians for millennia. Some are improved versions of the Babylonian method of computing square roots, a method that is now understandable by using algebra. This clever method uses the fact that the average of a guess of the square root of a number and the number divided by this guess is a better approximation of the square root; this is repeated using the last result as the new "guess" until the approximation is "good enough" for you.[65] This type of progression is an *iteration* (repetition of the method) that approaches or *converges* to the answer, here *converging* to the exact value, and is an example of *fixed-point iteration*. For $2^{1/2}$, with initial guess of 1, the next better approximation is $(1 + 2/1)/2 = 3/2$. The next better approximation is $(3/2 + 4/3)/2 = 17/12$, and then $(17/12 + 24/17)/2 = 577/408$, and then $(577/408 + 816/577)/2 = 665{,}857/470{,}832$, and so on. The fraction $665{,}857/470{,}832 \approx 1.41421356237$ is about a millionth of a millionth away from $2^{1/2}$. The square root of 2 has been calculated in decimal form to at least 2 trillion digits.[clxxiii]

Irrational numbers can also be expressed as continued fractions that never terminate, with a repeating pattern in some cases, but not in others. Those that are ultimately periodic are the sum of one integer and the square root of a second integer that is not a perfect square (and so other

[65] For a number n, which need not be integral, a guess of its square root of x leads to $(x + n/x)/2$, which becomes the next value of x.

than 1, 4, 9, …), and are called quadratic irrationals. As seen in Figure 9.1b, the continued fraction for $3^{½}$ starts with 1 and then has a repeating sequence of 1 and 2.[66] For $2^{1/2}$ (not shown in this figure) it starts with 1 and then has a repeating 2 and for the golden ratio $(1 + 5^{½})/2$ (also not shown) it also starts with 1 and has a repeating 1. Such periodic patterns are seen for solutions of quadratic equations (those containing sums of x^2 and x and constants[67]). The continued factions representing some other irrational numbers also have some type of patterns, such as that for e (Figure 9.1c),[clxxiv] while those for others do not, such as that for π (Figure 9.1d).[clxxv,clxxvi]

One conceptual way to see how rational numbers can be used to approach (but never arrive at) an irrational number is by considering two sets of rational numbers, A and B. Each number in A is smaller than every number in B, and A does not have a largest number. If B has a smallest number, then one can say that there is a *Dedekind cut* at this smallest rational number in B (as introduced by German mathematician Richard Dedekind ca. 1860). If it does not, then the "number space" between A and B can be thought to give an ever-improving approximation to the irrational number characterized by this "cut" (with ever increasing lower limits in A and ever decreasing upper limits in B). The sequence of increasing 1/1, 7/5, 41/29, 239/169, 1391/985, … approaches $2^{1/2}$ from smaller numbers (set A), while the sequence of decreasing numbers 2/1, 3/2, 17/12, 99/70, 577/408, 3363/2378, … approaches it from larger numbers (set B) (Figure 9.3). 99/70 is about one ten-thousandth larger than $2^{1/2}$.[clxxvii]

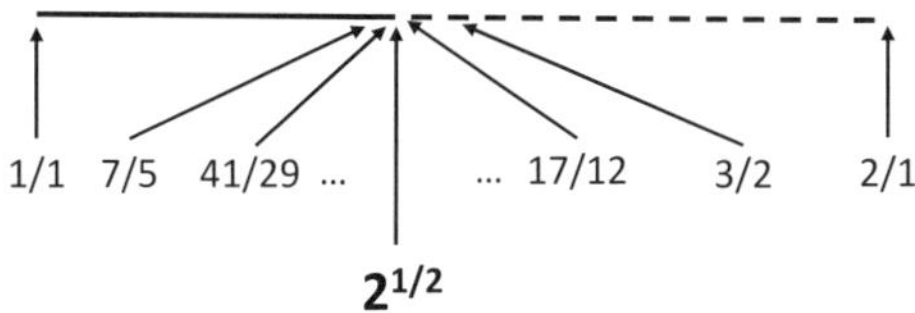

Figure 9.3: Dedekind cut.

[66] This continued fraction for $3^{½}$ can also be expressed as $1 + (1 + (2 + (1 + (2 + (1 + (2 + \ldots)^{-1})^{-1})^{-1})^{-1})^{-1})^{-1}$ or more simply as [1; 1, 2, 1, 2, 1, 2, …].

[67] These are $x^2 - 3 = 0$, $x^2 - 2 = 0$, and $x^2 - x - 1 = 0$ for these three examples, respectively.

Some series and approximations of irrational numbers are good to various degrees of accuracy, but are in fact merely math curiosities and coincidences. For example, $\pi \cong 22/17 + 37/47 + 88/83$, which happens to be accurate to 9 digits. Moreover, it is remarkable, but not really significant, that e and π can be combined to produce a number that is very closely an integer, $e^{\pi} - \pi = 19.999099979\ldots \cong 20$.[clxxviii]

The square root of a positive number is another real number. A positive number and the negative of it have the same square. However, the square root of a negative number is not a real number, and so it is called an imaginary number. The numerical symbol denoting the solution to $x^2 = -1$ is called i (or j by some engineers), and so $i \times i = i^2 = -1$,[clxxix] and also $i^3 = -i$ and $i^4 = 1$. So the cube root of i is $-i$,[68] while that of a negative real number is a negative real number. There are actually two solutions to $x^2 = -1$: $+i$ and $-i$ (or $\pm i$), and to $x^2 = -4$: $\pm 2i$. The product $(2i)(3i) = 6i^2 = -6$, while $(2i)(-3i) = -6i^2 = +6$. Also, i taken to the 0th power equals 1 ($i^0 = 1$), as it is true for all numbers aside from 0 itself.[clxxx]

The sum of a real number, a, and an imaginary number, bi where b is real, is called a complex number, $a + bi$. Imaginary and complex numbers are harder to comprehend than real numbers because they are not ordinary counting numbers and are not used in measurements, but they serve bone fide mathematical functions. Much in mathematics could not be done without them. The 18th-century work of the Swiss mathematician Leonhard Euler and the early 19th-century work of German mathematician Carl Friedrich Gauss led to the use of imaginary and complex numbers in mainstream mathematics. Prior to this, they were viewed with suspicion. In fact, the term “imaginary” was first assigned to them to be derogatory.[clxxxi]

The rules for the arithmetic of complex numbers are largely the same as those for real numbers, as initially set down by Italian mathematician Rafael Bombelli in 1572. Complex numbers are added and subtracted by adding or subtracting the real and imaginary parts separately, as with $(-2 + 3i) + (4 - 5i) = (-2 + 4) + (3i - 5i) = 2 - 2i$ and $(2 - 3i) - (4 - 5i) = (2 - 4) + (-3i - (-5i)) = -2 + (-3i + 5i) = -2 + 2i$. Complex numbers are multiplied

---

[68] and so the cube root of imaginary number $a^3 i$ is $-ai$, for real number a

by considering each term separately by using the distributive principle, such as with $(-2+3i)\times(4+5i)=(-2)(4)+(-2)(5i)+(3i)(4)+(3i)(5i)=-8-10i+12i-15$ (because $15i^2=-15$), which equals $-23+2i$.

A complex number has a magnitude, which is the square root of the sum of the squares of the real and imaginary parts, which for $3-4i$ is $(3^2+4^2)^{½}=25^{1/2}=5$. The *complex conjugate* of a complex number is the same number with the sign of the imaginary part changed from + to – or from – to +, so for $3-4i$ it would be $3+4i$. Multiplying a complex number by its complex conjugate gives a real number that is the square of the magnitude of the number (and also of its complex conjugate), or here $(3-4i)\times(3+4i)=(3)(3)+(3)(4i)+(-4i)(3)+(-4i)(4i)=9+12i-12i+16=25$.

The golden rule for fractions, that multiplying a fraction by x/x does not change the value of the fraction, also applies to complex numbers. Division by complex numbers proceeds by multiplying both the numerator and denominator by the complex conjugate of the denominator, so the denominator becomes the square of the magnitude, a real number. The division problem represented by $6/(1-i)$ proceeds by multiplying the numerator and denominator by the complex conjugate of $1-i$, which is $1+i$, so $6/(1-i)=[6/(1-i)][(1+i)/(1+i)]=(6)(1+i)/[(1-i)(1+i)]=(6+6i)/(1+1)=(6+6i)/2=3+3i$.

We showed earlier that e^x is $(1+x/N)^N$ in the limit that N goes to infinity.[69] Choosing x to be $i\pi$ leads to the surprising and well-known Euler's formula $e^{i\pi}=-1$.[clxxxii]

Even within the realms of the more mundane whole and rational numbers, there are numbers outside much of everyday life that still need to be put under our control and then used comfortably.

---

[69] With x set equal to iy, where y is a real number, this gives the very important Euler's formula, that e^{iy} equals the cosine function evaluated at y plus i times the sine function evaluated at y, or $e^{iy}=\cos y+i\sin y$.

Chapter 10

Really, Really Big and Really, Really Small Numbers

The math in our lives uses scientific notation and logarithms to bring very large and small numbers under our control. But, sometimes numbers are bigger or smaller than we can imagine.

10.1 Are There Enough Big Numbers?

One concern about really big numbers, those that are still within the realm of imagination, is whether there are enough of them to satisfy our needs, such as enough telephone numbers for, say, everyone in the U.S. (and others within the North American numbering plan). Telephone numbers have 10 digits, the first three are called the area code, the next three are the central office code or the prefix for a switch in that area code, and the final four are the station code or the suffix for the particular phone. If the numbers 0, 1, 2, … 9 could be used for each of them then there would be $10 \times 10 \times \ldots = 10{,}000{,}000{,}000 = 10^{10}$ or 10 billion numbers. But, there are fewer useful numbers because of signaling restrictions on some of the numbers. The first number of the area and office codes can only be the eight numbers from 2 through 9, so the total number of phone numbers can be $(8 \times 10 \times 10) \times (8 \times 10 \times 10) \times (10 \times 10 \times 10 \times 10) = 6{,}400{,}000{,}000$ or 6.4 billion. This corresponds to 10,000 numbers for each of the 800 office codes in a given area code, and 8,000,000 phone numbers[70] in each of the 800 area codes (mostly in the

[70] To avoid confusion with area codes whose second and third numbers are 1, the second and third digits of nine-digit phone numbers (the second and third digits of the central office code) cannot both be 1. So, only 99 of the $10 \times 10 = 100$ of this set can be used,

U.S.).[clxxxiii] With over 300 million people in the U.S. (ever since ~2007), say each with one personal and one business phone, and with future increases in population and the hesitation in reusing numbers, this appears to be plenty (after accounting for the ~100 million people outside the U.S. who are also within the North American numbering plan).

There are a plethora of available numbers because an earlier anticipated shortage of new phone numbers was avoided by an increase in numbers that became available with the new, current signaling schemes. This increased the number of available numbers from 1 billion[71] to the "current" 6 billion, but not all of these are actually available at present.[clxxxiv] Four to seven new areas codes are assigned each year and these should not be depleted in the thirty years starting in 2017. Often a new area code is not needed because all of the 8 million numbers in a given area code have been used, but because all of the 800 office codes or prefixes in that area code have been assigned and more are needed. This occurs because all of the station codes or suffixes in many of the office codes have not been used, and the remaining ones cannot be used at present in other switching areas.[clxxxv]

Similarly, there need to be enough passwords in security systems because the more the allowable possible passwords, the harder it is to crack them. For an eight-digit password, using the ten numbers from 0 to 9 for each digit and no restrictions, there are $10 \times 10 \times 10 \times 10 \times 10 \times 10 \times 10 \times 10$ or a hundred million, $100{,}000{,}000 = 10^8$ possibilities. Your code could become more secure by either increasing the choices for each digit or the number of digits in the password. Using number digits, letters in the alphabet, with upper and lower case being distinct, and eight types of punctuation {including . , ; : ? ! ( ) }, there are $10 + 26 + 26 + 8 = 70$

---

reducing the number of phone numbers per area code to $(8 \times 99) \times (10 \times 10 \times 10 \times 10) = 7{,}920{,}000$. Furthermore, the telephone numbers 555-0100 to 555-0199 are officially reserved for fictional use, reducing the available number per area code to $7{,}920{,}000 - 100 = 7{,}919{,}900$.

[71] With the older signaling methods in the 1960s, there were additional restrictions on phone numbers. The first number of the area code could be 2 through 9 but the second number had to be one of the two numbers 0 or 1, and the first two numbers of the office code had to be 2 through 9. So, only $(8 \times 2 \times 10) \times (8 \times 8 \times 10) \times (10 \times 10 \times 10 \times 10) = 1{,}024{,}000{,}000$ or approximately 1 billion numbers were available then.

possibilities for each code digit, and 70^8 or almost 580 trillion or 5.8×10^{14} possibilities. Alternatively, to increase the number of passwords by a factor of a million by using numbers only, you would need a password with six more number digits, for a total of 14, and 10^{14} possibilities.

Again, restrictions could reduce the number of possibilities. If you cannot use the same digit more than once, then for the eight-digit numerical password there would be 10 choices for the first digit, 9 for the second, eight for the third, and so on until you have only three choices for the last one. The total number of possibilities would then be $10 \times 9 \times 8 \times 7 \times 6 \times 5 \times 4 \times 3 = 1{,}814{,}400$, corresponding to 10!/2! permutations or arrangements for passwords (Section 4.10), or more than 50 times fewer than before. If for some reason you were not allowed to repeat digits and now their order was not important (another uncommon password restriction), with the 8! = 40,320 ways of ordering these numbers (Section 4.10), there would be only 10!/(2!8!) = 1,814,400/40,320 = 45 combinations or choices.

10.2 Infinities and Infinitesimals

The animated toy character Buzz Lightyear led charges with the proclamation "To infinity… and beyond!" in the movie 1995 *Toy Story* and its sequels.[clxxxvi] Does this make sense? Can you reach infinity, much less beyond it?

When we count positive integers, we know there is no end and that the largest number we count may "approach" *infinity* in our minds, but it never reaches it. In this sense, infinity is the limit in counting to the biggest number. Likewise, the limit of this largest number is also the total number of integers, in the collection or set of positive integers, and so the number elements in this set, called the *cardinality*, approaches infinity the same way. When we choose a small positive number (not an integer) x, and look at its reciprocal $1/x$ as we let x get smaller and smaller, approaching zero, this reciprocal also approaches infinity. Similarly, when x gets larger and larger, $1/x$ becomes smaller and smaller (and so *infinitesimal*), and approaches 0.

Though infinite, the set of positive integers is *countable*, because we can always try to count them. But, are there more integers than just positive integers? Are there more rational numbers than integers? Are there more numbers, rational and irrational, than rational numbers only? We will see these answers are no, no and yes, and so there is more than one type of infinity.

You can compare sets of numbers that never end by trying to "map" one set into the other "in a countable way." If you can do this they are the same type of infinity and if you cannot, they are not. You can map all non-negative integers (positive integers and 0) to all even positive integers, so 0 to 2, 1 to 4, 2 to 6, 3 to 8, and so on, and all negative integers to all odd positive integers, and so -1 to 1, -2 to 3, -3 to 5, -4 to 7, and so on. So, the infinity describing all integers and that of all positive integers are the same. They are the same level of infinity of countable numbers and are *equinumerous* (have the same "numbers" of members) or have the same *cardinality* (or cardinal type). This approach was developed by German mathematician Georg Cantor in the 1870s. Cantor denoted the first infinite cardinal, this "lowest" level of infinity, $\aleph_0$, pronounced "aleph-zero." (The aleph is the first letter in the Hebrew alphabet.)[clxxxvii]

The positive and negative integers have the same cardinality, but they have different *order-types* (or ordinal types), because they are different when you order them from their "smallest" to "largest" elements. The positive integers have a first element (which is 1), but no specific last one, and are therefore classified by the order-type denoted by the lower-case Greek letter $\boldsymbol{\omega}$ (omega). The set of negative numbers have no first element, but a last one (-1), and so belongs to the order-type $\boldsymbol{\omega}^*$. The set of all integers is of order-type $\boldsymbol{\omega}^* + \boldsymbol{\omega}$. The order-type $\boldsymbol{\omega} + \mathbf{1}$ describes the positive integers plus one element that is larger than any of them.

Intuitively, you would think there are more rational numbers than integers, because in the range spanned by an integer and that integer plus 1 there many numbers. But, you would be wrong, because you can map one set onto the other. Rational numbers can always be expressed as fractions of integers. There is 1 way of writing a fraction whose numerator and denominator sum to 1 + 1 = 2, namely 1/1. There are 2 ways of writing a fraction whose numerator and denominator sum to 2 + 1 = 3, namely 2/1 and 1/2. There are 3 ways of writing a fraction whose numerator and denominator sum to 3 + 1 = 4, namely 3/1, 2/2 and 1/3, and so on. So, you can map the set of all rational numbers into the set of integers, even after removing the duplicates such as 1/1 and 2/2, and so the rational numbers and integers are of the same cardinality of infinity. They are all *countable* or *numerable.* Including negative rational numbers does not change this counting argument.[clxxxviii]

In contrast, the sets of real and irrational numbers are uncountable and so they correspond to a "higher" level of cardinality and infinity. You can prove this by making a never-ending list of "all" real numbers between 0 and 1 expressed as decimals, labeling the first one as Number 1, the second one as Number 2, and so on (Figure 10.1). Select one digit in each number and change it by adding 1 to it if it is from 0 to 8 and changing it to 0 if it is 9. In the first number do this to the first digit, in the second to the second digit, in the third to the third digit, and so on, and then combine the selected digits together. If the first four numbers are 0.**6**387…, 0.4**9**21…, 0.03**2**7…, and 0.188**4**…, with the digit to be changed in bold, the number resulting by combining these "diagonal" digits would be **0.6924…** before the change and **0.7035…** afterward. The number resulting from this *Cantor diagonalization* method cannot equal any of these chosen numbers and so the resulting number cannot be mapped onto the set of integers (1st one, 2nd one, 3rd one, …) or rational numbers. So, the real numbers and therefore also the irrational numbers, are not equinumerous with the integers and rational numbers and must represent a larger cardinality of infinity.[clxxxix,cxc,cxci]

0.**6**387…	Before diagonalization
0.4**9**21…	**0.6924…**
0.03**2**7…	
0.188**4**…	After diagonalization
…	0.7035…

Figure 10.1: Cantor diagonalization.

The set of real numbers is therefore uncountable. The cardinality of the real numbers set of infinity is thought to be 2 raised to the power aleph-zero ($\aleph_0$), or $2^{\aleph_0}$. It is also the infinity of the sets of all imaginary and complex numbers because of mapping. According to the famous so-called *Continuum Hypothesis*, $2^{\aleph_0}$ equals the next level of infinity called $\aleph_1$ or aleph-one.[72] Cantor thought that these were the only two sets of infinities, which is also part of this hypothesis, but could not prove it. In 1940 the Austrian-American mathematician Kurt Gödel showed that this hypothesis cannot be disproved from standard principles and in 1963 American mathematician Paul Cohen showed that it also cannot be proved using them. Not everything is well understood!

This second level of infinity is also the infinity of the numbers of points in a line of a finite length. If the line is 1 inch long and you call the leftmost point 0 and the rightmost 1, the distance of any point from the left is described in inches by a real number between 0 and 1. Though possibly shocking, the type of infinity of points within a square or in a cube is the same as that in a line.[73]

So we know now that when Buzz Lightyear led charges with "To infinity… and beyond!" he was being both correct and incorrect. There are different levels of infinity, so you can imagine something being beyond or larger than infinity, but physically you cannot approach an infinite distance, much less surpass it.

The method of forming numbers developed by English mathematician John Horton Conway ca.1969 provided new insight into how we think about numbers. It characterizes numbers by the set of other numbers that are either smaller than it or larger than it. So, 7 can be defined as {… 3, 4, 5, 6 | 8, 9, 10, 11, …}, where the set on the left are

---

[72] One could wonder whether it could follow that $\aleph_2 = 2^{\aleph_1}$, $\aleph_3 = 2^{\aleph_2}$, and so on, as in the *Generalized Continuum Hypothesis*.

[73] You can describe all points in a 1-inch square by two numbers between its leftmost and rightmost values 0 and 1, and uppermost and lowest values 0 and 1. (In analytic geometry these would be called the two coordinates of the point.) For every point in a line, say 0.36298745…, you can write a real number using either the odd or even digits, here 0.3284… and 0.6975…, which then maps to a point in the square. You can also map in reverse with 0.5901… and 0.8223… mapping to 0.58920213… in a line. In addition, you can map points in a line to points in a cube and show that they also correspond to this second level of infinity, by mapping the point into to the three numbers needed to describe a point in the cube, by choosing the first, fourth, seventh, … number for the first number, the second, fifth, eight, … for the second number, and the third, sixth, ninth, … for the third number.

those that are smaller than 7, and those on the right are those that are larger. 7 is an integer and we have provided only integers in the two sets, but this approach is easily generalized to rational numbers and irrational numbers, both positive and negative, and to 0. In the last chapter, we saw that Dedekind cuts "localize" irrational numbers. This approach expands that characterization of real numbers by providing ways of determining numbers that are infinite or infinitesimal. An infinite number can be defined as being larger than any integer, with no number larger than it, or {1, 2, 3, 4, 5 … | }. This number is called by its order-type $\boldsymbol{\omega}$**.** With this approach you can also define other infinite numbers that are larger, including $\boldsymbol{\omega} + \mathbf{1}$ and $\mathbf{2}\boldsymbol{\omega}$, and smaller, including $\boldsymbol{\omega} - \mathbf{1}$ and $\boldsymbol{\omega}/\mathbf{2}$. You can define an infinitesimally small number, that is larger than 0 but smaller than any positive number, {0 | 1, 0.1, 0.001, 0.0001, 0.00001, …}, which is called $\boldsymbol{\varepsilon}$. Though still technically real numbers, these more general numbers were dubbed *surreal numbers* by Donald Knuth.[cxcii,cxciii] As noted by mathematician Gian-Carlo Rota, "Surreal numbers are an invention of the great John Conway. They will go down in history as one of the great inventions of the (20th) century. … Thanks to Conway's discovery, we have a new concept of number."[cxciv]

Concepts such as countable and uncountable numbers extend beyond our experience base. The possibility of doing arithmetic with whole numbers and no fractions or decimals may also seem unusual. However, it is less esoteric than it may seem and is closer to our everyday life experiences.

Chapter 11

The Whole Truth of Whole Numbers

Integers impact your everyday life "behind the scenes," as with prime numbers and their importance in encryption and coding. These numbers help keep our records secure, and so they control our lives in a positive sense. We will see this in the next chapter where we will explore the world of arithmetic with only whole numbers, called modular arithmetic. For now, we amuse ourselves with whole numbers, specifically how to express powers of some integers as the sums of powers of other integers and whether and how integers can be *factored* into products of other integers, and consider potential implications and applications later.

11.1 Integers Leading to Other Integers

As 18th-century British author Samuel Johnson noted "Round numbers are always false."[cxcv], which meant that you should be suspicious of measurements or statistics that are presented exactly as whole numbers. But integers are employed properly often, as in ordered lists, #1, #2, #3, …, in house addresses, and so on.

Relations between positive integers are amusing diversions to some, often more obscure than obvious, and sometimes they are very significant. Every positive integer is special in some way. In 1918, when mathematician G. H. Hardy suggested that the number of a taxi, 1729, was "dull," mathematical genius Srinivasa Ramanujan showed off his spontaneity and genius by replying, "No, it is a very interesting number; it is the smallest number expressible as a sum of two cubes in two different ways, the two ways being 1^3 and 12^3, and 9^3 and 10^3." The brilliant, unusual, and unfortunately very short life of Ramanujan was the subject of the 2016 movie *The Man Who Knew Infinity*, based on the

1991 book by Robert Kanigel of the same name.[cxcvi]

When we add the squares of integers, we naturally get an integer. But, can we represent every integer as the sum of the squares of some two integers, even including zero as one of these integers? No. Clearly, $1 = 1^2 + 0^2$ and $2 = 1^2 + 1^2$, but we cannot express 3 this way. How about as the sum of three squares? We see, $1 = 1^2 + 0^2 + 0^2$, $2 = 1^2 + 1^2 + 0^2$, $3 = 1^2 + 1^2 + 1^2$, $4 = 2^2 + 0^2 + 0^2$, $5 = 2^2 + 1^2 + 0^2$, and $6 = 2^2 + 1^2 + 1^2$, but we cannot express 7 this way. How about as the sum of four squares? Yes, every integer can be expressed as the sum of four squares.

More generally, in 1770 Cambridge mathematician Edward Waring conjectured that every number can be expressed as the sum of no more than 4 squares, or 9 cubes, or 19 fourth powers, or 37 fifth powers, and so on for all powers. In most cases, it is not obvious why this is true. By the way, the only number that needs as many as 37 fifth powers is 223, which equals the sum of thirty-one 1^5s and six 2^5s ($= 31 \times 1 + 6 \times 32$).[cxcvii]

There are many examples of the sums of the squares of two integers being equal to the square of another integer: $3^2 + 4^2 = 5^2$, $5^2 + 12^2 = 13^2$, $8^2 + 15^2 = 17^2$, …. These are examples of the Pythagorean Theorem (as in Chapter 9) for sides of integral lengths. These are known as *Pythagorean triples* and *primitive triples* if the integers have no common factors. Of course, you can multiply each number in a triple by the same integer (2, 3, 4, 5, …) and get another triple, so the primitive triple 3, 4, 5 from $3^2 + 4^2 = 5^2$, also gives $6^2 + 8^2 = 10^2$, $9^2 + 12^2 = 15^2$, and so on. Jenny's number, 8,675,309, is the hypotenuse of a primitive Pythagorean triple: $8{,}675{,}309^2 = 2{,}460{,}260^2 + 8{,}319{,}141^2$.[cxcviii]

However, there are no examples of the sums of the cubes of two integers being equal to the cube of another integer, the sums of the fourth power of two integers being equal to the fourth power of another integer, and so on. This is the famous *Fermat's Last Theorem*,[74] by 17th-century French mathematician Pierre de Fermat. It was either first proved by him in 1637, when he noted in the margin of a book that the margin was too small to hold his proof so his proof was lost to the world or, as many think, first proved only after numerous efforts over three centuries later,

---

[74] There is no solution for $x^n + y^n = z^n$ for integers x, y, and z, for any integer n larger than 2.

by Andrew Wiles in 1994 (and published in 1995).

More generally, equations involving polynomials (sums of terms of different orders of one or more variables) are known as Diophantine equations when the two or more variables and other numbers in it are constrained to be integers. They are named after Diophantus of Alexandria of the 3rd century AD. Examples relate to Pythagorean triples ($x^2 + y^2 = z^2$) and Fermat's Last Theorem (involving $x^n + y^n = z^n$). Far simpler are linear Diophantine equations, such the one that addresses how many ways you can form 50 cents from x quarters, y dimes, and z nickels (which can be expressed by $50 = 25x + 10y + 5z$). You can do this with 2 quarters; 1 quarter and either 5 nickels, 3 nickels and 1 dime, or 1 nickel and 2 dimes; or 0 quarters and either 10 nickels, 8 nickels and 1 dime, 6 nickels and 2 dimes, 4 nickels and 3 dimes, 2 nickels and 4 dimes, or 0 nickels and 5 dimes, for a total of 10 solutions.

(Another Diophantine equation and its solution came to me during a visit to the dentist:[75] $x^y - y^x = x + y$. The "trivial" solution is $x = 1$ and $y = 0$ (with $1^0 - 0^1 = 1 + 0$, because $1^0 = 1$ and $0^1 = 0$). The nontrivial integer solution is $x = 2$ and $y = 5$ (with $2^5 - 5^2 = 2 + 5$, because $2^5 = 32$ and $5^2 = 25$, leading to $7 = 7$).)

The number 1729, the so-called Hardy-Ramanujan number, is an example of a Diophantine equation in which the sum of two cubes equals a sum of two other cubes (as in $w^3 + x^3 = y^3 + z^3$), with the smallest non-trivial solution being $1^3 + 12^3 = 9^3 + 10^3$. It has been used in popular media as an "insider" math joke, as by a mathematician in the 2005 film *Proof*, in the animated television series *Futurama* as the spaceship Nimbus BP-

---

[75] I always have trouble remembering how many teeth adults with a full set of teeth have. I know this number is divisible by 4 because there are an equal number of teeth in the upper and lower jaws and in the right and left sides of the mouth, so it could be …, 20, 24, 28, 32, 36, … . However, I always remember that I have had 7 teeth extracted (to prepare for orthodontal work and to remove impacted wisdom teeth). Recently at the dentist, I noted the tooth diagram showed 32 teeth, so then I knew that I have 25 teeth. Because it is easy for me to remember that 7 is the difference between the powers of two integers, and given the possibilities for this, from now on it will be easy for me to remember that a full jaw must have 32 teeth, because $2^5 - 5^2 = 32 - 25 = 7$ is the only nontrivial way this can occur. When I told my dentist about this, he pointed out that 7 is also the sum of 2 and 5, and this led to the Diophantine equation $x^y - y^x = x + y$.

1729 and as the robot character Bender's serial number, in a Christmas card,[cxcix] and in the television show *The Simpsons*. It is also known as the second taxicab number, because it is the smallest number that can be expressed as the sum of two cubes in "2" different ways. In 1657 French mathematician F. de Bessy found the third taxicab number, 87,539,319, which is the smallest number that can be represented as a sum of two cubes in three different ways: $167^3 + 436^3$, $228^3 + 423^3$, and $255^3 + 414^3$.[76] The fourth, fifth, and sixth taxicab numbers are also known, the fourth one being "uncovered" in 1991 and the others more recently.[77,cc]

Some relations between powers of integers might seem daunting, but are actually well known. For example, the sums of the terms in the sequence $1^1 + 2^1 + 3^1 + \ldots + n^1$,[78] $1^2 + 2^2 + 3^2 + \ldots + n^2$,[79] and so on are obtainable for integers raised to any power by using Faulhaber's Formula, developed in 1631 by Johan Faulhaber from Ulm (in what is now called Germany).[cci]

11.2 Factoring and Prime Numbers

All numbers can be written as a product of factors. For example, 6 can be written as 6×1, so 6 and 1 are factors, and as 3×2, where 3 and 2 are factors. So, 6 has divisors of 1, 2, 3 and 6. 36 can be factored in several ways, 6×6, 3×12, 2×18, $2 \times 3 \times 6$, $2 \times 2 \times 3 \times 3$ (which has only prime number factors, so is *prime factorization*) and so on, and so has divisors 1, 2, 3, 6, 12, 18 and 36. Numbers that have only 1 and themselves as factors are called *prime numbers* (or primes).[ccii,cciii,cciv] After you begin to factor a number, you can continue by factoring the factors until all that remains are prime numbers, as with $36 = 2 \times 2 \times 3 \times 3$. The great geometer Euclid also made key insights into number theory over two millennia ago, including one that (essentially) said there is a unique way of factoring numbers into primes.[ccv] So-called *perfect numbers* are those

---

[76] This number was used as the taxicab number in the *Futurama* episode: "Bender's Big Score."

[77] A broader definition of taxicab numbers includes <u>all</u> numbers that are the sums of two cubes in two or more ways (and not just the smallest numbers), and includes 1729, 4104, 13,832, 20,683, 32,832, … .

[78] $1 + 2 + 3 + \ldots + n = (n^2 + n)/2$

[79] $1 + 4 + 9 + \ldots + n^2 = 1/3[n^3 + (3/2)n^2 + (1/2)n]$

whose divisors (other than the number itself, but including 1) sum to give you the number itself, such as $6=1+2+3$ and $28=1+2+4+7+14$.[ccvi]

Two positive integers have common factors or divisors, sometimes 1 and sometimes 1 and other numbers. The *greatest common divisor* of 8 and 12 is 4. It is easy to find the greatest common divisor by prime factorization. The great common divisor of $36=2^2 \times 3^2$ and $126=2 \times 3^3 \times 7$ is $2 \times 3^2 = 18$. When the greatest common divisor of two integers is 1, meaning they have no common factors other than 1, such as 8 and 15, the two numbers are said to be *relatively prime* (or mutually prime or coprime). If you divide two numbers by their greatest common divisor, the resulting two numbers are relatively prime. So for 16 and 36, the great common divisor is 4, with 16/4 = 4 and 36/4 = 9, which are relatively prime.[80] The number of integers smaller than and relatively prime to a given positive integer is called the phi-function of that number (named after the lower-case Greek letter phi, ϕ). This pops up in cryptography. There are 6 integers that are relatively prime to 9 (and smaller than it): 1, 2, 4, 5, 7, and 8, so the Euler phi-function of 9 is 6.[81] The numbers 1, 3, 5, 7, 9, 11, 15, 17, 19, 21, 23, and 25 are relatively prime to 26, so the Euler phi-function of 26 is 12. We will use this in our discussion of cryptography.

11.2.1 Fortuitous factoring

Sometimes factoring leads to interesting results, such as the factoring of 5,280 = 88 × 60, which means than 5,280/60 = 88. A car moving at 60 miles per hour (or 60 mph) moves how many feet per second? We change units by multiplying 60 miles/hour by three ways of expressing 1: (a) 5,280 feet/mile (because there are 5,280 feet in a mile), (b) 1 hour/60 minutes (60 minutes in an hour), and (c) 1 minute/60 seconds (60 seconds in a minute). After cancelling the units of miles, hours, and minutes, which both appear in the numerator and denominator, we are left with (60 × 5,280)/(60 × 60) or 5,280/60 = 88 feet per second due to

---

[80] (8,12) = 4 is standard notation that means that the greatest common divisor of 8 and 12 is 4. 8 and 15 are relatively prime, and so (8,15) = 1. (16,36) = 4, and so 16/4 = 4 and 36/4 = 9 are relatively prime because (4,9) = 1.

[81] and so $\phi(9)=6$

this fortuitous factoring.

11.2.2 More prime thoughts

Early 20th century British mathematician G. H. Hardy noted "317 is a prime, not because we think so, or because our minds are shaped in one way rather than another, but because it is, because mathematical reality is built that way."[ccvii] Prime numbers include 2, 3, 5, 7, 11, 13, 17, 19, 23, 29, 31, continuing forever. The others are called *composite numbers*. (1 is not included in either category.) All positive integers can be factored into prime numbers, such as $180 = 2 \times 2 \times 3 \times 3 \times 5$. There is only one way to do such prime factorization (as we noted), aside from the order of the factors; it is unique, and this is known as the *Fundamental Theorem of Arithmetic*.[ccviii]

The only prime even number is 2. Other even numbers are divisible by 2, so they are not prime and so the minimum separation between prime numbers (aside from 2 and 3) is 2. 2 and any other prime are separated by odd numbers, because with such a separation one of the two would be odd and the other even, and 2 is the only even prime. (The sum of two even or two odd numbers is even, while the sum of an even and odd number is odd.) Prime numbers other than 2 and 5 cannot be separated by 3. Numbers can be confirmed to be prime by dividing them by integers 2, 3, 4, …, to the largest integer smaller than $n^{1/2}$, and finding no remainder. (Any larger factor would have a co-factor that would be smaller than that square root, so these larger numbers need not be tested.) Even with this simple method, it requires much computer time to test suspected large primes.

One mode of decrypting messages depends on whether or not an integer is prime. If it is prime it is hard (or "impossible") to decode.[ccix] If it is not, it can be easily decrypted. If the number is not prime, but can be factored only by using state-of-the-art computer power and time, the message cannot be effectively decrypted. The larger the prime, the harder it is to crack a code based on primes (because it takes longer and longer time to try to factor it), which explains one motivation in finding large primes. Another is just for fun and the challenge of "exploration."[ccx,ccxi]

Nature may also "use" prime numbers. Several species of cicadas reappear every year, but others only every 13 or 17 years, both prime numbers. Paleontologist Stephen Jay Gould hypothesized that such multiyear cycles composed of these "larger" prime numbers of years could confer to them an evolutionary advantage in avoiding predators that appear every 2 or 3 years. For instance, cicadas with a 17-year cycle would encounter predators that appear every 3 years only every $17 \times 3 = 51$ years, because 3, 6, 9, 12, …, 45, 48, 51 first overlap a factor of 17 then.[ccxii]

There are a host of fascinating features of prime numbers. Over most ranges of numbers, almost half of prime numbers can be expressed as the sums of two squares and slightly more than half cannot.[ccxiii] Fermat showed that the only primes that can be expressed as the sum of two squares are 2 and those odd primes that have a remainder of 1 after being divided by 4 (*Fermat's Theorem on the Sum of Two Squares*).[ccxiv] For example, the prime number 144,169 equals the sum of the squares of 315 and 212. The next prime is 144,173, which is the sum of the squares of 338 and 173. The next prime, 144,203, is not the sum of squares. Dividing these three consecutive primes by 4, gives 36,043 with a remainder of 1, 36,043 with a remainder of 1, and 36,050 with a remainder of 3, respectively, as expected.

One reason such odd primes with a remainder of 1 are called *Pythagorean primes* is that, being the sums of squares, their square roots are the (nonintegral) hypotenuses of right triangles whose other sides are integers. Another, is that each such prime is <u>also</u> the (integral) hypotenuse of a <u>different</u> right triangle whose other sides are integers, and so the largest member of a primitive Pythagorean triple. The number 5 is a Pythagorean prime, with $5 = 1^2 + 2^2$ and also $5^2 = 3^2 + 4^2$. So is Jenny's number, with $8{,}675{,}309 = 422^2 + 2{,}915^2$,[ccxv] and $8{,}675{,}309^2 = 2{,}460{,}260^2 + 8{,}319{,}141^2$.[ccxvi]

It is not obvious (to me) that every positive even integer greater than 2 can be written as the sum of two prime numbers, sometimes in more than one way. This is *Goldbach's Conjecture*. For example, $4 = 2 + 2$, $6 = 3 + 3$, $8 = 3 + 5$, $10 = 3 + 7 = 5 + 5$, $12 = 5 + 7$, … $20 = 7 + 13$, …, $50 = 3 + 47$, …, $100 = 3 + 97 = 11 + 89 = 17 + 83 = 29 + 71 = 41 + 59 = 47 + 53$, … . This is still a conjecture because it has still not been formally proven, but has been shown to be true up to very large numbers, up to 4×10^8.

Mathematicians have also taken note of the integer 137 because it is the 33rd prime number and is also a *twin prime*. Twin primes are prime numbers separated by 2. 137 is the smaller member of the 11th twin prime pair. Physicists have also long thought that the integer 137 might be very special in nature. The strength of the force between any pair of electric charges depends on a special constant called the fine-structure constant. It has a value that is approximately 0.00729735257. This is also approximately equal to 1/137.036, which is enticingly close to 1/137, so it has long been speculation that it is exactly equal to this reciprocal of this integer—but experimental evidence says this is not so. By the way, Jenny's number, 8,675,309, is not only prime, but one of the twin prime pairs: 8,675,309 and 8,675,311.[ccxvii]

If you were one of the many people who tried to obtain Jenny's number 8,675,309 for your own area code, you may wonder whether the resulting 10-digit number would still be prime. There were 379 three-digit NANPA (North American Numbering Plan Administration) area codes being used by the general public in the U.S., Canada and elsewhere (as of June 16, 2017)[ccxviii] and 42 of them[ccxix] result in a 10-digit prime number. Each of these is a sum of two squares and so is also the hypotenuse of primitive Pythagorean triple.[82] Three of the area codes, 252, 563, and 804, lead to a prime in a twin prime pair, and two of them, 681 and 708, lead to *Sophie Germain primes*, meaning that twice it plus one is also prime.[ccxx] (The song's writers never realized they had chosen such an interesting number!)

Some primes have other curious properties. The prime number 144,169 looks like the square of 12 (which is 144) followed by the square of 13 (169). The number 1,234,567,891 looks like a counting sequence and is also prime.[ccxxi]

11.2.3 How many primes are there and how far apart are they?

Checking when large numbers are prime is both challenging and

---

[82] Each Jenny's number preceded by an area code is the sum of 8,675,309 and the three-digit area code × 10^7. This second term has no remainder when divided by 4, because it has 100 as a factor and 100 has no remainder when divided by 4. So, the remainder in the sum after division by 4 is 1 and, as with Jenny's number, each 10-digit Jenny's number that is an odd prime is also a Pythagorean prime.

important. The way to test if a number is prime is by trying to factor it. Shortcuts can help show that a given number is not prime. For example, *Fermat's Little Theorem* states that any integer raised to a prime number, minus that integer, is a multiple of that prime number.[83,ccxxii] For integer 4 and prime 5, $4^5 - 4 = 1{,}024 - 4 = 1{,}020 = 5 \times 204$, which is a multiple of 5. If this is not true for a trial prime number, you know it is not prime. If it is true, the trial prime may or may not be prime.

There are an infinite number of prime numbers. (For small and large primes, see Refs. [ccxxiii] and [ccxxiv].) They are not rare, but they become progressively rarer for large primes, with more composite numbers in between them. Paul Erdős noted: "I'll tell you once, and I'll tell you again. There's always a prime between *N* and 2*N*." (for every positive integer *N*).[ccxxv] This is a restatement of *Bertrand's Postulate* by French mathematician Joseph Louis Francois Bertrand from 1845, which was first proved in 1852, by Russian mathematician Pafnuty Lvovich Chebyshev.[ccxxvi] Still, the largest gap between primes is clearly much smaller than *N*.

Up to $100 = 10^2$, there are 25 primes so there is a prime for every (100/25 =) 4.0 numbers. Up to $1{,}000 = 10^3$, there are 168 primes or on the average of 1 every 5.96 numbers. Up to $1{,}000{,}000{,}000 = 10^9$ there are 50,847,534 primes, a prime every 19.8 numbers, and up to 10^{16} the average frequency is down to 1 in 35.8. This corresponds to an average of ~1 prime for every 2.3n numbers for numbers up to $N = 10^n$, and so the larger the prime number the larger the separation.[84,85,ccxxvii,ccxxviii] Several mathematicians, ranging from French mathematician Legendre to German mathematician Carl Friedrich Gauss to mid-19th century German mathematician Georg Friedrich Bernhard Riemann developed and

---

[83] For prime number p and integer a, this means that $a^p - a$ is a multiple of p. This also means that when a^p and a are divided by p they have the same integer remainder or residue (see below).

[84] The number of prime numbers up to and including N is called the prime-counting function $\pi(N)$; so $\pi(1) = 0$ (since 1 is not considered prime), $\pi(2) = 1$ (for the 1 prime 2), $\pi(3) = 2$ (for the 2 primes 2 and 3), ... For large N, it is between $N/\ln N$ and $(1.25506\,N)/\ln N$.

[85] Because n equals $\log N$, and so $\ln N \sim 2.3\,n$, on average there is roughly 1 prime for per $\ln N$ numbers up to N and so the Nth prime number is very roughly $N \ln N$.

improved such estimates.[ccxxix]

Moreover, the *Twin Primes Conjecture* (again, a conjecture because it is currently unproven) is that there are infinitely many pairs of primes that differ by exactly 2 (the twin primes), such as … 3, 5, 7, … 179, 181, … .

Related to the infinite number of primes, in 1938 Scottish mathematician Robert Alexander Rankin showed that for large numbers N, the largest gap between prime numbers smaller than N (i.e., the longest string of composite numbers) is of order of $\ln N$.[86,ccxxx] Some think this maximum separation may increase even faster, and as $(\ln N)^2$, as Swedish mathematician Harald Cramer noted in 1936.[ccxxxi] The many such expressions in number theory involving ln (pronounced as "log") have led to a "joke" among number theorists: "What does a drowning number theorist say? 'Log log log log …'", as noted by renowned number theorist Terence Tao.[ccxxxii]

Methods that find long strings of composite numbers can help set "weak" lower limits on the separation of primes. Such a string can be constructed starting from the factorial of a given number. Because $101! = 101 \times 100 \times 99 \times \ldots \times 4 \times 3 \times 2 \times 1$, the sum $101! + 2$ is divisible by 2 because both parts are divisible by 2—the 2 from the factorial and from the 2 itself, $101! + 3$ is similarly divisible by 3, and so on until $101! + 101$, which is divisible by 101. So, this string of 102 (= 101 + 1) numbers from 101! to 101! + 101 definitely does not include a prime.

One way to find primes is to eliminate numbers that are not prime. Two millennia ago Eratosthenes of Alexandria developed the simple yet ingenious prime number sieve (Figure 11.1). You list the integers 1 2 3 4 5 6 7 8 9 10 11 and so on. Leave 1 alone and then 2, which is the smallest prime number. Then cross out all multiples of 2 (all of the even numbers, 4 6 8 10 and so on), then all multiples of the next number that are not crossed, 3, and then those of the next one, 5, and so on. Those that are not crossed out are prime. (Show this!) This is, in fact, implicit factoring.[ccxxxiii]

Some deceptively simple ways that seem to generate primes turn out to be just plain wrong. Take 2 as the smallest prime number and add 1 to

---

[86] and, more precisely, at least $(\ln N)[\ln(\ln N)](\ln\{\ln[\ln(\ln N)]\})/\{3\ln[\ln(\ln N)]\}^2$.

it, to get 2 + 1 = 3, which is prime. Now add 1 to the product of the primes 2 and 3 to get $2 \times 3 + 1 = 7$, which is the next prime, and then with the next three in the sequence $2 \times 3 \times 5 + 1 = 31$, $2 \times 3 \times 5 \times 7 + 1 = 211$, $2 \times 3 \times 5 \times 7 \times 11 + 1 = 2311$, you get a prime number—though not the next largest prime number for each. However, the next number $2 \times 3 \times 5 \times 7 \times 11 \times 13 + 1 = 30{,}031$ is not prime (= 59×509), and this is true for many other numbers in the sequence.[ccxxxiv] The same is true for *Fermat numbers*, which are 2 to an exponent that is a power of 2, with 1 then added to it.[87] In 1640, Fermat guessed that all these numbers are prime. This is true for the first five of them, with $2^1 + 1 = 3$, $2^2 + 1 = 5$, $2^4 + 1 = 17$, $2^8 + 1 = 257$, and $2^{16} + 1 = 65{,}6537$, but in 1732 Euclid showed, that $2^{32} + 1$ is not prime because it equals $4{,}294{,}967{,}297 = 641 \times 6{,}700{,}417$. More recently, others have shown that $2^{64} + 1$, $2^{128} + 1$, and $2^{256} + 1$ are not prime either.[ccxxxv]

Cross out multiples of 1 2 3 4 5 6 7 8 9 10 11 12 13 14 15 ...
2: 1 2 3 ~~4~~ 5 ~~6~~ 7 ~~8~~ 9 ~~10~~ 11 ~~12~~ 13 ~~14~~ 15 ...
3: 1 2 3 ~~4~~ 5 ~~6~~ 7 ~~8~~ ~~9~~ ~~10~~ 11 ~~12~~ 13 ~~14~~ ~~15~~ ...
5: 1 2 3 ~~4~~ 5 ~~6~~ 7 ~~8~~ ~~9~~ ~~10~~ 11 ~~12~~ 13 ~~14~~ ~~15~~ ...
7: 1 2 3 ~~4~~ 5 ~~6~~ 7 ~~8~~ ~~9~~ ~~10~~ 11 ~~12~~ 13 ~~14~~ ~~15~~ ...

Figure 11.1: Prime number sieve.

The search for prime numbers has relied on computers for some time. The largest prime number found as of the time of this writing is $2^{82{,}589{,}933} - 1$ (found on Dec. 7, 2018),[ccxxxvi] which has 24,862,048 digits, but there are many smaller prime numbers that remain unknown. This particular prime number is called a *Mersenne prime*, so named after a 17th-century French theologian and mathematician who studied them. This 51st known Mersenne prime was found as part of the Great Internet Mersenne Prime Search (GIMPS)[ccxxxvii] and is among the much larger set of known primes. *Mersenne numbers* have the form $2^n - 1$, and they can be prime (and then called a Mersenne prime) only when the exponent *n*

---

[87] So, the Fermat numbers are $2^{2^n} + 1$ for $n = 0, 1, 2, 3, 4$ and so on, so the exponents are 1, 2, 4, 8, 16 and so on.

is prime—but as with previous methods, not all such generated numbers are prime.[ccxxxviii] It is prime for prime exponents up to 7, such as $2^2 - 1 = 4 - 1 = 3$ and so on, but not for 11 because $2^{11} - 1 = 2{,}047 = 23 \times 89$, and then again it leads to Mersenne primes for prime exponents from 13 up to 19, but not for 23 because $2^{23} - 1 = 8{,}388{,}607 = 47 \times 178{,}481$, and so on. Still, this approach is significant because it is a way of generating potential large prime numbers, which are then tested by trying to factor them. Apple Computer uses very large Mersenne primes for rapid message encryption and decryption.[ccxxxix]

Mersenne primes are also interesting because of how they are related to perfect numbers. Euler showed that one half of a Mersenne prime plus one, times the prime itself[88] is a perfect number. Eighteenth-century Swiss mathematics Leonard Euclid then showed that these are the <u>only</u> even perfect numbers. (All currently known perfect numbers are even.)[ccxl]

Many wondrous properties of numbers can be found online, including whether or not they are prime.[ccxli]

Armed with the knowledge that integers have curious properties and relate to each other in amazing ways, we enter the world where integers are used to control and assist many facets of our lives.

---

[88] This equals $(2^n)(2^n - 1)/2 = (2^{n-1})(2^n - 1)$.

Part III

The Second World of Math: The Math of Doing

We now enter the second world of math, the one that controls our lives, "the math of doing," first by addressing the subworld of the arithmetic of whole numbers, "the math of the digital world."

Chapter 12

The Math of the Digital World: Modular Arithmetic (or Using Number Leftovers)

We use whole numbers to present addresses, credit card numbers, and so on, but not when we measure object dimensions, temperatures, and the like, where using fractions and decimals is often essential. A distinct way of performing the arithmetic of whole numbers has been developed, modular arithmetic. It is identical to ordinary arithmetic, but its results are presented in a different manner, with integral leftovers after division and not fractions or decimals. Though not obvious, this approach can be very productive. It has close connections with the math of whole numbers of Chapters 7 and 11, especially that related to math performed in different base systems.

12.1 Modular Arithmetic: When Does 7 Divided by 3 Give 1?

You can always convert a certain number of dollars into 4× as many quarters, but you cannot always convert a certain number of quarters entirely into dollars. With 9 quarters, you can receive two dollars, but will have one quarter leftover. More generally, you could have 0, 1, 2 or 3 quarters leftover.

More generally, when you divide a whole number (the *dividend*) by a factor (the *divisor*), you could get an integer, as with 36/12 = 3.0. Another way of looking at this is you obtain the integer 3 with no integer remainder (or an integer remainder of 0). You can express 36/10 as 3.6 (which is the quotient), or the integer 3 (which in this type of math is also called the *quotient*) with an integer *remainder* or leftover of 6. In shorthand, for the quotient: 3 = 36 **div** 10 and for the remainder: 6 = 36 **mod** 10 (or 36 (**mod** 10) or 36 modulo 10). (Mod is the abbreviation for modulus or modulo and is pronounced in the standard way.) The answer to the first example can be written as 3 = 36 **div** 12 and 0 = 36 **mod** 12. The answer to our earlier example of converting quarters into dollars can be expressed as 2 = 9 **div** 4 and 1 = 9 **mod** 4. This way of doing arithmetic is called *modular arithmetic* (and is part of modular math), and was developed by Carl Friedrich Gauss. This analysis based on remainders is very important in unexpected ways, as in message encryption.[ccxlii]

When you subtract from 1, 4, 7, 10, 13, or 313 the largest multiple of 3 that does not exceed the number, the remainder or *residue* is 1. Consequently, each is *congruent to 1 modulo 3* (so it acts just as does 1 after removing the highest factor of 3). Any pair of these numbers are said to be congruent modulo 3 because their differences are divisible by 3, so modulo 3, 7 is congruent to 313 and this can be represented by $7 \equiv 313$. In modular arithmetic, 7 and 313 are the same number modulo 3, though $7 \neq 313$ in the usual sense.

We routinely use modular arithmetic. Our system of currency is modulus 100. $12.47 means 12 dollars and 47 cents, where 1 dollar equals 100 cents. It is also 1,247 cents. So, 12 (dollars) = 1247 **div** 100 and 47 (cents) = 1247 **mod** 100. In the game bingo, numbers 1-15 correspond to the B column, 16-30 to I, 31-45 to N, 46-60 to G, and 61-75 to O. If you take the number and subtract 1, and divide it by 15 (the divisor) the B column has quotient of 0, I has 1, N has 2, G has 3, and O has 4. Odometers on cars used to be modulus 100,000, so the odometer would read the same if a car had 95,000 miles or 195,000 miles. Odometers now often range from 0 to 999,999 miles, and so are modulo 1,000,000.

Our system of time is based on modulus arithmetic, on multiple scales. Modulo 60, the number 134 has a quotient 2 and remainder 14 ($134 = 2 \times 60 + 14$), so 134 seconds is 2 minutes and 14 seconds, and also 134 minutes is 2 hours and 14 minutes. So, seconds are modulo 60 (before the next minute starts), as are minutes (which turn into hours). Modulo 24, the number 77 has a quotient of 3 and remainder of 5 (because $77 = 3 \times 24 + 5$), so 77 hours is 3 days and 5 hours. In the 24-hour clock common outside the U.S., the 24 hours are labeled from 0 to

23 (as in 00:01 a minute after midnight), corresponding to hours 1, 2, 3, …, 24. In the 12-hour clock the hours are labeled 12, 1, 2, … 10, 11 AM (as in 12:01 AM a minute after midnight), and then 12, 1, 2, … 10, 11 PM (as in 12:01 PM a minute after noon), corresponding to hours 1, 2, 3, …12 in the AM and hours 1, 2, 3, …12 in the PM. So, the 12-hour clock is essentially the 24-hour one, but modulo 12.

Both clocks turn into days. The days of the week are congruent modulo 7. If you call Sunday the first day of the week, Tuesday is day 3 of the week, whatever the week. Because months have different numbers of days and weeks, there is no correspondence there, but when the month is over, we return to day 1 of the next month. Months of the calendar are modulo 12 (turning into years). Modulo 365, the number 757 has a quotient 2 and remainder 27 (because $757 = 2 \times 365 + 27$), so 757 days is 2 non-leap years and 27 days, and so days of the non-leap year calendar are also modulo 365 (turning into years).

The number 9 is special in modular math. We know that $1 = 1$, $10 = 9 + 1$, $100 = 99 + 1$, $1000 = 999 + 1$, and so on. So, for example, $4{,}173 = 4 \times 1000 + 1 \times 100 + 7 \times 10 + 3 \times 1$ also equals $4 \times (999 + 1) + 1 \times (99 + 1) + 7 \times (9 + 1) + 3 \times 1$. Because 4×999, 1×99, and 7×9 and are multiples of 9, the remainder of 4,173 modulo 9 is just the sum of those other terms that do not have 9, 99, 999, and so on as a factor, and this is $4 + 1 + 7 + 3$, and this is exactly the same as summing the digits in the number. Both sums equal 15, which is congruent to 6, modulo 9. This means that you can find the remainder of a number modulo 9, by adding its digits, which is called *casting out nines*.[ccxliii]

Casting out the nines is a good, though not perfect, check of addition, subtraction and multiplication (meaning that it can find many errors, though not all of them). For multiplication, you cast out the nines of the numbers you want to multiply and then multiply these remainders, and this result will have the same remainder modulo 9 as the product. (You can prove this by writing out each number in terms of $10 = 9 + 1$ and so on, as above.) For example, for 23×48 the remainders are 5 for 23 and 3 for 48; their product 15 has remainder 6, which is also the remainder for final product $23 \times 48 = 1{,}104 = 9 \times 122 + 6$. This method will tell you that the product 23×48 is not 1,105 (which has a remainder of 7), 1,106, 1,107, 1,108, 1,109, 1,110, 1,111, or 1,112, but it does not tell you it is not 1,113, which also has remainder 6.[89] Also, casting out nines will alert you to a mistake caused by only one digit being replaced by another (as for the 1,105, 1,106, 1,107, 1,108, 1,109 listing here), except for a change of a 0 for a 9 or a 9 for a 0 (which do not change the remainder). We will see below that similar approaches are used to check for errors when using ISBN book identification numbers.

12.2 Using Modular Arithmetic to Make the World Work and Stay Safe

Modular arithmetic is prevalent in a vast range of activities, including some of your frequent interactions, so it is important to understand how this subworld applies its fundamental concepts. It also underpins much of the digital world.

12.2.1 Making solar calendars work and does anyone really know what day of the week it is?

It is obvious that if March 10 is a Wednesday, so is March 3. But, determining the date of the prior Wednesday requires you to know the number of days in the prior month, and, in this case, whether it is a leap year. Modular arithmetic provides shortcuts in determining the day of the week for any date, a perpetual calendar (aside from the now even faster way of finding the day online). To do this, say for January 10, 1925, we first need to account for leap years.

The Egyptian calendar had the same numbers of days per month as we have now for a non-leap year, with 365 days per year. Every year is about a quarter of a day longer than the calendar would indicate, $\approx$365.2422 days long, so after approximately 4 years you need an extra day (Section 8.1.2). Julius Caesar (supposedly) simplified earlier calendar corrections with the Julian calendar by adding February 29 every fourth or leap year. This new calendar was still too long by $\approx 365.25 - 365.2422 = 0.0078$ days per year. After 1,500 years, the accumulated error amounted to calendars being $\approx 0.0078 \times 1{,}500 = 11.7$ days too slow. In 1582 Pope Gregory adjusted for the slippage of 10 of

[89] By the way, we know that the answer 1,113 is wrong for a different reason (aside from just being wrong); this product must end in a 4 because $3 \times 8 = 24$.

these days by redefining Oct. 5 to be Oct. 15, and removed some leap years to give us our current, Gregorian calendar. If every 100th year were not a leap year, then there would be 24 leap days per century or 24/100 = 0.24 days per year for a total of 365.2400 days per year, which is better but still too few. But, if every 400th year were still a leap year, then there would be 97 leap days per 400 years or 97/400 = 0.2425 days per year for a total of 365.2425 days per year, which is too long by only ≈0.0003 days a year or ≈1 day every 3,000 years, so changes will need to be made some time. (Furthermore, leap seconds are added periodically to keep clocks synchronized with the rotation of the Earth.[ccxliv])

To calculate the day of the week for any date, you need to know the day for any one date in the Gregorian calendar, say an earlier date, and count the number of days from that reference date modulo 7, accounting for the different lengths of the months in a non-leap year, and for leap years. If your answer is 0, the day of the week is the same and as the reference. If it is 1, the day is one day later in the week and so on.

Because it is easier to do this if the leap day is made the last day of the year, for the purpose of the calculation January and February are considered the 11th and 12th months of the previous year, and March is considered the 1st month of the current year, April the 2nd month, … and December the 10th. So the calendar year for the analysis for January 10, 1925 is 1924 and the month is 11. The century used for this date in the calculation is 19 (even though it was technically the 20th century) and the year of the century is 24, and the day is of course 10. The day of the week is then the sum of the following, modulo 7, where 0 means Sunday, 1 means Monday, … 6 means Saturday: (1) the day of the month, (2) the integer smaller than or equal to 2.6 × the month minus 0.2 (which surprisingly accounts for the different number of days in each month), (3) minus twice the century, (4) the year of the century, (5) the integer smaller than or equal to the year of the century divided by 4 (to account for leap years), and (6) the integer smaller than or equal to the century divided by 4 (to account for leap years).[90]

---

[90] Algebraically, $w = [k + (2.6m - 0.2) - 2c + y + (y/4) + (c/4)]$ **mod** 7, for the day of the week w, where k is the day of the month, m is the month, c is the century, and y is the year in the century, which are all defined in the text.

What day of the week did the first person step on the moon? This occurred on July 20, 1969 (EDT). The sum is 20 (for the day of the month) + 12 (for 2.6 × 5 (for July) – 0.2 = 12.8) – 38 (for 2 × 19) + 69 (for the year) + 17 (for the year 69/4 = 17.25) + 4 (for the century 19/4 = 4.75) = 84. Because 84 = 12 × 7, it is 0 modulo 7, so the first person landed on the moon on a Sunday.[ccxlv] By the way, January 10, 1925 occurred on a Saturday because 10 + 28 – 38 + 24 + 6 + 4 = 34 and 34 **mod** 7 = 6.

12.2.2 Scheduling round robin tournaments

You want to conduct a tournament with N teams or people, say to determine the Little League Baseball or chess champion, and need to schedule one contest between each pair of teams, having all teams compete in round 1, all then against different teams in round 2, and so on. Since each team will play $N - 1$ teams, you need to schedule $N - 1$ rounds in this round robin tournament. This is easy to do using modular arithmetic because it helps cycle through each pairing once. For simplicity, we first assume there are an even number of teams, and then number the teams 1, 2, … N. In each round, all pairs of teams whose team numbers sum to the number of the round modulo $(N - 1)$ play each other, with one exception. Team N plays the one team each round for which twice the team number modulo $(N - 1)$ is the number of the round. After round $N - 1$, each team will have played each other exactly once. When there are an odd number of teams, say 7, one dummy team is added and called the last team, here 8, and the team playing that team has a bye that given round.

This is clearer with an example. Suppose you need to set up 3 rounds for 4 teams. In round 1, team 1 plays team 3 because (1 + 3) **mod** 3 = 1 (because it is round 1). Obviously, team 2 must play team 4 because it is the remaining pairing, but also because team 2 is the one that would need to play the last team, team 4, because (2 × 2) **mod** 3 = 1 (for round 1). In round 2, team 2 plays team 3 because (2 + 3) **mod** 3 = 2 (for round 2) and it is team 1 that would need to play team 4 because (2 × 1) **mod** 3 = 2. In round 3, the round number **mod** $(N - 1)$ is 3 **mod** 3 = 0. Team 1 plays team 2 because (1 + 2) **mod** 3 = 0 (for round 3) and team 3 would have to play team 4 to complete the tournament; also (2 × 3) **mod** 3 = 0 (the round number (**mod** 3)) (Figure 12.1). This method is overkill for only 4 teams because it is obvious who plays whom, but it is very helpful for tournaments with many teams. It provides a way of scheduling teams without favoritism and without concern of the relative strengths of teams,[ccxlvi] especially when the team numbers are assigned randomly.

	Team 1 plays	Team 2 plays	Team 3 plays	Team 4 plays
Round 1	3	4	1	2
Round 2	4	3	2	1
Round 3	2	1	4	3

Figure 12.1: Round Robin tournament example with 4 teams.

12.2.3 File storage - modular

Let's say you want to assign the files of say several hundred people to a unique location in memory with unique addresses. You want to use their nine-digit number identification numbers, in years past it would have been their Social Security number in the U.S. It would be foolhardy to buy a memory system large enough to have a location associated with each of the billion possible ID numbers, so you would want to reduce their ID numbers to say a thousand or so different address numbers. If you used only the last three or four digits of their ID numbers, the system would be suitably small but pairs of people could end up with the same location numbers. The trick is to assign the location number to be the ID number (called the *key* or *k*) modulo a given modulus *m*. This assigning code is called the *hashing function*, such as $h(k) = k$ **mod** m. Choosing the modulus larger than the number locations is one step to unique memory locations. If you used the (invalid) Social Security Number 666-12-3456 and modulus 113, the location assignment would be 666,123,456 **mod** 113 = 95 for this particular hashing function. However, some types of moduli lead to nonunique assignments. If you have 350 ID numbers and used the modulus 1,000, both 038-45-4783 and 052-10-9783 would match to 783. Prime numbers, such as 113, are chosen as the modulus to avoid this problem.[ccxlvii]

12.2.4 ISBN numbers

Modular arithmetic is used to check that the ISBN (International Standard Book Number) number used to identify a given book is a valid type of number. Books published before 2007 have ten-digit ISBN numbers; more recently assigned numbers, for newer books and editions, have 13 digits. These series of numbers include a set identifying the publisher and the specific book number assigned by the publisher, so they are unique. In both systems, the last digit is a check number (or digit) chosen so that the two most common errors introduced in copying ISBN numbers can be checked with modular arithmetic. These are a single incorrect digit and the transposition (switching) of neighboring digits. This approach cannot identify most other types of errors.

For the older 10-digit ISBN number, this check is performed by first multiplying each digit by a specific number, by 1 for the first number, 2 for the second, … 10 for the tenth, and then summing them. The 10th number, the check number, is assigned so this weighted sum is 0, modulo 11, for the valid ISBN. This is effectively base 11, so there are 11 possibilities for this number, 0, 1, 2, …8, 9, X (where X is 10 in base 10). For example, the ISBN-10 number 0-12-342070-9 is valid because $1 \times 0 + 2\times 1 + 3\times 2 + 4\times 3 + 5\times 4 + 6\times 2 + 7\times 0 + 8\times 7 + 9\times 0 + 10\times 9$ is 198, which is 18×11, so it is 0, modulo 11. If exactly one of the digits were incorrect, the modular sum would not be 0. If a 3 were changed to 7, it would be congruent to $7 - 3 = 4$. Whereas this type of error would be detected with either a simple, unweighted sum or this weighted sum, the transposition of adjacent digits or another pair of digits is caught only with this weighted sum. For instance, if the 7th digit were 5 and the 8th digit were 2 the weighted sum would increase by the difference of $2\times 7 + 5\times 8 = 54$ and $5\times 7 + 2\times 8 = 51$, each modulo 11, or by 3; the weighted sum of 3 indicates that it is an invalid number.

In the weighted sum of the current 13-digit ISBN number, the odd-numbered digits are multiplied by 1 and the even-numbered digits by 3. The 13th number, the check number, is assigned so this weighted sum is 0, modulo 10, for the valid ISBN. For example, the ISBN-13 number 978-3-319-23930-9 is valid because $1\times 9 + 3\times 7 + 1\times 8 + 3\times 3 + 1\times 3 + 3\times 1 + 1\times 9 + 3\times 2 + 1\times 3 + 3\times 9 + 1\times 3 + 3\times 0$ is 101 and adding the final 9 gives 110, which is 11×10, so it is 0, modulo 10. The use of the 13-digit number and modulus 10 avoids the confusion of sometimes using X as the last digit, but not all transpositions are caught. For instance, no error is detected if two adjacent numbers that differ by 5 are transposed because the difference in the multiplying factors $3 - 1 = 2$, and the sum will be larger or smaller by 10, so a congruence of 0 remains 0. By the way, valid 10-digit ISBNs are converted to valid 13-digit numbers by using the prefix 978 and a new correct check digit.[ccxlviii]

12.2.5 Codes and encryption

When you deliver a message in secret, you first encrypt or encipher it using a specific cipher. You then transmit it to the intended receiver who needs to decrypt or decipher it by using the same specific cipher, used in reverse. Many ciphers are based on modular arithmetic.[ccxlix]

If you write a message that uses the 26 English letters (only) and replace each original letter with a unique code letter, a receiver who knows how this was done can decode it. How many ways can this be done? You can replace every A in the *plaintext* by one of the 26 letters, then replace B by any one of the remaining 25 letters, and so on until you choose the remaining letter when you reach Z. There are $26 \times 25 \times 24 \times 23 \ldots \times 2 \times 1 = 26!$ or 4×10^{46} ways to convert the plaintext to the coded *ciphertext* by using this *character or monographic cipher*. Because of the *one-to-one correspondence* (or mapping), the receiver who knows the chosen code can simply convert the ciphertext to the original plaintext.

The downside of this general cipher is the need to define it with 25 numbers that need to be transmitted between the encrypter and decrypter (with the 26th one being the remaining one). It is simpler to define the cipher by using directions. The first cipher, supposedly developed by Julius Caesar, replaces letters by simply shifting the alphabet, by replacing the 26 English letters A, B, C through Z by the positive integers 0, 1, 2, … 25, and then adding 3, the *shift number*, to it modulo 26. So, the 0 of A becomes 3, which corresponds to D, the 1 in B becomes 4 or E, and so on until the 22 for W becomes 25 for Z. The 23 for X become 26, but 26 **mod** 26 is 0, the X is replaced by A, and the Y by B, and Z by C. Using the remainders in modular arithmetic allow the alphabet to loop once, with each letter uniquely corresponding to another letter. In the encryption phase, the letters are grouped in series of 5 letters, ignoring the spaces in the original words, so the code cannot be guessed on the basis of the number of letters per word. In the decryption phase, 3 is subtracted from each number for the letter, again modulo 26.

In the more general shift transformation, you encrypt by adding any one number from 1 to 25, instead of 3, as the shift number, and decrypt it by subtracting this number, both modulo 26. (Adding 0 leads to no change and any number 26 or larger gives no new possibilities modulo 26.)

In the more general *affine* transformation (take affine to mean linear), you first multiply the number for a letter by a *multiplier number* and then add the shift number, which can now range from 0 to 25. This works only if the multiplier number is chosen so it uniquely maps each of the 26 letters to one distinct letter. With 2 as the multiplier and 0 as the shift number, the 3 for D becomes 6 for G, but the 16 for Q becomes $32 \textbf{ mod } 26 = 6$, which corresponds to G. D and Q cannot both map to the same letter, so 2 cannot be a multiplier. The mapping works well if the multiplier and 26 have no common factors other than 1. There are 12 choices for this multiplier number and with the 26 choices for the shift number, these two numbers define $12 \times 26 = 312$ distinct transformations; this includes the one with multiplier of 1 and adder of 0, which leads to no change. You can get the plaintext from the ciphertext by subtracting the shift number and then multiplying by what is called the inverse of the multiplier number and then evaluating it modulo 26. Though it is not obvious at all, the inverse of the multiplier is the multiplier raised to a power 11, and then the remainder is found modulo 26. (This uses the Euler phi-function from Section 11.2. The power 11 is one less than the Euler phi-function of 26, which is 12.)

These codes are far too simple to be of use today because all possible decryptions are quickly evaluated by using even very modest computer speeds. Moreover, cracking *such single character or monographic ciphers* is made much easier by analyzing the frequency of letters in longer codes. ~13% of all letters in common English text are E, 9% are T, 8% are I, N or R, 7% are A or O, and so on.

Advanced encryption methods overcome these limitations, as with those *block or polygraphic codes* composed of blocks of letters. The Vignere cipher from the 16th century is a modification of the shift transformation that uses a key of a given length to modify the plaintext in blocks of that length. For example, the very simple key XKT (with X letter 23, K 10, T 19) operates on blocks of three letters; the first three letters in the plaintext are respectively shifted by 23, 10, and 19 (modulo 26), and this is repeated for the next three letters, etc. So, with key PHYSICAL (or 15 7 24 18 8 2 0 11) becomes (38 17 43 41 18 21 23 21), which modulo 26 is (12 17 17 15 18 21 23 21), and this corresponds to MRRPSVXV. Note that repeating letters in this ciphertext do not correspond to repeating letters in the plaintext. Decryption uses the same code to shift the ciphertext in the opposite direction. The blocks are correspondingly longer with longer keys.

The Hill cipher is a *diagraphic cipher* that modifies sequential pairs of letters (with an X added at the end for an odd number of letters) using a variation of the affine transformation with no shift. The ciphertext letters are mixes of the plaintext letters. Each pair of sequential numbers, with number *plain1* and *plain2*, is converted to the pair *cipher1* and *cipher2* by using a key with four code numbers between 0 and 25, labeled as *code1*, *code2*, *code3*, and *code4*. *Cipher1* equals (*code1* × *plain1* + *code2*×*plain2*) **mod** 26. *Cipher2* equals (*code3*×*plain1* + *code4* × *plain2*) **mod** 26. (This type of routine is formally called matrix multiplication.) If you know the key, you can find the inverse key and decrypt the ciphertext.

In the *Vernam Cipher*, a simple type of a *stream cipher*, each letter in the plaintext is converted into a series of bits, to give the *plainstream* bit stream. If you wanted to use the ASCII code to do this, for example, "a" would be 97 in decimal and the eight bits 01100001. The code could be chosen totally randomly, as with using dice, or as the ASCII code for a number, letter or symbol, also chosen randomly, such as "+", which is 43 in decimal and 00101011 in binary, to give the *keystream bit stream*. The corresponding bits in the plainstream and the keystream bit streams are added in binary, ignoring any carryover. So, a key bit of 0 does not change the 0 or 1 in the plaintext and a key bit of 1 changes a 0 to 1 and a 1 to 0. This is the same as the "exclusive or," XOR, operation. For this *plainstream* bit stream 01100001 and keystream bit stream 00101011, the *cipher bit stream* is 01001010 (which is 74 in binary and ASCII character "J").[ccl] The ciphertext is converted back to plaintext stream by adding the key stream to it in the same way. Other stream ciphers can use other bases and key streams to shift letters. In one type, the key for one letter is the plaintext of the previous letter.

Ciphers based on (non-affine) exponentiation, *exponentiation ciphers*, are more difficult to decrypt. The plaintext letters are expressed from 0 to 25, but with single digits written as two digits, such as the 7 for H expressed as 07. The plaintext is expressed as even blocks of digits. The ciphertext block is this plaintext block raised to a positive integer called the enciphering key, and this is done modulo an odd prime. The *enciphering key* and *odd prime modulus* are the two cipher keys. For this to work well, they can have no common factor others than 1. For odd primes between 2,525 and 252,525, the block length needs to have 4 digits, representing 2 letters. (Once again primes are important!)

The *RSA cryptosystem* is well known and is used widely; it is named after Ron Rivest, Adi Shamir, and Leonard Adleman, who first publicly presented it in 1978.[ccli] It is a version of the exponentiation cipher, with the modulus being the product of two large prime numbers. The phi-function of this modulus (which is the number of integers smaller than and relatively prime to it) and the exponent must have no common factors other than 1. The keys for the other cryptosystems we discussed need to be private because once the encryption key is known, the decryption key can be found "quickly" using computer analysis. In contrast, the RSA encryption key of the modulus and the exponent constitute a *public key* known to all so all can encode with it, but the actual prime numbers used are secret. Furthermore, the decryption key is kept private and cannot be determined "fast enough" to be useful even with modern computation for large enough secret primes, because of the difficulty in factoring large integers. One motivation to find larger primes is to keep up with ever faster computational capability of factoring large integers.

Likely, new modes of encryption will be needed as new computing methods evolve.[cclii]

12.2.6 Generating pseudorandom numbers

In Chapter 8 we saw an early method of generating pseudorandom numbers. One better way uses modular arithmetic and the *linear congruential method*.[ccliii] To obtain a short series of pseudorandom numbers, you can choose a very small *modulus*, say 5. You need three more numbers: a *multiplier* not smaller than 2 or larger than the modulus minus 1 (which is $5 - 1 = 4$ here), say 3, and an *increment* and *seed* (the initial number) that are both not smaller than 0 or larger than the modulus minus 1, say 1 and 4 here. The first number comes from multiplying the multiplier by the seed, adding the increment, and then finding its remainder for the modulus given.[91] Here, the product is $3 \times 4 = 12$ and adding the increment 1 gives $12 + 1 = 13$ and 13 **mod** 5 is 3 (or $((3 \times 4) + 1)$ **mod** $5 = 3$). This is the first pseudorandom number and also the seed for the next step, to get the second number is $((3 \times 3) + 1)$ **mod** $5 = 0$. The third number is $((3 \times 0) + 1)$ **mod** $5 = 1$. The fourth number is $((3 \times 1) + 1)$ **mod** $5 = 4$, which is the seed here. So, the fifth number $((3 \times 4) + 1)$ **mod** $5 = 3$ repeats the sequence and so this particular sequence (3, 0, 1, 4) repeats after 4 numbers in this simple example. The maximum number of terms that can be generated without repeating using a given modulus is the modulus itself (0, 1, 2, …, $m - 1$ for modulus m) so, for this example only 4 terms of the 5 potential terms (0, 1, 2, 3, and 4) are generated. Computers can use this method with much larger moduli and often with an increment of 0 to generate very long series of pseudorandom numbers; they are often *normalized* to be between 0 and 1 by dividing them by the modulus. A modulus of $2^{31} - 1 = 2{,}147{,}483{,}647$ and multiplier $7^5 = 16{,}807$ are often used. They can generate $2^{31} - 2 = 2{,}147{,}483{,}646$ numbers before the pattern repeats.

Frequent generation of random numbers is central to modern encryption and computer security, but problems occur because these numbers can be less random than advertised."[ccliv]

To rule the math in our lives and be comfortable with it, we need enter the world of the math of "doing" and its first subworld of the math of the "what will be."

---

[91] For modulus m, multiplier a, increment c and seed x_0: with $2 \leq a < m$, $0 \leq c < m$, and $0 \leq x_0 < m$, the $n + 1^{st}$ term is obtained from the nth term by using $x_{n+1} = (ax_n + c)$ **mod** m for $n = 0, 1, 2, \ldots m - 1$; x_n/m gives numbers between 0 and 1. In the given example, $x_{n+1} = (3x_n + 1)$ **mod** 5 with $x_0 = 4$.

Chapter 13

The Math of What Will Be: Progressions of Growth and Decay

We have encountered different sets and combinations of numbers. Some are listings of different numbers, such as 4, -10, 37, …, which can be called *sequences*. Others are the sums of different numbers, such as 4 – 10 + 37 + … or different functions, which may approximate or define a number or function, as we saw for e and e^x, and this sum is a *series*. Sequences and series can contain a finite number of terms or an infinite number of them, as in an infinite series. Sometimes, the limiting terms and sums in sequences are quite important. Because these sequences are often projections from present values to those expected in the future, they are common in the world of the math of "doing," within the subworld of predicting "what will be."

13.1 Sequences and Progressions

1, 2, 3, 4, 5, …. is a sequence of integers, while 1, 4, 9, 16, 25, … is a sequence of squares of integers, and 1, 8, 27, 64, 125, … of cubes. 1, 1/2, 1/3, 1/4, … is a sequence of reciprocals. 2, 4, 8, 16, 32, 64, 128, … is the sequence of 2 raised to the powers 1, 2, 3, 4, 5, 6, 7, …, and 3, 9, 27, 81, 243, 729, 2,187, … is the analog for 3. The sequence 1, 2, 6, 24, 120, 720, … is the sequence of the factorials 1!, 2!, 3!, 4!, 5!, 6!, …. Databases currently have over 200,000 different integer sequences.[cclv] Most sequences consist of numbers other than integers. In any case, when the terms in a sequence are determined by a formula, such as the ones presented here, the sequence is often called a *progression*.

There are also sets of numbers that do not relate to each other and sequences that are trivial. Consider strings of 5 numbers, with String 1 being 6 72.2 160.9 12132014 280.98 and String 2 being 17 65.8 130.5 01102016 120,421.34. These strings could mean that the person with ID number 6 is 72.2 inches tall, weighs 160.9 pounds and on December 13, 2014 has $280.98 in the bank, while the person with ID number 17 is 65.8 inches tall, weighs 130.5 pounds and on January 10, 2016 has $120,421.34 in the bank. Digital images of photos and facial scan information are also strings of numbers.

Some sequences are meant to be amusing or whimsical. What is the next term in the sequence 1, 2, 2, 3, 3, 3? It is clearly 4, because 1 is given one time, 2 is given two times, and so on. What is the missing term in the sequence 16, 06, 68, 88, __, 98? (Spoiler alert: If you look at this page upside down, you will see it is 87.) What is the next term in the sequence 1, 11, 21, 1211, 111221, 312211, 13112221? (Spoiler alert: 1113213211.) This is "The Look-and-Say Sequence": one, one one, two ones, one two one one, … that mathematician John Conway helped popularize.[cclvi] You can start such a sequence with any number, such as 55555, which is followed by 55, 25, 1215, 11121115, … Other sequences have more significance.

In the detective novel *Cat of Many Tails*, after being recruited by the police to find a serial killer, fictional sleuth Ellery Queen noticed that the ages of the 7 victims (at the time) monotonically decreased from one murder to the next by the sequence 44, 42, 40, 37, 35, 32, 25.[cclvii] He realized that there was a very low probability that their ages would decrease every time if the victims had been chosen at random (Chapter 15) and that it was surprising that this decrease in age was nearly the same each time. However, it was not exactly the same (whether or not the ages were rounded to the lower whole number), so there was a clear pattern in the sequence but no precise formula; this proved to be a useful clue!

When the next term in a sequence depends on the value of the "current" term or on the current term and previous terms, it is a progression determined by using a relation defining it called the *recurrence relation*, and this is called *recursion*. Such progressions start at an initial value or condition, and may or may not have explicit units (such as dollars, meters, days or pounds).

If you start an interest-free bank account with $300 and add $100 to it the first of every month, the balance on the first of every month will be \$300, \$400, \$500, \$600 and so on. Say you fill up your car gas tank so it has 15.2 gallons and the car travels 40 miles per gallon. If you monitor how much gas you have every 20 miles, you will obtain the sequence 15.2 14.7 14.2 13.7 and so on gallons. These are examples of *arithmetic progressions*. They have a recurrence relation in which a fixed number is added to or subtracted from the current term to get the next term.[92] The *arithmetic average* or mean of such progressions is the sum of all terms divided by the number of terms, which equals the middle value when there are an odd number of terms.

If instead of fixed monthly deposits, 10% of its current value is added to a bank account at the first of every month, the balance will be \$300, \$330, \$363, …, changing by a factor of 1.1 each month. This is the basis of compound interest. In the sequence 10, 20, 40, 80, 160, … the recurrence relation is that the next term is twice the current term. These are examples of *geometric progressions*, because the recurrence relation is multiplication or division by a fixed number.[93] If you start with 3 and multiply each term by 2, the progression is 3, 6, 12, 24, 48, …, from 3, 3×2, 3×2^2, 3×2^3, 3×2^4, … If you divide each term by 5 or multiply it by 0.2, the progression is 3, 0.6, 0.12, 0.024, 0.0048, … The geometric average or mean of such a progression of n terms is the nth root of the product of all terms. For two terms it is the square root of the product of the two terms. For an odd number of terms, it equals the middle value.

Figure 13.1 contains bar graphs that compare the first few terms in an arithmetic progression that starts with 2, with 4 added to each subsequent term, and that of a geometric progression that also starts with 2, but with each subsequent term multiplied by 3.

In an arithmetic progression, you can make the added or subtracted terms very small, perhaps increasing the total number of terms at the same time. In a geometric progression, you can make the multiplied or divided factor 1 plus or minus a very small number (much smaller than

---

[92] If the nth term in the progression is called a_n (with the subscript n), then the recurrence relation giving the $n+1^{\text{st}}$ term is $a_{n+1} = a_n + d$, with d being a number.

[93] The recurrence relation is $a_{n+1} = ra_n + d$, with r being a number, and n being the number of the term.

1), also perhaps increasing the total number of terms at the same time. In performing this you are performing a standard operation in calculus.

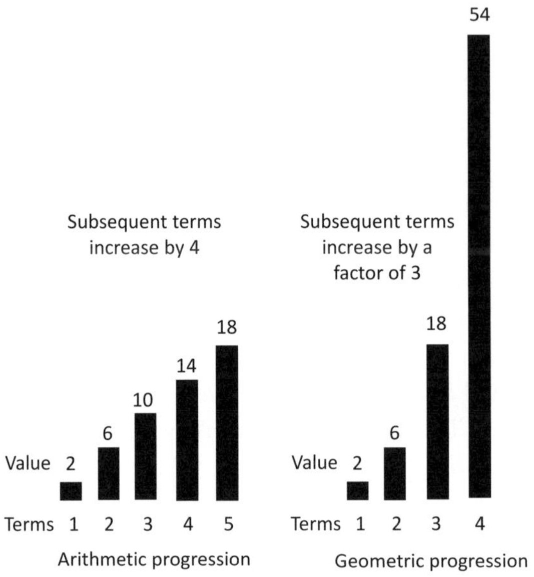

Figure 13.1: Examples of progressions.

Sometimes the evolution of a regular sequence is the goal; at other times the sum or the product of its terms is the goal. Take the arithmetic progression 110, 114, 118, 122, 126, 130, which has an even number of terms. Summing the first and last terms gives $110 + 130 = 240$. This is the same sum as adding the second and next-to-last terms $114 + 126 = 240$, and the third and third-from-last terms $118 + 122 = 240$. The total sum is $240 + 240 + 240 = 720$, which equals the sum of the first and last term times the number of pairs, $240 \times 3 = 720$, and the average of the first and last terms, $(110 + 130)/2 = 240/2 = 120$, times the number of terms, or $120 \times 6 = 720$. If there are an odd number of terms, the middle term is then added to the sums of the pairs. This is a general procedure for determining the sum of a finite arithmetic progression, and it equals the number of terms times the arithmetic average.[94] Whether or not it is finite, this sum is also called an *arithmetic series*. This leads to the question: *Is it easier to negotiate one agreement among 10 countries with widely differing interests or a series of bilateral agreements for each pair of countries, each of which may be easier to negotiate?* The deciding factor may be the sheer number of bilateral agreements, which is the sum of an arithmetic progression. The first country needs to negotiate 9 treaties. The second country needs to negotiate with the 8 remaining countries, and so on, until the 9th country needs to negotiate with only the 10th one. The number of needed bilateral agreements would be $9 + 8 + 7 + 6 + 5 + 4 + 3 + 2 + 1 = [(9+1)/2]9 = 45$.

There is an analogous procedure for summing a finite number of terms in a *geometric progression*, similarly called a *geometric series* whether or not it is finite. Take the geometric progression that starts with 1, with the next term multiplied by 3, that ends after 5 terms: 1, 3, 9, 27, 81. This same sequence with each term multiplied by 3 is: 3, 9, 27, 81, 243. The sum of this second sequence is clearly 3× that of the first sum. It is also equal to that of the first sum after you add its last term, 243, and subtract the first term of the first one. This means that three times the first sum equals one times it plus $243 - 1 = 242$. So twice the first sum equals 242, so it equals $242/2 = 121$.[95]

If the multiplying factor has a magnitude less than or equal to 1, so it is between -1 and 1 but equal to neither, each term has a magnitude that is smaller than the next, and the sum remains finite even when approaching an infinite number of terms. This same summing method is used but the "last" terms in the first and second sums are now 0. For a first term of 2.0 and multiplying factor 0.1, the sequence is 2.0, 0.2, 0.02,

---

[94] More generally, if the first term is a, the number added to each term is d, and the total number of terms is n, then the last term is $a+(n-1)d$ and the average of the first and last terms is $(a+a+(n-1)d)/2=(a+(n-1)d/2)$. The sum of the arithmetic progression is $(a+(n-1)d/2)n=na+(n(n-1)/2)d=na+((n^2-n)/2)d=(n/2)(2a+(n-1)d)$.

[95] More generally, if the first term is a, the multiplying factor is r, and the total number of terms is n, then the last term is ar^{n-1}. The sum is $a+ar+ar^2+\ldots+ar^{n-2}+ar^{n-1}$, $r\times$ this sum is $ar+ar^2+ar^3+\ldots+ar^{n-1}+ar^n$, and their difference is $(1-r)\times$ the sum. This also equals $a-ar^n=a(1-r^n)$, because the other terms cancel. So, the sum of the geometric progression is $a(1-r^n)/(1-r)$. This is valid for finite n, when r is not equal to 1.

0.002, 0.0002, …. In a second infinite series, with the same multiplying factor, 0.2, 0.02, 0.002, 0.0002, 0.000002, … The sum of the second sequence equals 0.1 times the sum of the first and also the sum of the first sequence minus 2.0. So, 1.0 – 0.1 = 0.9 times the sum of the first sequence equals 2.0, and this first sum equals 2.0/0.9 = 2.22222…, or 2 2/9. This is a general approach.[96]

13.1.1 Progressions impact everyday life

We are directly and indirectly affected by both types of progressions. What affects some of our lives more than how long we have to wait on line to get coffee? If a coffee shop can serve an average of 2 coffee drinks each minute, but 5 customers enter the line each minute each requesting one coffee drink, the line will grow at a rate 5 – 2 or 3 customers per minute. If the line started with 0 people, then the length of the line after each minute will be 0 (at the start), 3 (after 1 minute), 6 (after 2 minutes), 9, 12, 15, …, an arithmetic progression. This growth will continue unless more staff is added to serve the customers faster or people (finally) decide to leave the line or not enter it. If this coffee shop can serve an average of 5 customers each minute, but only 2 customers enter the line each minute (each requesting one coffee drink), any existing queue shortens at a rate 5 – 2 or 3 customers per minute. If the line started at 30, then the length of the line after each minute will be 30 (at the start), 27 (after 1 minute), 24 (after 2 minutes), 21, 18, 15, … This arithmetic progression terminates after 10 minutes at 0 customers, because negative queues do not make sense. In real life, as customers never enter at one exact rate (these variations can be called fluctuations), some drinks take longer to prepare than others, and so on, and this becomes a less regular sequence. This specific problem is an example in the theory of lines or queues, which we will revisit in Chapter 18.

The rapid advance in computer (and cell phone, …) hardware in recent decades has impacted our lives immensely. In 1965 Gordon E. Moore, the co-founder of Intel and Fairchild Semiconductor, noticed that the number of transistors in an integrated circuit (IC) had doubled

---

[96] The r^n term in the $a(1-r^n)/(1-r)$ sum goes to 0 as n approaches infinity when $|r| < 1$, so the sum of the infinite geometric progression approaches $a/(1-r)$.

approximately every year from 1959 to 1965, and projected that there would be great progress in production methods and lowered costs in the next decade. He made an educated guess that the transistor count per IC would double each year and therefore increase by a thousand in the coming decade (because $2^{10} = 1{,}024$); this geometric progression is widely called Moore's Law. From 1971 to 2011, the transistor count increased by about a million, from ~2,300 to ~2,600,000,000, as would be predicted by the 20 expected doublings over 40 years by Moore's law ($2^{20} = 1{,}048{,}576$), which we will see can be called *exponential* growth. Even though this is not a scientific or engineering law in any sense, it has been surprisingly accurate and became a driving force in this highly competitive industry, and in effect a self-fulfilling prophecy. It is now being "violated" more and more as integrated circuit dimensions have become so small, approaching the fundamental atomic limit.[cclviii]

The diameters of wires used in electrical circuitry are denoted by their "gauges." The gauge system is a geometric progression in which the next higher gauge number characterizes a wire that is smaller in diameter in a specific way. In the American wire gauge (AWG) system, the diameter of the next larger gauge is ~0.89053× as large as the current one.[97,cclix]

The geometric progression is also the math behind the quintessential multi-level marketing practice known as the "pyramid scheme," in which members recruit others to their team to obtain resources from them, rather on marketing products, services or investments to the public.[cclx] The initial member recruits, say, 6 new members (for a total of $1 + 6 = 7$ members), each of whom recruits 6 new members, so at this second level $6^2 = 36$ new members need to be recruited (for a total of $1 + 6 + 6^2 = 43$ members), who in turn recruit 6 new members for total of $6^3 = 216$ new members at this third level (for a total of $1 + 6 + 6^2 + 6^3 = 259$ members), and so on. In the 11th level, $6^{11} = 362{,}797{,}056$ new members need to be recruited, which is more than the entire U.S. population in 2019, and in the 13th, round $6^{13} = 13{,}060{,}694{,}016$ new members need to be recruited,

---

[97] In the AWG system, the wire diameter decreases by the same factor in the 39 steps between the 40 gauge numbers: 0000, 000, 00, 0, 1, 2, 3, 4, …, 35, 36, and overall by a factor of 92. So, per step, it decreases by a factor of $92^{1/39} \approx 1.12293 \approx 1/0.89053$.

which more than the population of the entire world. Clearly, this scheme cannot be sustained for many levels. We will see below how we need to understand the math of progressions to understand and handle legitimate investments.

There is a clever analog of progressions in the world of geometry, called *fractals*. Though they do not affect your everyday life, they amuse many. *Fractals* are geometric shapes that approximate the large and small scales views of mountains and coastlines. They are formed using geometric analogs of progressions, in which a geometric "rule" is successively applied to one shape to obtain the next one in the series. They look pretty similar when viewed with successively higher levels of magnification and so the view at one spatial scale determines that at the next finer one. Figures 13.2 and 13.3 show two examples of fractals as they are formed, both starting with a line segment. [98,cclxi,cclxii,cclxiii]

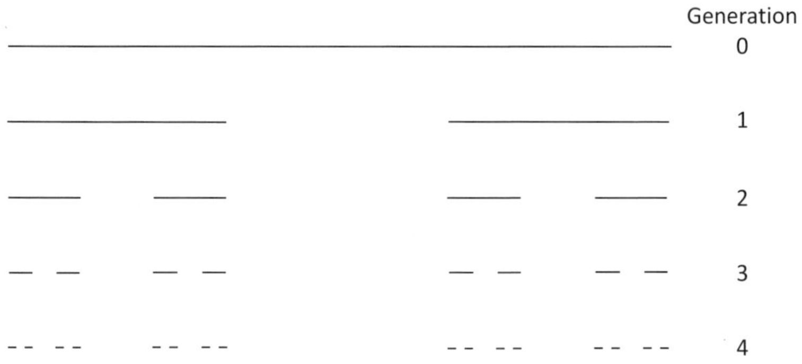

Generation
0
1
2
3
4

Figure 13.2: Cantor set fractal diagram.

[98] The rule in the Cantor Set is to remove the middle third of remaining line segment, while in the von Koch curve the middle third is the base of an equilateral triangle that is replaced by the other two sides of the triangle.

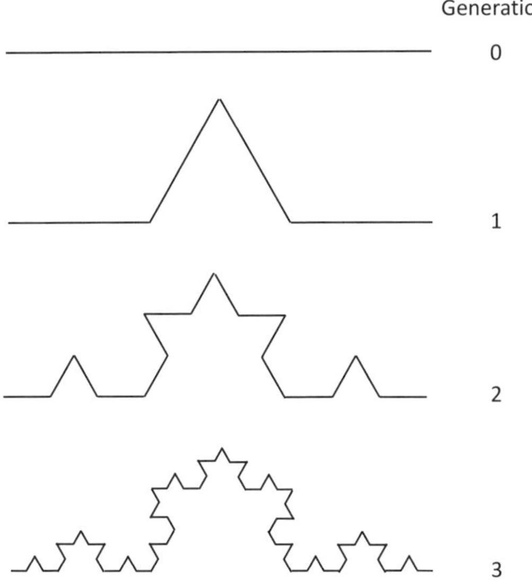

Figure 13.3: von Koch fractal diagram.

13.2 Compound Interest

Regular progressions describe the application of the same continual change at the beginning or end of every distinct section, such as the end of every 3 years or after traveling another 10 miles. Sometimes these changes can be applied so frequently that mathematically they are essentially being applied continuously. You have \$10,000 in an account that accrues 6.0% interest per year. If the interest is applied only once a year, for the first 364 days you have \$10,000 and on the 365th you will have \$10,600. If interest is compounded every month, then at the end of the first month, you earn 6.0%/12 = 0.5% interest and will have \$10,000 × (1 + 0.005) = \$10,050. At the end of the second month, you will have \$10,050 × (1 + 0.005) = \$10,000 × (1 + 0.005)(1 + 0.005) = \$10,000 × (1 + 0.005)^2 = \$10.100.25, which is a quarter more than the 1% simple interest expected after two months (6.0%/6). After twelve months you will have \$10,000 × (1 + 0.005)^{12} = \$10,000 × (1 + 0.06/12)^{12} = \$1,0616.67 (after rounding down), or \$16.67 more than with simple interest, corresponding to ≈6.17% with compounding. Compounding daily gives you \$10,000 ×

$(1 + 0.06/365)^{365}$ = \$1,0618.31 (or ≈6.18%). If you would compound more frequently, every hour, every minute, or every second and so on you reach a limit of \$10,618.36 (rounding down), so you gain only a little extra. So the nominal interest rate of 6.0% effectively compounded continuously lead to an effective interest rate of \$6.18%.

13.2.1 Mortgages and annuities

The math behind continually-compounded interest also underpins mortgages and annuities, but with additional factors included. Consider a 20-year fixed rate mortgage for \$240,000 by first examining the extremes. If there were no interest on the loan, you would pay \$240,000/240 = \$1,000 for each of the 240 months. (This is not going to happen!) If you have a loan with 6% interest per year and pay the monthly interest but never pay off the principal, your monthly payment would be (0.06/12) × \$240,000 = \$1,200 or 0.5% per month of the fixed principal of \$240,000. At the end of 20 years, you would owe a balloon payment of \$240,000. (You do not want this to happen!) If you do not pay back any of the interest or the principal, your loan amount will continue to increase, as we have seen with compound interest. (This will not be allowed to continue for long!) With the fixed rate mortgage, each month you pay interest of 0.5% of the remaining principal plus a part of the principal, which lowers the principal for the next month, for a total monthly payment that is calculated so it is the same each month;[99,cclxiv] the principal owed is zero after 240 equal monthly payments. Before calculating the exact monthly payment, let's estimate the monthly payment. The lower limit of what it could be is the larger of the two limiting payments we just noted, \$1,000 and \$1,200, and so \$1,200. The upper limit is the sum of these two payments, and so \$2,200.

More exactly, the first month for this *amortized* loan you pay 0.5% interest on the \$240,000 total, \$1,200, plus a principal of \$510.43, for a total payment of \$1,719.43 (which is in between our two limits). In the second month, you pay 0.5% interest on the remaining balance of \$240,000 – \$510.43 = \$239,480.57, which is \$1,197.40, plus the slightly

---

[99] The fixed monthly payment P for a loan of L over n months at monthly interest rate r, such as a fixed-rate mortgage, is $P = L(r(1+r)^n)/((1+r)^n - 1)$.

larger principal of \$522.03, for the same monthly payment of payment of \$1,719.43. Near the beginning, almost all of your payment is interest and the principal decreases slowly. Every month you pay less interest and more of the principal until the interest and principal payments become equal in month 102 in this example. In month 239 you pay 0.5% interest on the remaining \$3,413.25 principal, which is \$17.07, plus \$1,702.37 principal, for a total of \$1,719.43. In the final month 240 you pay 0.5% interest on the remaining balance of \$3,413.25 – \$1,702.37 = \$1,710.88, which is \$8.55, plus this \$1,710.88 principal to zero out the principal, for the final \$1,719.43 payment. The total interest paid is \$172,664.29, for the principal of \$240,000.

For higher interest rates, say 12%, the monthly payment is \$2,642.61, the total interest paid is \$394,225.61, and the interest and principal payments become equal much later, in month 172. For lower interest rates, say 3%, the monthly payment is \$1,331.03, the total interest paid is \$79,448.22, and the principal paid is always greater than the interest paid. (I obtained these numbers from an online mortgage amortization calculator.) In adjustable-rate mortgages (ARMs), the interest compounds at a rate that can change in time. This is a potential opportunity because rates start lower than those for fixed rate mortgage, but a risk, because the loans may not be affordable if the interest rate increases much, even within preset cap limits. We will assess annuities when we consider the math of risk in Chapter 20.

The math behind credit card debt and payments is the same as that for fixed-rate mortgages; however, sometimes it is presented in a less transparent way. If your minimum allowed monthly payment equals the interest that accrued for that month and no principal, and you decide to pay only this amount, you will be paying this amount forever. If it is just a bit above it, and you pay that amount, you will be paying off only a little principal each month and you will be making payments for a long time. Neither is to your advantage mathematically or financially. Understand the math![100,cclxv]

[100] If your effective monthly interest rate is 1% and your loan is for \$5,000, and your minimum allowable payment is \$5,000 × 1% = \$50, you will be paying this amount forever. By the way, 1% compounded monthly, corresponds to an effective annual rate of

13.2.2 Arithmetic and geometric rates of investment returns

Different ways of presenting the sequence of investment holdings are used to reflect the potential variability of returns and to account for portfolio withdrawals and deposits.[cclxvi] Say, during year 1 your investment increased from \$100 to \$110, in year 2 it decreased from \$110 to \$105, and in year 3 it increased from \$105 to \$130. Over the three years, your portfolio increased by \$30 or by 30%, so a simple (and incorrect) view would be that the interest was 10% per year. This is not the rate that is presented to you on your balance sheets (and this is good). During the first year, your investment increased by a factor of 1.10 (= (110 – 100)/100), and 100 and 110 could be the first two terms of a usual geometric progression. In the parlance of finance, this factor minus 1 is presented. So, in year 1, your account increased by 10.00% (= (110 – 100)/100 – 1.0 = 1.10 – 1.0), in year 2 it decreased by ≈4.55% (= the negative of (105 – 110)/110 – 1.0 ≈ 0.9545 – 1.0), and in year 3 it increased by ≈23.81% (= (130 – 105)/105 – 1 ≈ 1.2381 – 1.0). The *arithmetic mean return* is the simple average of these three percentages, or ≈9.75% here. Investment professionals do not provide this, because they consider the *geometric mean rate of return* a better measure since it includes the effect of compounding (as in compound interest) and market variations or *volatility*. It is the interest that would be needed during each of the three years to arrive at the same overall outcome.[101] Here, the outcome is changed by a factor of 1.30 (≈1.10 × 0.9545 × 1.2381), which is $\approx(1.0914)^3$. So, the *geometric mean rate of return* is ≈9.14%, which is the constant rate needed for a 30% return in a geometric progression.

Investments are more complex than this because you may add money to your account at times during this period, as when you contribute to your 401K account, and/or remove some, as when you make withdrawals. Tracking of the geometric mean performance of a given

---

$(1+0.01)^{12}-1 \approx 1.126825-1 \approx 12.68\%$. This compounded annual rate is the same as that using the annual simple interest rate of 11.94235% compound daily (for 365 days), because $(1+0.1194235/365)^{365}-1 \approx 1.126825-1 \approx 12.68\%$.

[101] If this interest rate is r, the final amount is $(1 + r)^3$ times the initial amount, or $100.00(1+r)^3 = 130.00$ here.

initial investment, portfolio or fund given these changes is called the *time-weighted rate of return*. The *dollar- or money-weighted rate of return* reflects changes in the market rate, and it weights rates more heavily when you have more money in the account; this is the rate that is often presented in investment reports.[102]

13.3 Exponential Growth

Compounding interest becomes relatively more important at higher interest rates, which is good when you earn interest on savings and bad when you owe money on a loan with compounding interest. While the simple rate of 6.0% becomes ≈6.18% with continuous compounding, the simple rate of 60% becomes ≈82.21%, which is much higher than 10× the compounded rate for 6.0%, 10 × 6.18% = 61.8%.[cclxvii] In fact, while applying simple interest at x% gives the initial investment × $(1 + x\%)$ for an increase of x%, with continuous compounding the investment increases by a factor of $(1 + x/N)^N$, where N increases without bound, to infinity. This factor is e^x (as in Section 7.2), which leads to an increase by an amount $e^x - 1$.[103,cclxviii]

The exponent here is the simple interest rate per unit time × the time

[102] If in our example you added $20 to your account at the very beginning of the third year, you would then have $125, which would grow by the same 23.81% to $156.76 at the end of the year. The dollar-weighted rate of return for your account is that needed for your account to end at $156.76, by investing $100 for three years and $20 for one year, all at the same constant rate. This is the value of r for which $100(1 + r)^3 + 20(1 + r) = 156.76$, or ≈10.43% here, which is higher than the time-weighted return of 9.14% because you added money just before the rate increased to the highest level in the three years. If instead, you removed $20 from your account then, you would have $85 at the beginning of the third year, which would grow by 23.81% to $105.24 at the end of the third year. Your dollar-weighted rate of return would then be that needed for your account to end at $105.24, by investing the entire $100 for three years and subtracting the amount that $20 would change to in one year, all at the same rate. This is the value of r for which $100(1 + r)^3 - 20(1 + r) = 105.24$, or ≈8.26% here, which is lower than the time-weighted return because you had less money when the market rate was the highest. When investment periods are shorter than a year or longer than a year, these rates are recast as being those per year, and called the effective annual return or EAR.

[103] This is what we just saw, with $e^{0.06} = 1.0618\ldots$ (an effective increase of 6.18...%) and $e^{0.6} = 1.8221\ldots$ (an effective increase of 82.21...%).

or, more generally, the simple rate of change per unit variable × the variable. If the variable is the time t, exponential growth is e^{Rt}, where R is the growth rate, or $e^{t/\tau}$, where τ (the lower-case Greek letter, tau) is a time constant that is equal to $1/R$. When the exponent is positive, the function *increases or grows exponentially* (Chapter 7, Figure 7.3a).[104] This means it increases from 1 at time 0 to e at time $1/R$ or τ, to e^2 at time $2/R$ or 2τ, and so on, and by the same $e^{T/\tau}$ or e^{RT} factor over the time span T. The exponential function increases to infinity, so it has limitations in modeling real circumstances. When the exponent is negative, as for e^{-Rt} or $e^{-t/\tau}$, where R and τ are still positive, the function *decreases or decays exponentially* to zero (Figure 7.3b). So, it decreases from 1 at time 0 to $1/e$ at time $1/R$ or τ, to $1/e^2$ at time $2/R$ or 2τ, and so on, and decreases by the same $e^{T/\tau}$ or e^{RT} factor over the time span T. Of course, exponential decay can also be represented by e^{Rt} where R is negative. The rates of change in these increasing or decreasing exponentials are proportional to their current values, so such models are called *proportional models*.

Exponential growth is encountered in many ways. For example, past age 30, the risk of death (probability per unit time) approximately doubles every seven years, and so this risk can be said to increase by a geometric progression or exponentially with time.[cclxix]

The *Malthusian population growth model*, by Thomas Robert Malthus in 1798, assumes that the growth of a population increases with a rate proportional to the current population. This would lead to exponential population growth, unless checked by factors that limit growth, such as limited resources of food and shelter, famine and disease (just like Moore's Law is being limited by the atomic limit). These factors can be included in the mathematical model. Malthus thought the future would be grim because a geometric increase in population with a presumably arithmetic increase in food production would lead to

---

[104] In e^x, x can be either positive or negative. With $x = Rt$, R is a positive rate constant (or sometimes called γ, the lower-case Greek letter gamma) (with units per unit time, such as 1/second) and t is time (such as seconds). Similar to the derivative $de^{ax}/dx = ae^x$, $de^{Rt}/dt = Re^{Rt}$ and $de^{-Rt}/dt = -Re^{-Rt}$, and with $x = t/\tau$, these are $de^{t/\tau}/dt = (1/\tau)e^{t/\tau}$ and $de^{-t/\tau}/dt = -(1/\tau)e^{-t/\tau}$.

starvation unless the birth rate was controlled, but food production has generally increased faster than he had expected due to improved technology.

The impact of continuously compounded interest is sometimes estimated by using the *Rule of 72*. The natural log of 2, $\ln 2 \approx 0.693$, so $e^{0.693} \approx 2$. This means that if your money was invested and compounded continuously it would double in one year if the annual rate of increase were ≈0.693 or 69.3%, or in two years at half that rate 69.3%/2 = 34.7% or in 69.3 years at 1%—or for any product of years and annual interest rate with a 0.693 product. In using the common, approximate form of this, the Rule of 72, divide the interest rate in % into 72 to obtain the approximate number of years it takes to double your money. It is widely used to estimate investment returns because it easy to calculate 72/1, 72/2, 72/3, 72/4, ….[105] This rule has been around for more than 500 years and is often wrongly attributed to Albert Einstein.[cclxx]

13.4 Exponential Decay

Exponential decay shows how much of an item remains after a given time or distance, and so on, and is widely observed and used.

If a set of N people die at a rate R, the population decays as Ne^{-Rt}. Say the decay rate R is 0.0125 per year, so the characteristic lifetime, $1/R = \tau$, is 80 years. In the coming year, 1.25% or 1/80th of the people at the beginning of the year will die on average that year and the other 79/80 = 98.75% remain. In reality, statistical studies show the mortality rate depends very much on gender, age, environmental factors and so on, so different population groups need to be tracked separately. The death rate increases with age in adulthood, so R for a group of people of the same age is not a constant, but increases with time as $R(t)$. Accordingly, the population decrease is nearly but not exactly exponential, as it would be

---

[105] The more exact form of this geometric progression, for annual rate r compounded annually over y years, gives $2 = (1 + r)^y = (1 + R/100)^y$, where here R is r expressed in % (so $R = 100r$). Taking the natural log of both sides gives $\ln 2 = y \ln(1 + R/100)$, so with annual compounding the doubling time y is more nearly equal to $\ln 2/\ln(1 + R/100) \approx 69.3/R$ than that given by the Rule of 72. This uses $\ln(1 + x) \approx x$ for $|x| << 1$ and $\ln 2 \approx 0.693$.

for a constant R; still, over short periods R is approximately constant and the decay can be assumed to exponential.

The U.S. Center for Disease Control and National Center for Health Statistics tracks mortality rates (which vary with time, as well as with age).[cclxxi] Such rates can give you an ever-changing snapshot of health. From 1950 to 2016, the overall age-adjusted death rates (that are adjusted for changing population distributions) per 100,000 each year decreased from 1,446.0 in 1950 to 728.8 in 2016, and deaths attributed to disease of the heart decreased from 588.8 to 165.5, to cerebrovascular diseases (strokes) decreased from 180.7 to 37.3, and to malignant neoplasms (cancer) decreased from 193.9 to 155.8. From 1950 to 2016, the overall crude (total, but not adjusted) death rates per 100,000 each year decreased from 963.8 in 1950 to 849.3 in 2016, and deaths attributed to disease of the heart decreased from 356.8 to 196.6, to cerebrovascular diseases (strokes) decreased from 104.0 to 44.0, and to malignant neoplasms (cancer) increased from 139.8 to 185.1. Deaths from cancer increased mostly because people lived longer without dying from heart disease and strokes, and at older ages they eventually contracted cancer and died from it. This shows that statistics must be interpreted with care![106,cclxxii]

The term *half-life* is frequently used when discussing radioactivity. The "amount" of remaining radioactivity in a substance decays exponentially in time because the rate of radioactive decay is proportional to the amount of remaining material. The characteristic

[106] Using crude death rates, rather than age-adjusted rates for $R(t)$ would be better, and using the data for specific age groups to follow deaths in those groups would be even more illuminating. In 2016, the overall death rate per 100,000 per year was 583.4 for those under 1 year of age, decreased down to 25.3 for those 1-4 years old and 13.4 for those 5-14 years old, and it steadily increased, to 405.5, 883.8, 1,788.6, 4,474.8, and 13,392.1, respectively for those 45-54, 55-64, 65-74, 75-84 years old, and those 85 years and older, respectively. Each rate was a large decrease from the rates from 1950 (respectively, 3,299.2, 139.4, 60.1, 853.9, 1,901.0, 4,104.3, 9,331.1, 20,196.9). The death rate for each age group decreased from 1950 to 2016 for heart disease, strokes, and cancer (aside for that for heart disease for those under 1 year and for cancer for those 85 and older). Twenty years past 2016, the death rate of those currently 45-54 (who survive) will likely not be that given for those 65-74 years old in 2016, but smaller.

decay time is given as the half-life, $T_{1/2}$, which is the time for radioactivity to decrease by a factor 2 (Figure 13.4). So, $e^{-T_{1/2}/\tau} = 1/2 = 0.5$ and so $\tau = T_{1/2}/\ln 2 \approx 1.44\, T_{1/2}$ (because $e^{\ln 2} = 2$). This tells you how long you can expect the radioactivity to last, which is important in assessing how nuclear power and nuclear waste affect society in general and you more personally. It also tells you how fast radiation is being released. The longer the half-life, the fewer alpha, beta, and gamma rays it emits in a given time, but the longer the radioactivity lasts. The radioactive form of hydrogen, tritium, formed in several natural and human-made ways, has a half-life of 12.3 years. Tritium can also appear in the human body in water, but it lasts in the body for much shorter times than its true half-life because it has a "biological" half-life of ~10 days, which is good for us if we happen to ingest a little.[cclxxiii] If there is a question about how well your heart is working, you may be asked to take a cardiac stress test to image your heart muscles during exercise. You ingest chemicals containing the radioactive element technetium-99m, which emits the gamma rays used to make the image, both before and during your exercise. Technetium-99m has a half-life of 6 hours, so it is well matched to your needs for two reasons. Clear images can be made because the emission is strong in the very roughly hour-time scale of the measurement and then the residual radioactivity in your body decreases very fast. Each day is four half-lives long, so it gets weaker by a factor of $2^4 = 16$ each day.[107]

[107] The same math is used to rack how light travels through objects. The transmission of light through a given uniform medium decreases exponentially with its thickness because the amount of light absorbed is proportional to the light level at that point, which is known as the Beer or Beer-Lambert law.

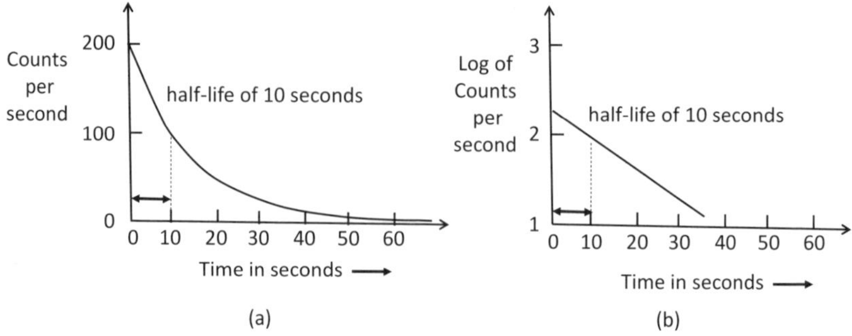

Figure 13.4: Exponential decay and half-life.

13.5 Limiting Growth and Promoting Chaos: Numerical Analysis

Nonlinearities in the math (which means expressions with other than linear terms and constants) are needed to treat problems as different as limiting population growth due to limited resources and chaotic events. Linear effects are much simpler to understand than nonlinear ones because they can be subdivided into components that can be solved and then combined.[cclxxiv] Still, the difficulty in solving nonlinear problems does not diminish the need to address and ultimately solve them. They are often addressed by using numerical methods, which we now explore.

Say that in a proportional or exponential model, the rate of change in a variable x with time t is $4x$.[108] x could represent the population and its initial value could be 0.2, perhaps representing 0.2 million people. The units of time could be seconds, years, or centuries. We know that the exact, analytic solution to this model is x equals its initial value (which is 0.2) times e^{4t}. After a time period 0.1, x equals 0.2 times $e^{4\times0.1} \approx e^{0.4} \approx$ 1.4918, and so $\approx$ 0.2984. After a second time period it equals 0.2 times $e^{4\times0.2} \approx e^{0.8} \approx (1.4918)^2 \approx 2.2255$, so $\approx$ 0.4452. Including the initial value, the sequence of values initially and then after each time period is (approximately, and to 10 terms) 0.2000, 0.2984, 0.4451, 0.6640, 0.9906, 1.4778, 2.2046, 3.2889, 4.9065, 7.3196, …, which shows the expected,

[108] So $dx/dt = 4x$, which has the solution $x(t)$ = (initial value, here 0.2) × e^{4t}.

exactly exponential increase.

This same problem can be solved using numerical or computational analysis, which we do to illustrate how this approach can be used to address nonlinear problems. Using this method, the value of x increases each time period by the rate of increase, 4 times its current value (which is $4x$), times the time period (which is 0.1 here), so after the time period its new value is its old value (which we still call x) plus this increase of $4x \times 0.1$. At time 0.1 after the start, the new value of x is $0.2 + 4(0.2) \times 0.1 = 0.2 + 0.08 = 0.28$. At a time 0.1 later, the new value is $0.28 + 4(0.28) \times 0.1 = 0.2 + 0.112 = 0.392$, and so on. Again including the initial value, the sequence of values after each time period is (approximately, and still to 10 terms) 0.2000, 0.2800, 0.3920, 0.5488, 0.7683, 1.0756, 1.5059, 2.1083, 2.9516, 4.1322, … The increase is slower than the previous exponential because the time period was chosen to be too large for this calculation to be even nearly exact. (With suitably small periods, these results would correspond to the exact solution results, of course.) Still, even this approximate numerical approach is sufficient to illustrate the nonlinear effects in a model of population growth in a world of limited resources.

The simplest way to show nonlinear effects is to subtract a squared or quadratic term from a linear term, such as x^2 from a linear term such as x, as in the expression $x - x^2$. Say that in the limited growth model the population change with time is $4x - 4x^2$, instead of $4x$, so it is the same as the above model with a subtracted quadratic term. Such a nonlinear model known is a type of the *logistic model* for population growth, developed by Belgian mathematician Pierre Verhulst in 1838.[cclxxv] In the numerical analysis, the value of x increases each time period by the rate of increase, which is now 4 times its current value (which is $4x$) minus 4 times the square of its current value (which is $4x^2$), times the time period (which is still 0.1). So, after a time period its new value is its old value (which we still call x) plus this increase of $(4x - 4x^2) \times 0.1$. At time 0.1 after the start, the new value of x is now $0.2 + 4(0.2 - 0.2^2) \times 0.1 = 0.2 + 0.064 = 0.264$. At a time 0.1 later, the new value is $0.264 + 4(0.264 - 0.264^2) \times 0.1 = 0.264 + 0.0777 = 0.3417$, and so on. Again including the initial value, the sequence of values after each time period is (approximately and now to 20 terms) 0.2000, 0.2640, 0.3417, 0.4317,

0.5298, 0.6295, 0.7228, 0.8029, 0.8662, 0.9126, 0.9445, 0.9655, 0.9788, 0.9871, 0.9922, 0.9953, 0.9972, 0.9983, 0.9990, 0.9994, ... With this subtracted term, the population no longer increases with time forever, but stops at the value 1.0.[109]

Sometimes introducing nonlinearities leads to a steady state, as here, and sometimes it "seems" to make the math go "haywire," as in *chaos*. Chaos describes situations where small changes in initial conditions and parameters can lead to remarkably large and unexpected changes in the outcome. One manifestation of chaos has been popularized as the *butterfly effect*, in which a butterfly flapping its wings on one side of the planet supposedly can induce wild weather events, such as hurricanes, on the other side. This is due to mathematical *nonlinearities* in the science of weather and the potentially large changes in outcomes that can occur with even small changes in initial conditions. Though they appear to be random and may be called random events in a colloquial sense, such chaotic events do not occur because of the random or stochastic, random processes we have described; they are fully *deterministic*.[cclxxvi] Such *nonlinear* effects in so-called nonlinear dynamical systems are described by functions that do not change strictly proportionally to their current value. Unexpected and extraordinary effects can occur with small changes in a model and with different initial conditions.

In 1975 Australian ecologist Robert May noted this expression $4x - 4x^2$ used for the limited growth model can also predict chaotic, unstable events (i.e., chaos) with very small deviations from predictable, periodic, stable events (such as "orbits" in physical space).[110] If you start with an initial value for x, then insert it in $4x - 4x^2$, and then use the result as the new x and insert it in $4x - 4x^2$, and so on, you can get a range of different sequences. If you begin with 1, you get the rather boring sequence 1, 0, 0, 0, 0, ... With input $(5 - 5^{½})/8 = 0.3454915...$, you get $(5 + 5^{½})/8 = 0.9045085...$ for the first output. The next output is $(5 - 5^{½})/8 =$

[109] This is not surprising because as x approaches 1.0, the term multiplied each time, $4x - 4x^2$, approaches zero.

[110] The function we used $4x - 4x^2$ can be written as $4x(1 - x)$, because of the distributive principle. It is often written in this latter form because it is closer to the form of a more general model.

0.3454915…, which is exactly the same as the initial input. Continuing, you see a repetitive "orbit" or stable sequence with period 2 (two states trapped at these points): 0.3454915…, 0.9045085…, 0.3454915…, 0.9045085…, 0.3454915…, …. It is surprising that if you start with a slightly different first input, the result is not repetitive, but a very unstable sequence. With input 0.34, the outputs are 0.8976, 0.3677…, 0.9299…, 0.2606…, 0.7708…, 0.7068…, 0.8290…, 0.5671…, and so on. This transition from a stable periodic to an unstable, chaotic trajectory is remarkable.[cclxxvii,cclxxviii,cclxxix] Modeling chaotic events using nonlinear equations is important in weather forecasting and trajectories for space probes, as well as in May's interest in population dynamics in ecology.

13.6 Spreading of Disease

Problems involving growth and decay can also have more than one parameter and rate constant. If you start with a fixed number of sick people, who never become well and die at a given rate, you have a one parameter or variable problem. Mathematically, this is like radioactive decay, with the decrease of the sick population per unit time being proportional to the current sick population times a rate. But life is more complicated than that and so the math is correspondingly more complex, with variables representing different types of populations and the changes in them depending on each other. Such mathematical modeling[cclxxx] is widely used to model infectious disease propagation and to track it.[cclxxxi]

This type of proportional model can explain endemic diseases that are restricted to a place or people or more widely spreading epidemics and pandemics. In the SIR model, developed by W. O. Kermack and A. G. McKendrick in 1927, the numbers of those who are susceptible to a disease (the S in SIR), infected and infectious (I), and recovered (and immune) or dead (R) are tracked as a function of time. Those in the S category can transform to the I category, and then can transform to the R category, as indicated by $S \to I \to R$. At any time the total number in the three categories is fixed at N.[111] The susceptible population decreases

[111] So, $S(t) + I(t) + R(t) = N$. The parameters in this model that describe the rates of the steps are the lower-case Greek letters beta, β, and gamma, γ.

with time proportional to the number who are susceptible, but here the rate is not a constant but is proportional to the number who are infectious. So, the disease spreads faster when there are more infectious people, and the decrease of the susceptible population is a product of the susceptible and infectious populations.[112] The infectious population changes with time, due to two additive contributions: it increases at the same rate the susceptible population decreases and decreases with time at a rate proportional to the infectious population that becomes the group of recovered or dead people.[113] This second contribution also describes the increase with time of the recovered and dead populations.[114] This model is a series of differential equations (Section 4.7) that is solved on a computer to track each group vs. time. The disease can spread quickly when the *reproduction number*, the ratio of the rate constant describing the rate susceptible people become infected to that characterizing how fast infectious people recover or die, exceeds 1.

The constant associated with infecting the susceptible may change with time, perhaps reflecting the greater prevalence in winter of the common cold or the increased contact of children during the school year. The once fast-spreading of measles and mumps among school children was largely eliminated with vaccination. Nowadays, with >95% local vaccination rates (for measles in many places in the U.S.), children *appear* to be protected even when they are not vaccinated.[cclxxxii] When a larger fraction is not vaccinated, the number who are infected increases and the disease spreads more quickly and widely among the susceptible, and they are no longer protected. The mathematics shows this, as has real-life experience.

Other situations are described by using modified models. In one modification, mothers can confer immunity to their newborn to some infections, and so for their first few months of the life infants belong to the new maternally-derived immunity category (M); they then leave this to join the susceptible group (S) in the MSIR model. In another, those who have had tuberculosis may still be carriers and infect others and also

---

[112] This is described by the differential equation $dS/\mathrm{d}t = -\beta SI/N$.

[113] So, $dI/\mathrm{d}t = \beta SI/N - \gamma I$.

[114] So, $dR/dt = \gamma I$.

get sick again, and these carriers can be treated as a fourth carrier group, and this would modify the SIR model in a different way. In a third modification, those who recover from the common cold are susceptible to re-infection but are not carriers, so there are really only the susceptible and infectious populations. There is no distinct recovered population, and those who recover rejoin the susceptible group in the SIS model.[cclxxxiii]

The spreading of epidemics from one place to another can also be addressed with mathematical models.[cclxxxiv] For instance, Bergsman, Hyman, and Manore modeled the bird-mosquito-bird infection cycle for the West Nile Virus, in which people are infected directly from the mosquitos and the birds include migratory and resident birds.[cclxxxv] Similar models are being used to model newer generations of mosquito-borne diseases, such as the Zika virus. They can be used to help plan mosquito spraying.[cclxxxvi] Transfer between population groups exacerbated the already devastating 1918-9 Spanish Flu Pandemic, which killed as many as 50 million people. The second wave of the pandemic in the U.S. was caused by one population, the military infected in Europe in WWI, returning home after the war and transferring the disease to another group, the general public, who mingled with these soldiers in celebratory parades and parties.[cclxxxvii]

The math spreading of rumors uses similar models.[cclxxxviii]

13.7 Other Sequences: How Rabbits Reproduce and More

In the examples so far, the next term in a sequence modeling growth or decay has depended only on the previous (or current) value, but modes of growth can also depend on earlier values as well. For example, people and animals need to reach a given age before they can reproduce in reality and therefore in population models, so updated populations cannot depend only on the most recent population; such models also need to include our limited lifetimes.

This first feature is included in the classic model of the population explosion in "breeding like rabbits," that occurs in a closed environment, such as on an island with plenty of food and no predators. Each male/female pair gives birth to a new pair every month once they are two months old and it is assumed that rabbits never die. The first pair is introduced at the beginning of the first month, so at the end of the first month there is 1 non-reproducing pair and 0 reproducing pairs, for a total of 1 pair. At the end of month 2 there is still the 1 non-reproducing pair born two months earlier, which is about to start to breed, and 0 reproducing pairs, for a total of 1 pair. By the end of month 3 the pair from month 1 begins to reproduce and has given birth to 1 new pair, so there is 1 non-reproducing pair (the new one) and 1 newly reproducing pair, for a total of 2 pairs. By the end of month 4, the 1 pair from month 2 has given birth to 1 additional pair, but the second pair has not started to reproduce yet because it is only one month old, so the total number of pairs is 2 (all pairs in the previous month, month 3) + 1 (newly born pair, which represents 1 pair for each pair alive two months earlier, month 2 here, because they are all fertile) = 3 pairs. So the number of pairs at the end of any month is the sum of the numbers of pairs from 1 month earlier and those from 2 months earlier; this is the recurrence relation.[115] The sequence is 1, 1, 2, 3, 5, 8 (= 5 + 3), 13 (= 8 + 5), 21 (= 13 + 8), 34, 55, 89, 144, and so on, so after a year there are 144 pairs of rabbits. This is the well-known *Fibonacci sequence.*[cclxxxix] This sequence is usually presented starting with 0 (term 0), so as 0, 1, 1, 2, 3, 5, 8, and so on. These numbers often pop up often in popular culture.[ccxc] They were mentioned by math genius Charlie Eppes in a 2005 episode of the TV show *Numbers* (*NUMB3RS*), and used by murderers in a 2008 episode of the TV show *Criminal Minds* and in Dan Brown's novel *The Da Vinci Code.*[ccxci]

The *Lucas sequence* has the same recurrence relation, but has 2 and 1 as terms 0 and 1. This results in a quite different sequence: 2, 1, 3, 4, 7, 11, 18, 29, 47, 76, 123, ... However, the ratio of each term to the prior one approaches the same limit for both the Fibonacci and Lucas sequences, 1.6180339..., the famous golden ratio $(1 + 5^{1/2})/2$ (Chapter 9).[116,ccxcii]

In real life this growth cannot continue forever, so the model

---

[115] This recurrence relation is $a_n = a_{n-1} + a_{n-2}$, starting with $a_1 = 1$ and $a_2 = 1$.

[116] The ratio of successive Fibonacci numbers are ... 55/34 = 1.61764..., 89/55 = 1.61818..., 144/89 = 1.61797..., ... and for the Lucas numbers ... 47/29 = 1.6206..., 76/47 = 1.6702..., 124/76 = 1.61842...., both approaching 1.6180339.... The ratios of the corresponding Lucas and Fibonacci numbers approach $5^{1/2}$.

assumptions need to be more realistic to account for the lifespan of rabbits and limited food. The lifetime is about a year for wild rabbits and about 10 years for pet rabbits. So, for example, to include the lifespan of wild rabbits, the number of pairs at the end of any month would be the number of pairs from 1 month earlier plus the number from 2 months earlier minus the number of pairs from 12 months earlier.[117]

13.8 Sequences to Very Large (or Small) Numbers, and Computing

Terms in sequences can increase very rapidly and may seem to approach infinity in one of several different ways. If the variable were an integer, n, it could increase linearly as n or "on the order of magnitude" of n. One type of slower increase is as the logarithm of n (with $\log n$ in any base; say in base e here, so as $\log_e n$ or $\ln n$). Progressively faster than n, are n times the log of n (which is $n \log n$), n to the power 2 (or n^2), 2 to the power n (or 2^n), and n factorial (or $n!$). n factorial increases are "slightly" slower than those of n to the power n (or n^n). "Increases exponentially" in this context usually means as Napier's number e to the power n (or e^n). This increases faster than 2 to the power n (or 2^n), but slower than 3 to the power n (or 3^n), which is still slower than n factorial. Similarly, terms can approach zero as the reciprocal of these functions. Decreasing as the reciprocal of n (so as $1/n$) is a relatively slow decrease, while decreasing exponentially (so as e^{-n}) is a relatively fast approach to zero.

This evolution is significant because it influences how fast you can solve particular problems with computers.[ccxciii] Mathematician Gian Carlo Rota noted: "the computer is just an instrument for doing faster what we already know how to do slower."[ccxciv] The question is whether or not the numerical analysis by computers can be done in a "reasonable" time. The time currently required to conduct a computer bit operation, the basic computing step, is approximately 10 millionths of a millionth of a second (10^{-11} seconds). To solve a problem of a given size, with n sections in time or positions, the number of bit operations needed is somewhere within these above orders of magnitude, ranging from $\log n$ to n, $n \log n$, n^2, 2^n and to $n!$.

If the number of bit operations varied linearly, such as with n, for 10

---

[117] This recurrence relation is $a_n = a_{n-1} + a_{n-2} - a_{n-12}$, with $a_1 = 1$ and $a_2 = 1$.

sections ($n = 10$), the computation time would be a hundred millionths of a millionth of a second (10^{-10} seconds = 10×10^{-11} seconds). Over the ranges from log n to $n!$, it would increase from 2.3×10^{-11} seconds (= $(\log_e 10) \times 10^{-11}$ seconds) to 3.6×10^{-5} seconds (= $10! \times 10^{-11}$ seconds), all quite short times. However, for 100 sections (10^2), it increases from 4.6×10^{-11} seconds for $\log_e n$ scaling to 10^{-7} seconds for n^2 scaling, and then to a whoppingly long 4×10^{11} years for 2^n scaling and to more than 10^{100} years for n factorial scaling, so algorithms that scale in this upper range would be "impossible" to implement. But, n is usually much larger than 100. In fact, for a million sections (10^6), the time increases from 2×10^{-10} seconds for log n scaling to 0.17 minutes for n^2 scaling, which is relatively fast, but becomes too long again with the more complex scaling rules and for larger n.

Computing mathematics tries to alleviate these problems in several ways. One is to find problems that computationally scale slower, as we will now see in discussing "complexity." Another way to accelerate computation is by developing faster computer switching electronics. Yet another is by developing and implementing parallel processing, in which operation sequences are done at the same time, with many smaller computers operating in "parallel." Each approach has been extremely successful in addressing computationally-intensive problems.

We have been assuming a computer step takes 10^{-11} seconds, so there would be 10^{11} steps per second. *But, actually how "fast" is a computer?* Its speed is characterized by the number of **flo**ating-**p**oint (arithmetic) operations, or flops or FLOPS, it can perform per second. The approximate hardware cost per gigaflop (10^9 flops) in 2018 U.S. dollars decreased from ~\$156.8 billion in 1961 to ~\$0.03 in October 2017. In 2018, the fastest supercomputer had a performance of ~120 petaflops (10^{15} flops).[ccxcv]

13.9 Onward to Very Large Systems: Complexity, and Computing

Our world is complex and the world we encounter is becoming evermore complex, be in physical, networking, computation, big data, biological, cryptography, environmental, social, or business and finance systems. Understanding complex systems will entail increasingly sophisticated mathematics and modeling, and computational methods to understand the increasing large numbers of interacting components and their impact, in this area termed *complexity*.[ccxcvi]

The properties and functions of common systems consisting of 10^{22}-10^{26} interacting atoms are usually complex, but at times they can be surprisingly simple. In a crystalline solid, the atoms are in very regular, periodic (repeating) structure, and this "symmetry" makes using physics and math to solve this "many-body problem" fairly easy. In a gas of molecules, the many molecules interact with each other only weakly and the net properties of the gas are well described by well-known "macroscopic" properties that provide an average view of the gas, such as pressure, temperature, and volume. However, living systems are exceedingly complex, and have defied understanding on several levels. They are often subdivided into many subsystems, such as organs (hearts, lungs, brains, and so on) and the interactions of these subsystems are examined. The human brain has roughly 100 billion or 10^{11} neurons (or nerve cells). Each neuron may be connected to as many as ten thousand or 10^4 other neurons, through as many as a thousand trillion = 10^{15} (= $10^{11} \times 10^4$) connections between them called synapses.[ccxcvii,ccxcviii] The exceedingly complex interactions between them have been an important topic of inquiry in complexity theory and practice, and this inquiry is expected to continue for some time. Still, there are some trends in the properties in complex organisms and their subsystems that follow surprisingly simple trends or scaling laws that depend on their size, such as their mass, as noted by Geoffrey West and others and noted in Section 8.2.5.[ccxcix]

Progressions illustrate how a network quickly becomes complex as it grows. If you a have group of people and each can interact with anyone else in the group, how many pair-wise interactions or interconnections will there be in the group? With two people, there is one interconnection. With three, there are three interconnections (two new ones), with four, there are six (three new ones), and so on. So, the nth person adds $n-1$ new interconnections to the group. The total number of interconnections is $1 + 2 + 3 + 4 + \ldots$ all the way up to $n-1$, an arithmetic progression that sums to half the square of the number of people minus half the number.[118] So, for large n, the total number of interconnections increases approximately quadratically, as the square of the number in the group.[119] This means that organizations in which everyone needs to interact with each other become very complex and unwieldy when the group becomes large.

The complexity of determining if a number with n digits is prime varies as a polynomial, and scales approximately as n^2. Such polynomial-type problems (of complexity class "*P*," meaning it uses a *polynomial-time algorithm*, with n raised to any fixed power) are distinguished from those whose complexity is not polynomial and whose solution time may increase exponentially with the number of elements (such as 2^n). For n digits, a class *P* problem would require very roughly $100^2 = 10^4$ or ten thousand steps, while an exponentially growing problem would entail $2^{100} \approx 10^{30}$, or $\sim 10^{26}$ = a hundred trillion trillion times as many steps. The chasm between these two extremes was well explained by Israeli cryptographer and co-inventor of the RSA cryptosystem Adi Shamir in response to a comment by someone attributing the difference between "easy" and "hard" computations to the different computer hardware that is needed: "The difference between us is that you consider one hundred steps in a computer to be an easy computation and one billion steps to be a hard computation. I define things in a different way. According to my definition, an "easy" computation is one which for inputs of size N asymptotically requires N, N^2, … N^{100}, … operations; a "hard" computation is one which, as N increases, requires 2^N steps."[ccc]

A class *P* problem is classified as being "easy" to *solve*, and its solution is even easier to *verify* because verifying is easier than solving (meaning it needs less computation time). "Nondeterministic polynomial" problems (class "*NP*") are those that might need to be *solved* with exponential scaling and so may be hard to solve, but they can be *verified* by using *only* a polynomially-increasing number of steps and

---

[118] So, $n(n-1)/2 = n^2/2 - n/2$.

[119] Seeing how the number of interconnections in a large group increases with the addition of a new member is an illustration of *differential calculus* in the limit of a large group, while the use of such increases to determine the total number of interconnections in the group is an illustration of the reverse process of *integral calculus*.

so their verification (only) would be easy. One of the great open questions, and one whose solution will be awarded a $1,000,000 prize by the Clay Mathematics Institute, is whether all problems with solutions that are easy to verify are also easy to solve, which is symbolically summarized as $P = NP$.[ccci] If this were true, public-key cryptography would eventually no longer be viable (which would be bad), but difficult scheduling problems would become more viable (which would be good).[cccii] In 2015, American mathematician and computer scientist László Babai appeared to have made a breakthrough in this area by proving that one major problem thought to be more *NP*-like, solving networks with many connections through *n* points or nodes, may be far simpler than had been thought, and may be more *P*-like.[120,ccciii]

Sometimes problems or figures may seem complex, but are simple to understand because they arise from simple rules, such as fractals (as discussed earlier this chapter). However, complex systems are usually very different on different scales and so they cannot be simplified in this manner. For example, on a coarse scale, a community of homes may look like a cluster of identical items. On a finer scale, each item consists of a house, lawn, garden, and driveway, and on an even finer scale, the garden is a set of different types of plants. Similarly, calculations can be made less complex by analyzing them separately on multiple scales in position or time because their computation needs differ at different scales; this is called *multiscale modeling*. The trick is to sample enough points in position and time at each scale to obtain accurate results, but not so many that the calculation would be too long.

We have seen that to understand and control our lives well we need to recognize that living in the subworld of the "what will be" is certain and sometimes this is sufficient, as in compounding interest, but when the evolution of progressions is not certain, as in how disease spreads, we

---

[120] Babai showed that the number of steps needed to solve graph problems with *n* vertices or nodes varies as $e^{(\log n)^{O(1)}}$, where $O(1)$ means on the order of 1, while the lower limit of its scaling behavior had been thought to be no faster than the much larger $e^{(O(n \log n)^{1/2})}$. For $n = 100$, these are 100 and 2×10^9, if "on the order of" is replaced by equal to, for log base e.

must step away from the deterministic math of the subworld of the "what will be" and enter the stochastic math describing the subworlds of the "what was" and "what might be."

Chapter 14

Untangling the Worlds of Probability and Statistics

Progressions represent the subworld of the "math of what will be," given assumptions and parameters that we know to be certain. Sometimes the future is uncertain and *random* and this calls for the math of *probability*, or the subworld of the "math of what might be." The determination and analysis of the probabilities of future events is key to the math that controls our lives. Educated guesses of probabilities often entail knowing the past. Gathering and analyzing this information is the math of *statistics*, or the subworld of the "math of what was." Statistics are used in many other ways as well. Many conflate the areas of probability and statistics so much that they do not understand them well enough to use them. After overviewing the features that describe and distinguish probability and statistics in this chapter, we will explore the math of both in the following two chapters.

Probability and statistics provide a *stochastic* view of the world in which variability or randomness is very natural, expected, and quantifiable. They are closely related, but distinct. As noted, statistical information is often used to determine probabilities. Because probability theory is used to predict the future and statistics are used to analyze the frequency of past events, some prefer to distinguish between the terms *probability* and *frequency*, with probability being the numerical outcome of knowledge or belief and frequency being the numerical result of repeated experiments. Measured frequencies used to help determine probabilities, are also called prior probabilities. We will not be concerned with such nuances, and will stick to the terms of probability and statistics.

One key principle in determining probabilities is that if a certain number of outcomes are equally likely, the probability that any one occurs is 1 divided by the number of potential outcomes. This has been ingrained in us and so it seems to be common sense. For an unweighted die, there is 1 chance in 6 of tossing and getting a 3, so this probability is $1/6 \approx 16.7\%$. The probability of getting a 2 in the first toss and a 5 in the second toss is $1/6 \times 1/6 = 1/36 \approx 2.78\%$, because the tosses are expected to be uncorrelated. So, the roll of a die is *random*, with probability of 1/6 for each possibility. The result of one roll and that of the next are *independent* of each other, i.e. they are *uncorrelated*, and these probabilities do not change with time. These are core concepts in probability. The probability of getting a 2 in one of the tosses and a 5 in the other is $2 \times 1/36 = 1/18 \approx 5.56\%$, because the probability is 1/36 for getting either a 2 and then a 5 or a 5 and then a 2. But, the probability of getting a 2 in one toss and a 2 in the other is only $1/36 = 1/18 \approx 2.78\%$. As we will see in the next chapter, this train of thought for several tosses puts us well on the way to obtaining the *binomial distribution*, which is a cornerstone in the subworld of probability. Extending this distribution to many tosses leads to the *normal or Gaussian distribution*, which is important and renowned in probability and statistical analysis.

We expect the probability that a tossed die will give an even number is $3/6 = 0.50000$ when we think it is unbiased. The frequency of getting an even number for a particular die might be measured to be 0.48 because the die is weighted. Alternatively, it could be unweighted, and with more trials the measured frequency would eventually reach 0.50000. This is part of statistical reasoning.

Gambling has been around for ages and so has been the development and use of probability theory to try to beat the odds. Ignoring any "cut" by the house, in gambling parlance, even odds, 1:1, correspond to a winning probability of $1/(1 + 1) = 1/2 = 50\%$; 2:1 and 1:2 odds respectively to $1/(2+1) = 1/3 \approx 33.3\%$ and $2/(2+1) = 2/3 \approx 66.7\%$ winning probabilities; 100:1 and 1:100 odds respectively to $1/(100+1) = 1/101 \approx 1.0\%$ and $100/(100+1) = 100/101 \approx 99.0\%$ winning probabilities.

Establishing odds is and can be made with different types of information. Presented probabilities can be based on hard information, such as the odds in tossing unbiased coins and dice, and spinning unbiased roulette wheels; the frequency of past events, assuming the statistics were gathered and analyzed properly and that circumstances have not changed since data acquisition; or on gut feelings. However, odds for gambling on sporting events and the like are based on the assessment of how the bettors will place their bets. Odds makers are successful when the "house" just about always wins. The "house" wins when the betting line is set so that equal money is bet on the two competing teams (for "even money"), say that team A will beat team B by 4½ points, because a winning bet pays a little bit less than the bet itself. In this case, the odds makers are not predicting a 50% probability that team A will win by 5 or more points and a 50% probability it will not, but that the betting public will bet with these probabilities.[121]

Probability is the mathematics used to help make decisions about uncertain future events. It is inherent in assessing risk and making decisions (Chapters 18-20). The corresponding and distinct math of uncertainty in logic and characterization is called *fuzzy math* (or *fuzzy logic* or *fuzzy set theory*), as advanced by the seminal work of mathematician and computer scientist Lotfi Zadeh in 1965. Both probability and fuzzy logic address uncertainty and may lead to multiple answers ranging from 0 to 1. The difference is that probability math addresses the likelihood that an event will occur and fuzzy math is concerned with how much an observation fits with a vaguely defined or fuzzy set, and the subsequent consequences. For a given temperature, with a continuous range of possible temperatures, the fuzzy set can correspond to whether something is cold, warm or hot. Some characterize probability as a model of ignorance (of outcome) and fuzzy logic as a model of vagueness (of characterization and

[121] If you bet \$100 with 2:1 odds and win, you receive your original bet plus \$100 × (2/1) = \$200, and so \$300. (If you lose, you lose your \$100.) If you bet \$100 with 1:4 odds and win, you win \$100 × (1/4) = \$25, and so receive \$125. In Moneyline betting, the odds are phrased differently. A line of +200 means that if you bet \$100 and win, you win \$200 (and your initial bet will be returned to you), and so these are in fact 2:1 odds. A line of -400 means that if you bet \$400 and win, you win \$100 (and also receive your initial bet), and so this is the same as 1:4 odds.

consequences).[122,ccciv]

The word *statistics* has been around since 1770, when it meant "science dealing with data about the condition of a state or community," and was derived from the Modern Latin words *statisticum (collegium)* "(lecture course on) state affairs."[cccv] The use of statistics is still pervasive in the public realm of government, but it is now also prevalent in the worlds of health, finance, sports, and so on. Gathering data for statistics to make decisions and educated guesses for probability analysis is much more valuable than evidence based on a few cases or anecdotes. The value of statistics depends on how the statistical data are collected, analyzed, and presented, and how actions are taken on them. Data must be sampled in a random manner. Statistics have little value and can be terribly deceptive if data are acquired in a biased or otherwise inaccurate manner, analyzed incorrectly, and/or presented with bias and deception.[cccvi]

If an extreme non-representative value is obtained on the first trial in data collection, on the second trial the value will likely be closer to the actual mean; this is called *Regression to the Mean.*[cccvii] Consequently, hasty, and often incorrect, generalizations are obtained after using too few samples for *statistical significance*; this is sometimes referred to as the *Law of Small Numbers*. The average of measured values should be near the actual mean when many trials are made and will be closer to the actual mean the more measurements that are made; this is called the *Law of Large Numbers.*[cccviii] These are three interrelated guiding principles in gathering and using data. In Chapter 16 we will see what constitutes a big enough sample. These principles are being invoked implicitly by baseball commentators who tell you to "look at the back of his baseball card" when referring to a mid-career major league player in a slump or on a hot streak. The card shows his career performance statistics and his current performance will likely progress toward his career averages in either case.

Cynical and skeptical witticisms about statistics abound because the

---

[122] In contrast, the Boolean math and logic of Section 7.1 are based on definitive outcomes of 1 or 0 for given inputs; this is the fundamental math of computing and also corresponds to the logical concepts of true or false.

use and presentation of statistics can be suspect or misleading, even when the math behind them is solid and exact (in a statistical sense).[cccix] Mark Twain quoted Benjamin Disraeli as declaring "There are three kinds of lies: lies, damned lies, and statistics."[cccx] Twain weighed in with "Facts are stubborn things, but statistics are pliable." Mathematician Gian Carlo Rota was quoted as saying "There is something in statistics that makes it very similar to astrology."[cccxi]

Uncertainties about their validity are not the only factors that cast doubt on statistics. Columbia University statistician Andrew Gelman has noted that in the days before the statistical link between smoking cigarettes and lung cancer was firmly established and in the public eye, several prominent people in the subworld of statistics were paid by the cigarette industry to point out even relatively inconsequential mistakes in the statistical analyses made by their opponents. This was done to obfuscate negative health data, including those linking smoking cigarettes and cancer and also those linking smoking to heart disease in the groundbreaking Framingham heart disease study (Chapter 16).[cccxii]

We now look at assessing the future by using the subworld of probability.

Chapter 15

The Math of What Might Be: Probability – What Are the Odds?

We need to enter the subworld of probability, the "math of what might be," to become comfortable with the uncertainty in our lives and to try to control it. Making educated guesses about the future often entails knowing the past, and this often uses the "math of what was," or statistics, which we encounter in the next chapter.

One pervasive example of the use of probability math is in gambling. In fact, gambling motivated the development of the math of probability. Its groundwork was established by French mathematicians Pierre de Fermat and Blaise Pascal in 1654 when they analyzed the rolling of dice. Analytic methods of probability were advanced by French mathematician Pierre-Simon de Laplace in 1812 and others. The odds you will eventually lose while gambling are pretty good, so some peoples' lives are indeed controlled by gambling and probability, and they seem to be comfortable with this.

15.1 Basics of Probability

Nobel-prize winning physicist Richard Feynman once noted: "You know, the most amazing thing happened to me tonight .... I saw a car with the license plate ARW 357. Can you imagine? Of all the millions of license plates in the state, what was the chance that I would see that particular one tonight? Amazing!"[cccxiii] Feynman's remark was in part probability analysis and in part a tease. There are 26 possible letters and 10 possible digits, and these are the equally possible choices for a letter or digit, from the Addition Principle in Section 4.10. So, using the Multiplication Principle (Section 4.10), there could be $26 \times 26 \times 26 \times 10 \times$

$10 \times 10 = (26)^3(10)^3 = 17{,}576{,}000$ possible "license plate numbers," because all letters and digits were potentially in use at the time. So, the probability of seeing this or any other particular one would be 1 out of 17,576,000 or $\sim 5.7 \times 10^{-8}$. The tease was his selecting this particular number, as opposed to say ARW 358. Of course, the probability of plate ARW 357 being issued since 1980 in his state of California is zero, because it started using three letters and four numbers then[cccxiv] to accommodate the exploding number of vehicles.

Sometimes choices are random, but with some restriction. In choosing a random 7-digit number, there are 10 ways of choosing the first digit, 10 ways of choosing the second digit and so on, so there are $10 \times 10 \times 10 \times 10 \times 10 \times 10 \times 10 = 10^7 = 10{,}000{,}000$ ways of choosing it, with each digit chosen randomly and uncorrelated to each other. The probability that a seven-digit number chosen this way is Jenny's number, 8,675,309 (*867-5309/Jenny* sung by Tommy Tutone), is 1 out of 10 million or 10^{-7}. The possible choices of 7-digit numbers with no repeating digits, as with Jenny's number, is more restrictive. There are 10 ways to choose the first number, 9 ways to choose the second, 8 the third, 7 the fourth, 6 the fifth, 5 the sixth and the 4 the seventh number, or $10 \times 9 \times 8 \times 7 \times 6 \times 5 \times 4 = 604{,}800$ such numbers, which we saw in Section 4.10 is the number of permutations or arrangements in a specific order, or 10!/3! here. Each digit is chosen randomly but from more restricted sets. The probability that you will randomly choose Jenny's number is larger now, but still tiny, 1 out of 604,800 or $\approx 1.65 \times 10^{-6}$.

You care about the order of the digits in a phone number, but not in the numbers drawn for a lottery. For a 7-digit number with no repeating digits, there are 7 ways of choosing one of them to be the first digit, 6 ways to choose the second, 5 for the third, 4 for the fourth, 3 for the fifth, 2 for the sixth and 1 for the seventh, or $7 \times 6 \times 5 \times 4 \times 3 \times 2 \times 1 = 5{,}040$ ways of choosing the "same" lottery number for this seven-digit sequence. So, the probability of choosing the ordered Jenny's number 8,675,309 from 0, 3, 5, 6, 7, 8, and 9, each chosen once, is 1 out of 5,040 or 2×10^{-4}. This also means that the number of ways of choosing in any order the seven non-repeating digits that constitute Jenny's number from the 10 digits, is $604{,}800/5{,}040 = 120$. We saw in Section 4.10 this is the number of combinations or choices without regard to order, or 10!/(3!7!)

here, and so the probability of choosing these or any other particular 7 digits is $1/120 \approx 0.0083$.[123]

15.1.1 The probability of single random and independent events in gambling

For single random events, the probability of getting heads or tails in tossing an unbiased coin is 1/2. In card games, you start with a deck of 52 cards that consists of 4 suits (hearts, diamonds, clubs, and spades) for each of the 13 values (2, 3, 4, 5, 6, 7, 8, 9, 10, Jack, Queen, King, and Ace), with $52 = 4 \times 13$. The probability of randomly choosing a specific suit is $13/52 = 1/4$, a specific value is $4/52 = 1/13$, and a specific card is 1/52.

In the American Roulette wheel, there are 38 numbers. 18 of the numbers from 1-36 are red and 18 are black. Numbers 0 and 00 are green. If you bet on red or black and it turns up the color you choose, you get back your initial bet plus an equal amount. The probability of winning in American Roulette is $18/38 \approx 0.4737$ and of losing is $20/38 \approx 0.5263$. In the European Roulette wheel, there are 37 numbers. Again, 18 of the numbers from 1 to 36 are red and 18 are black. There is only one extra number 0, which is green. If you bet on red or black, the probability of winning in European Roulette is $18/37 \approx 0.4865$ and of losing $19/37 \approx 0.5135$, which is slightly better for the bettor than in the American version. In American Roulette, you lose an average of ≈\$0.5263 per \$1 bet and win ≈\$0.4737; therefore, the fraction lost per bet is $\approx 0.5263 - 0.4737 = 0.0526 \approx 5.3\%$, so $\approx 100\% - 5.3\% = 94.7\%$ is returned on average per amount bet. For European Roulette, ≈2.7% is lost and ≈97.3% is returned per bet on average. In the 1960s and early 1970s, Richard Jarecki was able to consistently beat the house and win big in roulette by noticing wheel biases due to poor manufacturing and wear

[123] This is the same as the ways of choosing the 3 digits that are *not* included (when the 7 are being selected). There are 10 ways to choose the first, 9 way the second, and 8 ways the third, or $10 \times 9 \times 8 = 720$ (=10!/7!) ways. The order does not matter so the $3 \times 2 \times 1 = 6$ (= 3!) ways of choosing this triplet are the same, so there are $720/6 = 120$ (10!/(7!3!) = 10!/(3!7!)) distinct combinations. Often there is more than one way to get the same result!

and tear, and betting accordingly.[cccxv] Gambling establishments then took measures to rectify this situation.

Your intuition in gambling is often at odds with the actual odds (pun intended). At a gambling casino, you can engage in a game in which you randomly pick one card out of the deck. The house then randomly picks one of the remaining 51 cards. If your card has a value greater than that chosen by the house you win, as with your Ace beating their 5. If the house card has a value greater *or equal to* yours, the house wins. Since you pick first, the odds are in your favor. Right? Wrong! You have a $4/52 = 1/13$ chance of picking a 2, and then none of the remaining 51 cards will be less than yours, so the probability of winning if you pick a 2 is $1/13 \times 0/51$. You have a 1/13 chance of picking a 3, and only the four 2s, or 4 of the remaining 51 cards will be less than yours, so the probability of you winning because you pick a 3 is $1/13 \times 4/51$. You continue this for each of the 13 possible card values you can pick and then add the results. Your probability of winning is $1/13 \times (0/51 + 4/51 + 8/51 + \ldots + 44/51 + 48/51)$. This simplifies to 8/17, which is less than one half, and so on average the house wins.[124,cccxvi]

15.1.2 The probability of multiple random and independent events in gambling and other aspects of life

In a slot machine, where you win with five of a kind for any of the five possible symbols, the probability of randomly getting any one symbol in one line (or wheel) is 1/5, so the probability of getting five of one of the symbols, say cherries, is $1/5 \times 1/5 \times 1/5 \times 1/5 \times 1/5 = (1/5)^5 = 1/3{,}125 \approx 0.032\%$ because the lines are uncorrelated. Since you would win with any of the five sets of 5, the probability of winning is five times this or $1/625 \approx 0.16\%$. If you won \$625 for each \$1 winning bet, on average you would not win or lose any money, but the house would not want this. If the net payout averaged 95%, you should win $0.95 \times \$625 = \$594$ with

[124] There is a shortcut to this solution. For whatever card you pick, you know that 3 of the remaining 51 cards will always have the same value. On average, of the 12 of the remaining 13 values, 6 values (corresponding to 24 cards) will be greater than yours and 6 (24 cards) will be lower. So, on average 24 of the remaining 51 will have lower values, so you will win $24/51 = (3 \times 8)/(3 \times 17) = 8/17$ of the time.

five of a kind, and so you would be losing \$31/625 = 5 cents (= (100% – 95%) × \$1) per bet on average.[cccxvii] This is the typical payout of slot machines in Las Vegas, and is above the legal requirement of at least a 75% return.[cccxviii] Playing the same random, uncorrelated game multiple times is just one example of using the very important *binomial distribution* or binomial theorem, as we will see.

Each version of the dice game craps depends on the outcome in rolling two dice. You can get a total of 2, 3, 4, 5, 6, 7, 8, 9, 10, 11, or 12. There is one way getting a total of 2 (1 in "die #1" + 1 in "die #2") or 12 (6 + 6), so each has a probability of 1/6 × 1/6 = 1/36. There are two possibilities each for a 3 (1 + 2 and 2 + 1) and 11 (5 + 6 and 6 + 5), so their probabilities are each 2/36; three possibilities each for a 4 (1 + 3, 2 + 2, and 3 + 1) and 10 (4 + 6, 5 + 5, and 6 + 4), so their probabilities are each 3/36; four possibilities each for a 5 (1 + 4, 2 + 3, 3 + 2, and 4 + 1) and 9 (3 + 6, 4 + 5, 5 + 4, and 6 + 3), so their probabilities are each 4/36; five possibilities each for a 6 (1 + 5, 2 + 4, 3 + 3, 4 + 2, and 5 + 1) and 8 (2 + 6, 3 + 5, 4 + 4, 5 + 3, and 6 + 2), so their probabilities are each 5/36; and six possibilities for a 7 (1 + 6, 2 + 5, 3 + 4, 4 + 3, 5 + 2, and 6 + 1), so its probability is 6/36. (These sum to 1.) In the simple casino "field" bet in craps, you lose your \$1 bet when you roll a 5, 6, 7, or 8 and win when you roll a 2, 3, 4, 9, 10, 11, 12, so you win when you roll one of seven numbers and lose "only" when you roll one of the (fewer) four numbers, but the odds are still against you. If the house paid you \$1 when you won (even money), your average payoff from winning would be 1/36 + 2/36 + 3/36 + 4/36 + 3/36 + 2/36 + 1/36 = 16/36. If it kept your \$1 when you lost, the house would keep 4/36 + 5/36 + 6/36 + 5/36 = 20/36 for each \$1 you bet on the average. Your net average proceeds after a roll would be 16/36 – 20/36 = -4/36 = -1/9 = -0.1111... dollars, for a huge ≈11.1% house advantage. This "even money" or 1:1 odds arrangement is a very bad deal. You would likely lose all your money after relatively few rolls and then decide not to play this game again. To make it more attractive, the house decreases your average loss per bet by paying off a bit more than even money on some of your winning combinations.[125]

---

[125] For example, if when you roll a 2 you win \$2 (2:1 odds) and when you roll a 12 you win \$3 (3:1 odds), your payment from winning increases to 2×(1/36) + 2/36 + 3/36 + 4/36

What happens when you roll a die five times? In each independent roll, the probability is 1/6 of rolling a 2 (or any other number) and 5/6 of rolling a different number. In five rolls, the probability of rolling a 2 each time is $1/6 \times 1/6 \times 1/6 \times 1/6 \times 1/6 = (1/6)^5 = 1/7{,}776 \approx 0.0129\%$. The probability of not rolling a 2 in any roll is $5/6 \times 5/6 \times 5/6 \times 5/6 \times 5/6 = (5/6)^5 = 3{,}125/7{,}776 \approx 40.1878\%$.

The probability of rolling a 2 in the first four rolls but not in the fifth is $1/6 \times 1/6 \times 1/6 \times 1/6 \times 5/6 = (1/6)^4(5/6) = 5/7{,}776 \approx 0.0643\%$. This is the same probability as rolling a 2 in any four of the five rolls but not in the fifth one, so the probability of rolling a 2 only once is $5(1/6)^4(5/6) = 25/7{,}776 \approx 0.3215\%$. Similarly, the probability of not rolling a 2 in the first four rolls and rolling it in the fifth is $5/6 \times 5/6 \times 5/6 \times 5/6 \times 1/6 = (1/6)(5/6)^4 = 625/7{,}776 \approx 8.0376\%$. This is the same probability as not rolling a 2 in any one of the five rolls but rolling a 2 in the remaining one, so the probability of not rolling a 2 only once is $5(1/6)(5/6)^4 = 3{,}125/7{,}776 \approx 40.1878\%$.

There are two possibilities left. The probability of rolling a 2 in the first two rolls only is $1/6 \times 1/6 \times 5/6 \times 5/6 \times 5/6 = (1/6)^2(5/6)^3 = 125/7{,}776 \approx 1.6075\%$. If we do not care when the two rolls are, they could be in rolls 1 and 2, 1 and 3, 1 and 4, 1 and 5, 2 and 3, 2 and 4, 2 and 5, 3 and 4, 3 and 5, or 4 and 5, for a total of 10 possibilities, so the probability of rolling a 2 exactly twice is 10 times this value or $10(1/6)^2(5/6)^3 = 1{,}250/7{,}776 \approx 16.0751\%$. Similarly, the probability of not rolling a 2 in the first two rolls only is $5/6 \times 5/6 \times 1/6 \times 1/6 \times 1/6 = (1/6)^3(5/6)^2 = 25/7{,}776 \approx 0.3215\%$, and since there are the same 10 ways of doing this, the probability of not rolling a 2 exactly twice (and rolling a 2 exactly 3

$+ 3/36 + 2/36 + 3 \times (1/36) = 19/36$ dollars. Now, on the average, your proceeds after a roll would be $19/36 - 20/36 = -1/36 = -1/9 \approx -0.0278$ dollars, so the house advantage is now lower, $\approx 2.78\%$. The net probability of winning is $19/36/(19/36 + 20/36) = (19/36)/(39/36) = 19/39$ and of losing $(20/36)/(19/36 + 20/36) = 20/39$. You still lose on average with this version, but it usually takes more rolls for you to lose all your money. Unfortunately, most casinos pay only \$2 (2:1) for a 12, and your proceeds would be only $18/36 - 20/36 = -2/36 = -1/18 \approx -0.0556$ dollars, with the higher (and high) house advantage of $\approx 5.56\%$. Casinos pay out \$3 (3:1) on 2 and 12 only for limited times during promotions to attract new "players", because then the house advantage is zero (with net proceeds of $20/36 - 20/36 = 0$ dollars).

times) is 10 times this value or $10(1/6)^3(5/6)^2 = 250/7{,}776 \approx 3.2150\%$. The sum of the probabilities of rolling a 2 exactly 0, 1, 2, 3, 4 or 5 times in five tosses is 40.1878% + 40.1878% + 16.0751% + 3.2150% + 0.3215% + 0.0129% = 100%, as it must be (Figure 15.1a).

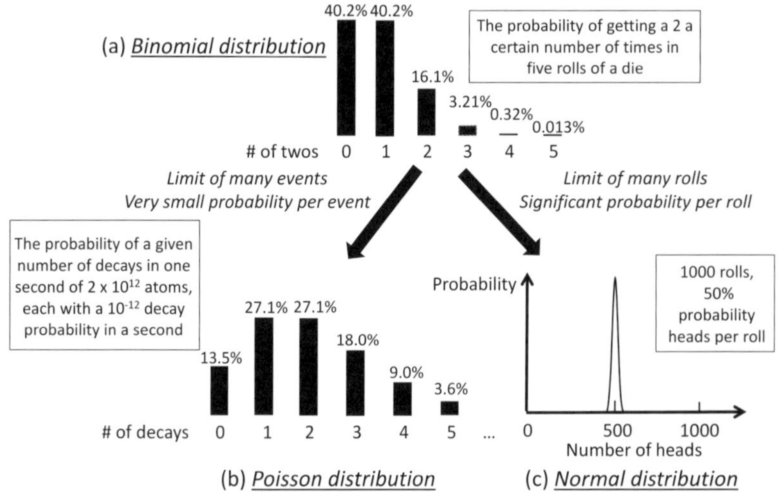

Figure 15.1: Examples of binomial, Poisson, and normal distributions (for three different cases).

Generalizing this analysis leads to the binomial distribution or binomial theorem, which gives of the probability of something random happening a given number of times in a certain number of independent trials. Say the probability in one trial is p, and so the probability of it not occurring is 1 minus p, or $1 - p$. As we have seen, the probability for any specific ordered outcome is (probability it occurs)$^{\text{times that it occurs}}$ × (probability it does not)$^{\text{times that it does not occur}}$. For rolling a 2 three times out of 5 rolls, so not tossing a 2 two times, we saw this is $(1/6)^3 \times (5/6)^2$, or $(p)^3 \times (1-p)^2$ with $p = 1/6$ and $1 - p = 5/6$. Again, there are several ways this can occur, and as we saw in Section 4.10, this number of combinations is (total trials)!/[(times it occurs)!(times it does not occur)!]. This is $5!/[(3!)(2!)] = (5 \times 4 \times 3 \times 2 \times 1)/[(3 \times 2 \times 1)(2 \times 1)] = 10$ here, as we just saw. Calling the total number of trials n (here 5) and the times it occurs r (here 3), the number of times it does not occur is $n-r$ (here $5-3=2$). When generalized to any p, n, and r, this is the binomial distribution (one type of probability distribution function).[126] The mean value is np, so the average number of times of getting a 2 in 5 rolls is $5 \times 1/6 = 5/6$ or in 50 rolls is $50 \times 1/6 = 50/6 = 8\ 1/3$. (Figure 15.1a gives a simpler example.)

The basic building block of this multi-event binomial distribution is the single event distribution, which tells whether the coin event is, say, a success by being heads, $p=0.5$, or a failure or not heads (and so tails), $1-p=0.5$, or whether the die event is, say a 2, with a success probability of $p = 1/6$, or a failure, and so it is not a 2, with $1-p = 5/6$. This probability distribution is called the *Bernoulli distribution.*[cccxix] So, the binomial distribution is the distribution of successes in a given number, say n, trials, with the result of each trial given by a Bernoulli distribution. The related, distribution of the number of failures before the first success is called the *geometric distribution.* The probability of not seeing a 2 in the first four trials is $(5/6)^4 = 625/1{,}296 \approx 48.23\%$, so the probability of seeing a 2 for the first time in the fifth trial is $(1/6)(5/6)^4 = 625/7{,}776 \approx 8.04\%$.[127,128,cccxx]

The binomial distribution is at the core of much of basic probability analysis, both as is and in two different important limits when there are many trials n. It becomes the *Poisson distribution* when there is a very small success probability per trial, p, so that the mean number of successes s, which is the product and n and p or $s=np$, is not very large (Figure 15.1b).[129,cccxxi] This distribution is useful in characterizing rare events, such as decay events in radioactivity. Because such events are

---

[126] The probability of seeing an event with probability p exactly r times in n chances is the binomial distribution, $\{n!/[r!(n-r)!]\}p^r(1-p)^{n-r}$ or symbolically as $\binom{n}{r}$.

[127] The probability of seeing an event with probability p for the first time after n failures is the geometric distribution, $p(1-p)^n$.

[128] This is one fifth of the probability we found for rolling a 2 exactly once in five tosses, $\approx$40.1878%, because the 2 could then be in rolls 1 through 5 in the binomial distribution, while in the geometric distribution it can occur only in the fifth roll.

[129] The Poisson distribution of r successful events occurring is $s^r e^{-s}/r!$, with the average number of successful events being s (= number of trials $n \times$ success probability per event $p=np$, for $n>>1$ and $p<<1$).

random and uncorrelated, the number occurring in a given time period, *r*, varies randomly, though the average number of decays occurring in that period, *s*, remains the same. You hear this variability when using a Geiger counter to detect radioactive decays because the events are not evenly spaced. If two trillion (2×10^{12}) atoms each have a probability of decaying in a trillionth (10^{-12}) of a second, on average 2 of their nuclei decay in a second. Poisson statistics show that no decays occur in 13.5% of the 1 second long intervals, one decay in 27.1% intervals, two decays in 27.1% of them, three decays in 18.0% of them, four decays in 9.0% of them, five decays in 3.6% of them, and so on.[130] As with the binomial distribution, the Poisson distribution specifies when an event occurs an integral number of times, so both are called *discrete* distributions.

The binomial distribution evolves into a very different limit for many trials when each trial has an arbitrary success probability. In this limit, it may be more interesting to learn, say, whether 1000 coin tosses produce heads between 49.9% and 50.1% of the time rather than exactly 499, 500 or 501 times. The form of this now *continuous* distribution is called the *normal or Gaussian* distribution (Figure 15.1c).[131] It is frequently used in probability and statistical analysis, as we will soon see, as well as in many other areas of science.

The binomial distribution gives the outcome after a specified number of random events. You could also be interested in how the distribution evolves with each event, such as after each roll of the die or tossing of a coin. This evolution is called a *random walk*, which we will discuss later.

15.1.3 Random, but with restrictions

Sometimes a choice may be random, but may still depend on history or other restrictions.

---

[130] For no decays this is $2^0e^{-2}/0! \approx 13.5\%$, for one decay it is $2^1e^{-2}/1! \approx 27.1\%$, for two decays it is $2^2e^{-2}/2! \approx 27.1\%$, for three decays it is $2^3e^{-2}/3! \approx 18.0\%$, and so on.

[131] In this limit of large *n*, *np*, and $n(1-p)$, the binomial distribution probability of an event with probability *p* occurring *r* times in *n* chances, $\{n!/[r!(n-r)!]\}p^r(1-p)^{n-r}$, evolves to the normal distribution $(1/\sqrt{2\pi np(1-p)})e^{-(r-np)^2/(2np(1-p))}$. In both distributions, the mean is *np*, and the standard deviation and variance (which are distribution widths that are described below and defined in Chapter 16) are $[np(1-p)]^{0.5}$ and $np(1-p)$. In the normal distribution, *r* is often called *x*.

Say 1/3 of the balls in a box are red and 2/3 are green, and 1/4 of them are big and 3/4 are small, and the probability of picking any single ball is equal. The probability of picking a red ball first is 1/3 and that of picking a small ball from the initial box is 3/4. The probability of picking a red ball and then a green ball would be $(1/3)(2/3) = 2/9 = 0.22222\ldots$ *only if you replaced the red ball after you first picked it*. If there were 60 balls at first, 20 would be red, 40 green, 15 big and 45 small. After a red ball is picked, there would be 19 red balls and still 40 that are green, and so the probability the next one picked randomly from the remaining 59 balls would be red is 19/59 and green is 40/59, which is >2/3. So, the probability of picking a red ball and then a green ball, *with no replacement*, would be $1/3 \times 40/59 = 40/177 \approx 0.226$, which is >2/9. Without replacement, the probability of picking a green ball after a red one (here $40/59 \approx 0.678$) is higher than that after picking a green one first ($39/59 \approx 0.661$), so the second pick is *correlated* with the first. The picks are still random, but the probabilities now depend on *history*. The issues of replacement and history are not relevant when tossing the same coin or die over and over again, but they can be in card games. Gambling institutions do not tolerate card counting, which means tracking history, because it can improve your odds.

Furthermore, the probability of picking a green ball and then a big ball from this initial box is $2/3 \times 1/4 = 2/12 = 1/6 = 0.166666\ldots$, if the first ball were replaced before picking the second one. But, what is the probability of picking a big green ball from the initial box? It would be the probability of the ball being green, 2/3, times that of it being big, 1/4, and so $2/3 \times 1/4 = 2/12 = 1/6 = 0.16666\ldots$, *only* if the color and size of the balls were uncorrelated. Then in our example, there would be $1/3 \times 3/4 \times 60 = 15$ small red balls, $1/3 \times 1/4 \times 60 = 5$ big red balls, $2/3 \times 3/4 \times 60 = 30$ small green balls, and $2/3 \times 1/4 \times 60 = 10$ big green balls, for the 60 balls. However, you need more information if the color and size of the balls are not uncorrelated. For this example, in one extreme limit all of the 15 big balls are green (with 25 being small), and in the other 0 big balls are green (with all 40 being small), so the number of big green balls can range from 0 to 15. Without more information, the best you could say is that the probability of choosing a big green ball first is between $0/60 = 0$ and $15/60 = 1/4 = 0.25$.

How a problem is phrased can affect the answer in ways that are not obvious, even for a sequence of coin tosses, when seemingly random events are linked by restrictions and the selection process. If you toss a coin four times, what is the probability that any time you get a heads in any of the first three tosses it is followed by a heads? The probability that the next toss is heads is always 50%, but the answer to this problem is not 50%, but ≈40.48% due to this subtle selection bias in selecting which data to analyze and the finite length of the sequence.[132,cccxxii] (Not fully understanding the phrasing of this problem has led to notable misinterpretations of this result.[cccxxiii])

An example of the subtleties in analyzing random events, and one with a red herring thrown in, occurred in the Robert Parker-character based 2009 television movie *Jesse Stone: Thin Ice.* "Paradise," Massachusetts police chief Jesse Stone, played by Tom Selleck, was searching for a local seven-year boy who had been born and kidnapped in New Mexico after birth because of a possible link to Paradise. He found evidence of a seven-year old who might be the same boy and was sure he had a match because both had the same birthday. Deputy Rose Gammon suggested that he should not be too optimistic because the

---

[132] There are 2 × 2 × 2 × 2 = 16 possible sequences for four tosses. However, we cannot consider the sequences TTTT or TTTH because for heads to follow another heads, one of first three tosses needs to be heads, and so only the 14 other combinations can be considered, and they are equally probable. For HHHH, the tosses after each of the first three heads are heads, so for this sequence, it is 100% probable a heads in the first three tosses is followed by a heads. It is also 100% probable for the two heads in the first three tosses for THHH and for the one toss in the first three for TTHH. These three sequences contribute 3/14 to the total. For HHHT, the toss after the first two heads gives heads, but after the third it is tails, so for this sequence, it is 2/3 = 66.66...% likely, and so it contributes 2/3 × 1/14 = 1/21 to the total. For HHTH, HHTT, HTHH, and THHT, one heads is followed by a heads and one by a tails, so it is 50% probable in each, and these four contribute 1/2 × 4/14 = 1/7 to the total. For HTHT, HTTH, HTTT, THTH, THTT, and TTHT, heads are never followed by a heads, so these six sequences contribute nothing to the total. Overall, when you toss a coin four times, the probability that a heads follows a heads that is seen in the first three tosses is 3/14 + 1/21 + 1/7 = 17/42 ≈ 40.48%, and is not 50% due this subtle selection bias in selecting which data to analyze and the finite length of the sequence. For longer and longer sequences of tosses, this probability does approach 50%.

probability that 2 out of 23 boys having the same birthday was greater than a half. The deputy was citing the classic *Birthday Problem*: What is the minimum number of people necessary in a room (or at a birthday party) so the probability of any two of them having the same birthday is greater than 50%?

Say one person in the room is joined by another, then another, and another until the probability of two not having the same birthday first becomes smaller than 50%; the number in the room then is the answer. Assume a leap year and say that being born on any one of the 366 days is equally probable and independent of each other. (Such imperfect assumptions affect the answer very little.) Whatever the birthday of the first person, the birthday of the second person could be on the other 365 days, so the probability of the second person not having the same birth as the first is 365/366. The third person would need to have been born in on one of the other 364 days and the probability of person 3 not having the same birthday is 364/366, for a given set of persons 1 and 2, which itself has probability 365/366. So, the probability of three people not having the same birthday is (365/366)(364/366). For four people, the probability of not having the same birthday is (365/366)(364/366)(363/366). This continues until this number dips below 50%. For 22 people this product is $\approx$52.75% and for 23 people it is $\approx$49.4%, so the probability is greater that two people will have the same birthday than not occurs with 23 (or more) people in the room (even in non-leap years, with 365 days). The catch is this same birthday could be any of the 366 days. Only with 367 people in the room can you be 100% certain that a pair will have the same specified birthday.[cccxxiv] So, though what the deputy said was correct, it was not relevant because the point was how probable it was that the located boy had the same specific birthday as the lost boy and not whether the two happened to have the same birthday. (In the end, Jesse Stone made the correct connection.)[133]

---

[133] Approximately 2,000,000 boys were born in the U.S. in 2002 (the birth year in question), with ~2,000,000/365 = 5,500 born on any particular day. The probability of the match was much greater than 1/5,500 because of the evidence. For example, narrowing the births to those in New Mexico and Massachusetts increases this probability by ~30,

We will look at another example of probability with restrictions, *conditional probability*, later this chapter.

15.2 Continuous Distributions

15.2.1 Distributions are often normal

In the limit of many trials, the binomial distribution evolves smoothly into a curve commonly called the normal distribution, which is known by physicists as a Gaussian distribution, after the mathematician Gauss.[cccxxv,cccxxvi,cccxxvii] Its shape is called a *Gaussian* or the *bell* curve (often by social scientists), because it looks like a bell (Figures 15.1c and 15.2). The variable is this case, x, could be the number of heads in many coin tosses or the fraction of these coin tosses that end up being heads. The distribution has an average value called μ, the lower case Greek letter mu, and a characteristic width called the *standard deviation* σ, the lower case Greek letter sigma, and whose square is called the *variance* σ^2. σ has a more precise definition, which we will see in Chapter 16. In a *standard normal distribution*, the mean is 0 and the standard deviation is 1. If you add all the probabilities over the range of possibilities in a normal distribution, you get 1, as you should for any for probability distribution. In calculus, you would say that you would "integrate" over the function, and its integral would be 1. Ensuring this is 1 is more generally called *normalizing* the function, making it a *normalized* distribution.

(An aside: Occasionally the same word in math has quite different meanings. Normalizing a function is yet another mathematical use of the word *normal*. Normal is also another way of saying perpendicular. In the normal distribution, "normal" denotes that which conforms to the norm. The nomenclature "normal distribution" appears to have been become standard when used by English mathematician and biostatistician Karl Pearson in 1894, though it apparently had been used earlier.[cccxxviii])

For a normal distribution, ≈68.26% or about two thirds of its values, and so of the probability, are within one standard deviation of the mean,

---

and narrowing them to those in their respective communities and by other commonalities increases this probability by even more.

≈95.44% are within two standard deviations, and ≈99.74% within three standard deviations (Figure 15.2a). This is sometimes called *the 68-95-99.7 rule*. So, there is a probability of ≈99.7% that a value will be within the range from 3σ below the mean to 3σ above it, or ≈0.3% or approximately 1 out of 300 that the values will be outside this range and these are called *outliers* for this "3σ" event. Half of them, or ≈0.15%, are above this upper limit and ≈0.15% are below the lower limit. In the normal distribution in Figure 15.2b, the average is 100 and the standard deviation is 5, so ≈68% of the probability is between 95 and 105, ≈95% of it is between 90 and 110, and ≈99.7% of it is between 85 and 115.

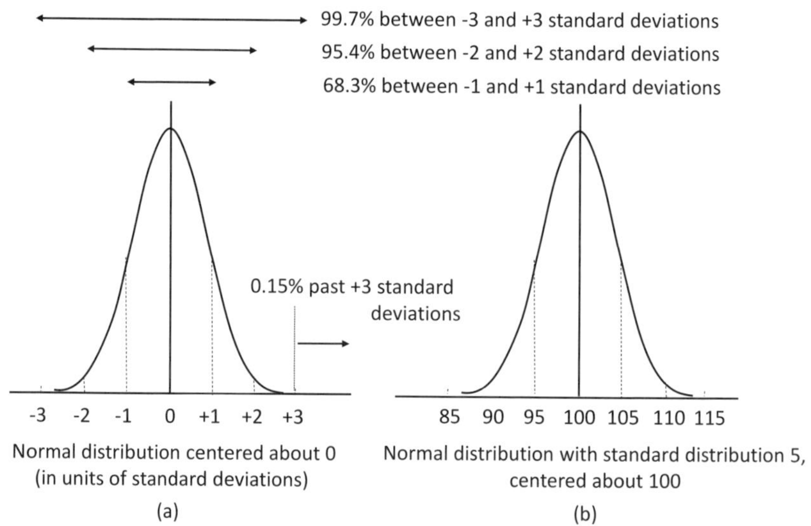

Figure 15.2: Normal distribution.

A 6σ event corresponds to a chance of about one outlier out of 500 million events, so it is exceedingly unlikely. The 7'6" height of former professional basketball player Shawn Bradley is 8.6 standard deviations above the average height of U.S. males, and so he is in the 99.99999th percentile of height. (Since Bradley has no known medical conditions, much of his extreme height may be linked to the statistics of genetic factors that are not very well understood at present.[cccxxix]) The occurrence of apparently "several-σ events" either more or less frequently than expected means this rare observation is a statistical anomaly or that the distribution is not exactly a normal distribution representing random events. The math analysis of rare events in general, *extreme value analysis*, is very important in analyzing weather (as in the occurrence of tornadoes), events (flooding, large wildfires), finance (stock market crashes, large insurance losses), and engineering (pipeline failures).

The normal distribution also arises in statistics in a profound way. One key result of probability theory is the *Central Limit Theorem*, first developed by Laplace in 1810, which says that when independent random variables are measured or sampled many times and summed, the result will usually have a normal distribution. This is true for any reasonable distribution of the random variables, and these do not have to be normal distributions themselves. This is why normal distributions can be used in statistics even if the underlying details of the distribution are not known, but only when the sample size is suitably large, usually 30 or more. Furthermore, the more samples you take, the more likely your average will be the average of the original distribution. (More on this in the next chapter.)

One curious, important, and unique feature of the normal distribution is that it is the only function whose Fourier transform (Section 4.4) has the same function form (and so is also a normal distribution).

A continuous probability distribution, such as this normal distribution, is often presented in two, seemingly different, yet essentially equivalent ways. The probability that an event x occurs is given is called the *probability density function (PDF)*, often written as $f(x)$. The normal distribution is in this form in Figure 15.2.[134] The continuous variable, x, can be the actual variable (ranging say from \$0 to \$600) or a normalized variable instead (from 0.0 to 1.0). It can be centered about 0 or the mean.

---

[134] For the normal distribution, the PDF (the profile itself) is $(1/\sqrt{2\pi}\sigma)e^{-(x-\mu)^2/2\sigma^2}$ and the CDF (the integrated form of this profile, which is defined below) is $(1/2)\{1+\text{erf}((x-\mu)/\sqrt{2}\sigma)\}$, which is also called the error function, or abbreviated as "erf", and is a tabulated function. The mean, median and mode are μ, the variance and standard deviation are σ^2 and σ, and the skewness and excess kurtosis (which are explained below) are 0.

In some distributions, x does not have finite bounds (and so it cannot be normalized), such as time, which could start at 0 (seconds) and continue forever.

For a discrete probability density distribution it makes sense to give the probability for something to occur, say, exactly 5 times, exactly 6 times, and so on. For a continuous PDF, it is meaningful to give the probability $f(x)$ over a small range, say 0.01, about a given x, say 0.7, and so from $0.7-(0.01)/2$ to $0.7+(0.01)/2$, which is $0.7-0.005$ to $0.7+0.005$ or 0.6995 to 0.7005. If the value of f for $x=0.7$ is 0.2, this probability is ~ the probability at x times the width of this range, or $\sim 0.2 \times 0.01 = 0.002 = 0.20\%$ here.[135]

You can also start at one extreme of a PDF ($f(x)$), say at $x=0$, and sum all the probability values from $x = 0.0$ to every value of x as you increase x, eventually to $x=1.0$. This sum starts at 0.0, becomes 0.5 at the peak of the distribution (only if the distribution is symmetric, as for a normal distribution, and x is defined this way), and approaches 1.0 as you approach the other range—because this is the total probability for all possible values of x (Figure 15.3). This cumulative probability for all values smaller than or equal to x, is called the *cumulative distribution function (CDF)* often written as $F(x)$, and is the second form of the probability distribution. The CDF and PDF contain identical information, but provide a different insight. In the jargon of calculus, $F(x)$ is the integral of $f(x)$, while $f(x)$ is the derivative of $F(x)$.[136] The cumulative form of the normal probability distribution is called the *error function.*

Normal distributions are frequently used in data analysis. They are also used to specify some distributions, even when there is no underlying statistical or scientific basis. The distribution of raw scores of IQ tests of a population is converted to a normal curve so that the mean, mode, and median scores map to an IQ of 100 and the standard deviation of the distribution of IQ scores is set equal to 15. This means that by definition

---

[135] For a small width of Δx, this equals $f(x)\,\Delta x$. (Δ is the upper case of the Greek letter delta. Δx is a symbol, not a product, as also in dx.) In calculus, this is equivalent to the integral of $f(x)$ from $x-\Delta x/2$ to $x+\Delta x/2$. For very small widths, in calculus, the width Δx is expressed as the differential dx or the small quantity h.

[136] In calculus notation these are $F(x) = \int f(x')dx'$ and $f(x) = \frac{dF(x)}{dx}$.

≈68.26% of the population have IQs within one standard deviation of 100, from 85 to 115; they are merely those with the median ≈68.26% of raw scores. The same is true for the median ≈95.44% with IQs with two standard deviations, from 70-130, and so on. The fraction of people with IQs in the tails below 70 and above 130 is then ≈100% – 95.44% = 4.66%, which means that, by definition, ≈4.66%/2 = 2.33% have IQs below 70 and ≈2.33% have IQs above 130.[137,cccxxx,cccxxxi] In another ranking application, baseball executive pioneer Branch Rickey developed a system of grading prospects that often resembled a normal distribution.[138,cccxxxii]

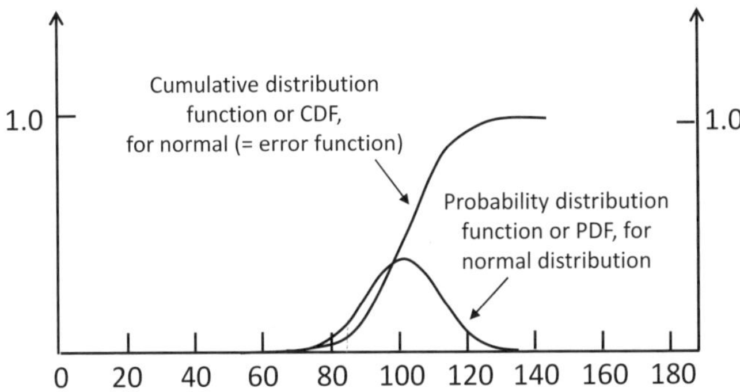

Figure 15.3: Normal distribution, centered at 100, showing the difference between probability and cumulative distribution functions.

---

[137] The normal profile is used to assign IQ scores from the ordered rankings of raw test data, and are not themselves statistical data *per se*.

[138] He graded prospects from 20 to 80 to assess how they would perform in the major leagues in each performance category (or tool, such as hitting with power, fielding and so on), compared to the sample set of current major leaguers. A 20 corresponded to a non-prospect, a 40 to a marginal major leaguer, a 50 an average major leaguer, a 60 above average, and 80 the best of the best. For many tools, it turns out that the distribution sample set of major leaguers happened to look like a normal distribution centered at 50 with a standard deviation of 10, though this was not likely Rickey's intention.

15.2.2 Some distributions are not normal

The normal function and other probability distribution functions are used to characterize shapes of functions in probability and statistics and in other fields. These other distributions do not follow binomial, Poisson, or normal distribution statistics, because they arise from different analyses and assumptions.[cccxxxiii] All such profiles are also characterized by their means, modes, medians, and widths, and by what fraction of the probability distribution falls within 1, 2 or 3 standard deviations about the average.[139,cccxxxiv,cccxxxv] Some profiles, such as the normal distribution, are symmetric; others are not and are called *skewed. Skewness* describes this profile asymmetry numerically, with the mean located above the median, which is itself above the mode (peak) (if skewed to the right) (Figure 15.4a) or with the mean located below the median, which is itself below the mode (peak) (if skewed to the left). You would expect the statistics of your travel times to or from work by car, bus, or train during rush hour to be asymmetric. Travel times would be somewhat shorter than the median (say ~20 minutes compared to 30 minutes) when traffic is relatively light, but either somewhat, much, or very much longer than the median (being ~40, ~50 or ≥~60 minutes), on days with somewhat heavier than typical, quite heavy, and terrible traffic, respectively. Some profiles decay rapidly away from the peak and some extend much further, and so have long tails. *Kurtosis* describes how dominant the extremes or tails are relative to those of the normal distribution, which is said to have 0 "excess" kurtosis. The faster the tails fall off, the more negative its excess kurtosis and the slower they fall off, the more positive its excess kurtosis (Figure 15.4b).

---

[139] The fraction of any normalized distribution, even fairly irregular ones, that is within a number z times the standard deviation about the average (and so $\pm z\sigma$ about it) is at least $1 - 1/z^2$; this was shown by the Russian mathematician Pafnuty Chebyshev. So, the fraction within $\pm 2\sigma$ of the mean is usually at least $1 - 1/4 = 75\%$ and that within $\pm 3\sigma$ of the mean is at least $1 - 1/9 \approx 89\%$. We saw for that for the normal distribution the actual fractions between these two bounds were larger than these values, ≈95%, and ≈99.7% respectively. These bounds set by *Chebyshev's Inequality* are "weaker," though more general results. Such bounds help in estimating results from distributions and assessing whether or not conclusions from statistical studies make sense.

One variation of the normal distribution occurs when you apply the Central Limit Theorem to the sums of the logarithms of variables, $\ln x$, instead of the random variables themselves; then you obtain a normal distribution of $\ln x$. Because the sum of the logs of two or more numbers equals the log of the product of the numbers, this is, in essence, using the Central Limit Theorem to look at the distribution of the product of random variables, and not their sum. This leads to a modification of the normal distribution called the *log-normal or Galton distribution*, which is merely the normal distribution with the x replaced by $\ln x$.[140,cccxxxvi] It looks very different than the normal distribution, when plotted vs. x (Figure 15.4a).

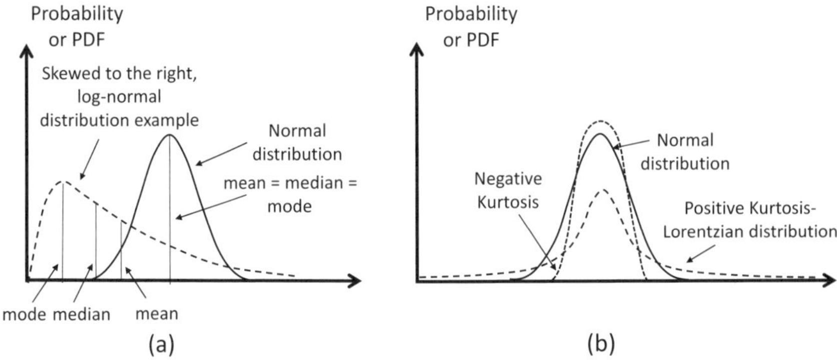

Figure 15.4: Skewness and kurtosis in example distributions.

Medical exam blood test results commonly list for each blood component your results and a reference or normal range, which is the middle 95% range of values of healthy people.[cccxxxvii] The distributions of many of the components in blood approximately follow either a normal or log-normal distribution. For the normal distributions, this range corresponds to two standard deviations about the mean. Because these

---

[140] The PDF of the log-normal distribution is obtained by replacing x in the normal distribution by $\ln x$, and by adding an x in the denominator (for a calculus reason---to keep the differentials the same), to give $f(x) = (1/\sqrt{2\pi}\sigma x)e^{-(\ln x-\mu)^2/2\sigma^2}$. The CDF is $F(x) = (1/2)\{1 + \mathrm{erf}\big((\ln x - \mu)/\sqrt{2}\,\sigma\big)\}$, where "erf" is the error function (Figure 15.3); the values of this function can be found in math tables.

values cannot be negative, of course, when distributions have low mean values and large variances, they are typically skewed and are often fit well by log-normal distributions. (x cannot be negative, though $\log x$ and $\ln x$ can be.) Such log-normal distributions also fit the latency periods of infectious disease, the distribution of minerals in the Earth's crust, and the distribution of sizes of the "grain" microstructure in metals.[cccxxxviii,cccxxxix]

Several other continuous distribution functions are also widely used. The *beta function* (denoted by a lower-case Greek letter beta, β) also arises in the analysis of random events and its variable x ranges between 0 and 1. Its shape is defined by two parameters. When both parameters equal 1/2, this distribution is the well-known and surprising *arcsine* ("*arc*" "*sine*") *distribution* (Figure 15.7 below).[cccxl] We will see that it describes the evolution of sequences of random probabilistic events, called *random walks* (Figures 15.5 and 15.6 below). In contrast, the binomial and normal distributions characterize the final states of such sequences.

Another often-used continuous probability distribution function is the *gamma distribution*[cccxli] (denoted by a lower-case Greek letter gamma, γ), which is commonly used to model aggregate insurance claims and rainfalls. One form of it, the *chi-squared distribution*[cccxlii] (or the chi-square or χ^2 distribution, where χ is the lower-case Greek letter chi, pronounced "ki"-with a long i), is used to assess the *goodness* of a fit to a model function, by using the *chi-squared test*, as we will see in the next chapter.

The *Weibull probability density function* has a quite different application. It is often used to characterize component, system and materials failure, and so in extreme value analysis, as we will see in the Chapter 20 discussion on risk. In one limit it is the usual exponential decay function, but in others, it can be used to describe a failure rate that either slows with time due to limited initial defects or increases with time due to aging.

The very different continuous probability distribution is the *Cauchy distribution*, known by physicists as the Lorentz, Cauchy–Lorentz, or Breit–Wigner distribution or simply as a *Lorentzian profile*. As seen in Figure 15.4b, its wings extend much further than those of the normal distribution, and so its excess kurtosis is large and positive.[cccxliii] It is significant and widely used in physics and engineering, because it is the Fourier transform (see Section 4.4) of the exponential decay function.

15.3 Where Do Probabilities Come From? And What Do They Mean?

As we have noted, some probabilities are determined objectively from mathematical analysis, such as from past frequencies and statistics, and analyses of events that seem to have equal probabilities, such as 1/2 for each side in tossing coins and 1/6 for each face in tossing dice. The projection into the future by using probabilities from statistical frequencies might be warranted if the conditions have not changed. Some probabilities are determined more subjectively, from seemingly reasonable expectations or personal beliefs, and these are called Bayesian probabilities.[cccxliv]

Probabilities can be determined by supply and demand, such as those of winning a housing lottery. In three lotteries in 2014 and 2015 for subsidized NYC public housing apartments, the probability of winning was 1 out of 490, 560 and 2,110 applications, with the last one drawing more than 80,000 applications for 38 units. These are pretty bad odds, but as has been noted, better than the 1 in 22 million probability of winning the New York Lotto jackpot (on a $1 play, in 2015).[cccxlv]

Probabilities can also be set by non-quantitative assessments that may be dubious. In late June 2016, reports were that basketball superstar Kevin Durant of the Oklahoma City Thunder basketball team in the NBA would sign again with his team of nine years. When on June 30, 2016, Marc Spears reported that Durant's friends and colleagues told him "His decision (to resign with Oklahoma City) is 90 percent made.",[cccxlvi] the probability of his resigning could be set at 90%, so that of his signing with another team was only 10% and so it was very unlikely. On July 4, 2016, Jacob McCleland reported that Durant would sign with the Golden State Warriors.[cccxlvii]

Some probabilities are derived more quantitatively, but may seem to some to be guesswork. For example, the meaning of probabilities in weather forecasting obtained from detailed computer codes is often misunderstood. Results are transmitted to you as a probability of rain, sun, or whatever. Say the forecast is for at least measurable precipitation

(0.01 inches or more) in 50% of the area specified by the forecast during at least some period in the time slot specified by the forecast. If the confidence in this forecast is 80%, it is reported that the chance of precipitation is $0.5 \times 0.8 = 0.4$, or 40%, at least according to the U.S. National Weather Service. So a 5% chance of precipitation could mean that there is 100% certainty that there will be precipitation in 5% of the stated area, a 5% certainty that there will be precipitation in 100% of the area, or many other things. There is also confusion about long-term weather forecasts (climate) of precipitation and temperature. They are often quantitatively classified relative to the distributions of historical data in terms of *terciles* or thirds. Predictions in the lower third of the distribution are denoted as being "below normal," those in the middle third as "near normal," and those in the upper third as "above normal."[cccxlviii]

Sometimes probabilities or odds are presented more whimsically to explain one's philosophy of life and observations about human behavior. Journalist and satirist H. L. Mencken observed: "The cynics are right nine times out of ten",[cccxlix] probably without the aid of statistical analysis. Comedian Henny Youngman once quipped, "I bet on a horse at ten to one. It didn't come in until half-past five."[cccl] Humorist Damon Runyon noted that "All life is six to five against.",[cccli] meaning that one is a bit more likely to lose (6/11 probability) than win (5/11) in all things in life. In the song *Guys and Dolls*, from the musical play of the same name based on two stories by Damon Runyon, Frank Loesser wrote "Call it sad, call it funny; But it's better than even money; That the guy's only doing it for some doll.", meaning that his stated motivation is expected to be true.[141]

[141] In the play and movie version of *Guys and Dolls*, a man sings this song in reference to a smitten man. In a very early version of the play, this song was sung by a woman about a smitten woman: "But you can give odds forever, that the doll's only doing it for some guy.", meaning only that you would be well served by giving odds in favor of it being true.

15.3.1 Assessing probabilities of future events using statistics from past events

In an early *Inspector Morse* TV episode[ccclii] homicide detective Morse expounded Morse's Law that "there is always a 50:50 chance that the man who found the body did the deed," presumably on the basis of his experience, and therefore on data.[142]

Statistical data can be useful in wisely assessing the future. If high school and college sports players want to rationally assess the odds that they will make it to the top professional leagues, they should know that the probability of a U.S. college player being drafted to the pros is only 1.6% for football (to the NFL), 1.2% for men's basketball (to the NBA), and 0.9% for women's basketball (to the WNBA).[143,cccliii] That major league baseball players were hit by a pitch during ~0.1% of their at bats in 2018, could be used to estimate this probability in the future.[144,cccliv]

---

[142] He later corrected himself and said that Morse's Law is really "There's always time for one more pint" of beer.

[143] Such data include the number of recent participants at each level, such as from a 2019 report by the NCAA, to obtain the (fractional) frequency that members of a given level enter a higher level. For example, the probabilities of high school participants in football, men's basketball, and women's basketball playing these college sports were 7.1%, 3.4%, and 4.0%, respectively. Again, the probabilities of a U.S. college player being drafted to the pros were 1.6% for football (to the NFL), 1.2% for men's basketball (to the NBA), and 0.9% for women's basketball (to the WNBA). So, the probabilities of high school players in these sports being drafted to the pros were the products of the high school to college and college to pros probabilities, or ≈0.1%, 0.04%, and 0.04% for football, men's basketball, and women's basketball, respectively. (These numbers include assumptions and variable factors, such as the probability of drafting international players and how many years high school players are on their teams, and so on. They also ignore participation in other pro leagues.) If the probability distribution of the likelihood of getting to the next level as a function of demonstrated talent or ability is described by a normal distribution, 99.9% or all but 0.1% or 1 out of 1000 would between $\pm 3.290\sigma$. Then $(100\% - 99.9\%)/2 = 0.05\%$ or ≈1 out of 2,000 would be outside this range and in the upper region, and so only those beyond 3.3 standard deviations above the mean of high school team talent have pro talent in these typical sports.

[144] In 2018, an average of 0.41 batters on one team were hit by a pitch per game (HBP) during the average of 37.96 plate appearances per game in the American League and an average of 0.38 batters on one team were hit by a pitch per game for the average of 38.19 plate appearances in the National League, so ~0.1% were hit by a pitch. However, this HBP rate has been as low as ~0.15 in the past, so you cannot be certain about this probability still being accurate after 2018.

Unfortunately, such past frequencies, are sometimes presented by television sports broadcasters as the cause of future events, such as when they claim that only 21% of National Football League (NFL) teams that win 2 of the first 5 of their 16 games will make the playoffs.[145,ccclv]

It seems rational to use past frequencies to assess the probability you will win a football pool that uses the last digit in the scores of the two teams at the end of the game or after each of the four quarters. Because the last number can range from 0 to 9, there are $10 \times 10 = 100$ possible pairs of digits. This is displayed as a big square with 10 boxes in each row and column, covering each ordered pair. Your bet may be placed randomly for 1 of these 100 boxes, but the probability that a given pair in this box pool is a winner is far from being random. The most common scores are from the sums of small multiples of 3 and 7, 3 being the points for a field goal and 7 being those for a touchdown (worth 6 points) plus an extra point kick conversion made (1 point). Scores of 8 points are possible when the relatively infrequent 2-point conversion is attempted and made. Scores of 6 points occur for touchdowns when either conversion fails. Scores or 2 points for safeties are rare. The most common last number for any final score was 7 (17.2% of the time), followed by 0 (16.6%), 4 (15.2%), and 3 (13.8%), for the 1,067 regular and post-season games from 2006-07 to 2013-14; each was greater than the 10% expected for random scores. 2 was the least common (3.3%).[ccclvi] Including the scores at the end of each quarter distinctly, the most common pair (regardless of order) was 7 & 0 (which is the sum of those for the ordered pairs 7 & 0 and 0 & 7) (13.16%).[146,ccclvii]

There are pitfalls in using statistical data to predict useful future probabilities, particularly when they are broadly averaged. Using data

---

[145] After playing five of their 16 regular season National Football League (NFL) games, the fraction of football teams that later made the NFL playoffs in the 1990-2015 seasons was 90% when they began the season with a record of 5 wins and 0 losses (5-0), 76% with a 4-1 record, 50% with 3-2, 21% with 2-3, 6% with 1-4, and 0% with 0-5 it was 0.

[146] In the 2015 season, the line of scrimmage for the extra-point snap was moved from the 2-yard line to the 15-yard line in professional (NFL) football, and the probability of a successful 1-point conversion decreased from 99.4% in 2014 to 94.2% in 2015, so scores of 7 became a bit less frequent and of 6 points a bit more frequent, and consequently these distributions changed. (The reference is in the text.)

averaged over a year and over the entire U.S. to predict the future occurrence of tornadoes can be faulty because the frequency depends on the region, the time of year, and the time of day. From 1991-2010, there were an average of 1,253 tornadoes in the U.S. each year, but there were none in Alaska, the state with the largest area, and relatively few in the West and New England.[ccclviii] Per 10,000 square miles, the average rate was 3.5 per year, but it was not uniform. It was largest in the Southeast and Midwest, peaking with 12.2 per year in Florida and 11.7 per year in Kansas. However, because not all tornadoes are observed, such presented frequencies of tornadoes are unintentionally lower than the actual frequencies.[147,ccclix]

15.3.2 Estimating probabilities without hard data: educated guesses and rare events

Estimated probabilities pop up frequently in popular media, with varying degrees of success in predicting even fictional events. In Isaac Asimov's classic *Foundation* series, statistician Hari Seldon developed psychohistory, which correctly predicted the end of the Galactic Empire.[ccclx] The probabilities for rescuing people from seemingly imminent death are often given, without any basis, in the TV show *Scorpion*, in which four geniuses use brilliant and innovative split-second thinking and implementation to prevent world-class catastrophes.[148,ccclxi]

A rational basis for estimating probabilities, and one you may emulate in your real-world analyses, is presented in *Last Bus to Woodstock*, the first Inspector Morse mystery novel (1975) by Colin Dexter.[ccclxii] The inspector attempts to estimate how many people could be murder suspects and have the identified red car. If the suspect were

---

[147] Tornadoes in the U.S. are rated by their intensity, from EF0 to EF6 corresponding to increasing observed damage to trees, houses, etc. For tornadoes encountering the same structures, this also correlated with increasing wind speed. However, because the most intense tornadoes may not encounter structures and so inflict great observed damage, they may be unintentionally and incorrectly rated as weaker tornadoes, or may not be reported at all.

[148] In one episode, the computational genius Sylvester, known a Sly, instantly calculated that Walter, the lead genius, had less than 0.01% probability of surviving his plunge into the La Brea Tar Pits in Los Angeles. Walter survived.

local to North Oxford and environs, this would be the number of people in this area, 10,000, times the probability that a person would be a suspect. He and his colleague Sergeant Lewis narrowed this down by assessing what fraction of them fit the presumed characteristics of the suspect by estimating that about a quarter would be men (presumably meaning adult men), about half of them would be between the targeted 35 and 50 years of age, four out of five would be married, and about half would regularly go out for a drink. Only the top 5% in intellectual capacity could have perpetrated the crime. Three out of five could be sufficiently attractive to women to have pulled it off, two out of three owned cars, and one in ten of these cars would have the identified color, red. So, the probability that a given North Oxford and environs resident would be the guilty party would be $0.25 \times 0.5 \times 0.8 \times 0.5 \times 0.05 \times 0.6 \times 0.666\ldots \times 0.1 = 0.0001$ or 0.01% and so there was an estimated $\sim 10{,}000 \times 0.0001 \sim 1$ person who fit the categories of the suspect. They found this person of interest, (spoiler alert) who turned out not to be the murderer. Though this procedure was admirable, the estimates were largely wild guesses without any basis. The analysis also assumed that all categories were independent (which does not seem likely); only if this were true would it be proper to multiply the probabilities.

A more serious and more controversial probability estimate uses the Drake equation. It was developed by Frank Drake in 1961 to help estimate the number of civilizations in the Milky Way galaxy that would be able to communicate with us. It is the product of (1) the rate of star formation in the galaxy, (2) the fraction of formed stars that have planets, (3) the average number of planets per star that could support life, (4) the fraction of these planets that actually develop life, (5) the fraction of such planets that developed intelligent, civilized life, (6) the fraction of these that developed communication that could be detected in space, and (7) the length of time these civilizations release such detectable signals. Terms (2)-(6) are fractions that are probabilities usual in such analysis. Terms (1) and (7) come from an approach called rate equations.[149] The assumption that the factors are not correlated with each other is

[149] The overall product gives a number without any units, because the units of the formation rate in (1), per unit year, and the length of time in (7), years, cancel each other.

reasonable. It is controversial because the values used in it are all very uncertain.[150,ccclxiii]

These Morse and Drake analyses are examples of the branching of decisions in *decision and fault tree analysis* and other ways of estimating probabilities in Chapter 20 (Figure 20.4 below). This includes the future probability of rare events, accidents, materials and parts failures, and unknowns such as the health effects of new technologies.

Rare events also pop up in discussions of how long it would take for monkeys at a typewriter (or keyboard) to reproduce Shakespeare's work by randomly hitting keys,[ccclxiv] of the likely fruitless approaches in assessing the odds of terrorism (aside from using past frequencies of such events), in extreme value analysis, of seemingly fluke occurrences and coincidences[ccclxv] (such as meeting someone in the hallway or Feynman seeing the particular license plate number ARW 357), and in assessing whether your vote can determine the outcome of an election.[151,ccclxvi]

In the disk of assessing risk in Chapter 20, we will revisit the use of past frequencies to determine the probabilities of future rare events.

15.3.3 Sure bets: one way and the other

Be suspicious if you are told that something is essentially certain to happen, with say 99.99999% probability, or that the event is essentially impossible, so it could happen with say <0.00001% probability. Often

---

[150] Using this, Drake and colleagues made the estimate of: (1 star is formed per year) × (20% to 50% of the stars have planets) × (1 to 5 planets per star can develop life) × (100% of these will develop life) × (100% of these will develop intelligent life) × (10-20% of these will be able to communicate) × (these civilizations will last between 1,000 and 100,000,000 years), or a range of between 20 and 50,000,000 of such civilizations in our galaxy. More recent estimates give the even wider, uncertain range of 2 to 280,000,000.

[151] For the U.S. presidential election, this equals the probability that your state is needed for a candidate to win a majority in the Electoral College times the probability the vote in your state would be tied without your vote. Statisticians Andrew Gelman, Nate Silver, and Aaron Edlin showed that for 2008 presidential election, the probability that a voter could decide the election was on average ~1 in 60 million, and was higher, ~1 in 10 million, for voters in New Mexico, Virginia, New Hampshire, and Colorado. (The reference is in the text.)

missile defense systems are characterized as providing perfect protection, which is nonsense. Sometimes outlandish probability claims prove to be pure conjectures. Ridiculous claims about DNA matching are common in TV shows. DNA mismatches found from testing are certain, but apparent DNA matches are not certain to 99.999% or so.[ccclxvii] (More on this below.) Claims that disinfectants kill 99.99% of germs are rarely true.[ccclxviii]

Still, some such claims are not outlandish, for example, the risk of some rare events (Chapter 20), such as airline deaths. The probability of you being safe when you next fly commercially is better than 99.9999%. The frequency of being on an airline flight from 1999-2012 that resulted in at least one fatality was 1 in 3.4 million for 78 major world airlines and 1 in 10.0 million for the 39 safest airlines, and that a given person was killed on these flights was 1 in 4.7 million and 1 in 19.8 million, respectively. (The two sets of numbers would be the same only if all died on board every time there was any fatality.)

Some things are certain, and so have probabilities of 100%. In gambling, it is certain that you will eventually lose all your money if you don't quit when you are ahead. Also, it is certain that you will lose if you delude yourself about the real odds of an event. In the movie *Guys and Dolls*, Nathan Detroit tries to con the accomplished and worldly gambler Sky Masterson into a sucker bet of 1,000 bucks (that the then-popular restaurant Mindy's sold more strudel than cheesecake the day before). Sky responds with advice his father once gave him: "One of these days in your travels, a guy is going to show you a brand-new deck of cards on which the seal is not yet broken. Then this guy is going to offer to bet you that he can make the jack of spades jump out of this brand-new deck of cards and squirt cider in your ear. But, son, do not accept this bet, because as sure as you stand there, you're going to wind up with an ear full of cider."[ccclxix]

15.4 Are Random Events Really Random?

Odds change only when the basis of the odds changes. As Leonard Mlodinow remarked in *The Drunkard's Walk: How Randomness Rules Our Lives*:[ccclxx] "Another mistaken notion connected with the *Law of Large Numbers* (Chapter 14) is the idea that an event is more or less likely to occur because it has or has not happened recently. The idea that the odds of an event with a fixed probability increase or decrease depending on recent occurrences of the event is called the *Gambler's Fallacy*. For example, if (probability mathematician) Kerrich landed, say, 44 heads in the first 100 tosses, the coin would not develop a bias towards the tails in order to catch up! That's what is at the root of such ideas as 'her luck has run out' and 'he is due.' That does not happen. For what it's worth, a good streak doesn't jinx you, and a bad one, unfortunately, does not mean better luck is in store."[152,ccclxxi]

In Tom Stoppard's Play *Rosencrantz and Guildenstern Are Dead* [ccclxxii] Rosencrantz and Guildenstern bet on the flips of apparently unbiased coins. Rosencrantz wins 92 flips in a row, each time betting on heads, which has a probability of $(1/2)^{89}$ or ~1 out of 10^{27} (a billionth of a billionth of a billionth). It is so unexpected that Guildenstern suggests that they may be "within un-, sub- or supernatural forces."

When you roll dice, the probabilities of outcomes in the next roll are independent of history, and so the outcome is uncorrelated with the results of the previous roll. As we have seen, the Gambler's Fallacy (or the Monte Carlo Fallacy) is that the probability of a random event that has been occurring more frequently than expected will become smaller in the future and that a random event that has been occurring less frequently than expected will become larger in the future. This latter case means falsely expecting that the probability of winning the next gambling event is higher than expected after a streak of bad luck. The *Hot-hand Fallacy* is that the probability of winning the next gambling event is higher than expected after a streak of good luck.

These are aptly called fallacies for events that are random, but some events are not. There has been much discussion about hot streaks in sports, relative to one's average performance, in the rates of hitting home runs in baseball and making field goals and free throws in basketball. There is evidence that even when one side, such as the offense, changes strategies, the other side, the defense, adapts and the individual events are still random (but with a bias based on the relative quality of the

---

152 John Kerrich and Eric Christensen followed the statistics of tossing a coin 10,000 times, while interned by the Nazis during World War II.

opponents), though some are not convinced about this.[ccclxxiii] More on this later.

15.5 Some Probabilities Depend on Information

Some probabilities are dependent other information or are "conditioned" on other factors (*conditional probability*), such as correlations and history. So, these are probabilities of random events, but with restrictions. Sometimes there is even additional information and the problem is more complex and affects how we assess situations, such as whether or not we can rely on tests such as eyewitness accounts, medical tests of disease, and DNA identification.

A company invites employees who have two children to dinner. For a given employee, what is the probability that both of these children are daughters? There are four possible pairs of children: both are sons, both are daughters, the older one is a daughter and the younger one a son, and the younger one is a daughter and the older one is a son. If you can assume the number of daughters and sons are equal, the probability for each pair is equal and is 1/4, and this is also the probability that both children are daughters. If for some reason the condition that the invited parent must have at least one daughter is added, there are only three possible pairs of children, the last 3 of the 4 given above. Each has equal probability, so the probability that both are daughters has now increased to 1/3.[ccclxxiv]

When you have more information you can make better decisions because some randomness is replaced with certainty. In the long-running TV show *Let's Make a Deal* the host Monty Hall would ask a contestant to choose door number 1, 2, or 3. Behind one of them was a grand prize and behind the other two were silly prizes, and so the probability of choosing the good prize was 1/3. But, after the contestant chose a door, say door #1, Monty Hall would always open one of the other two doors. If both were losers, he chose either one. If one were the winner, he would open the other one. Say contestant chose door #1 and Monty Hall opened door #3. He would then ask if the contestant wanted to change to the other, unopened door, here door #2. If the contestant switched, the probability of winning would increase from 1/3 to 2/3, because Monty

Hall did not randomly choose to open door #2 or #3. He had "inside" information.[ccclxxv]

Extra information could come from using more finely-binned statistics. You are told that approximately 5% of the general population cannot distinguish between red and green, so they are red-green color blind, which is the most common type of color blindness. Out of a thousand randomly selected people you would expect an average of 50 to be red-green color blind. (You should be suspicious if every time you chose a random set of 1000, exactly 50 turned out to be color blind, because of statistical fluctuations, as we will see in the next chapter.)

By binning the statistics more finely you can learn more. Approximately 9% of men are color blind (say 9.0%) and 1% (say 1.0%) of women, so assuming equal numbers of men and women the probability would still be 5.0% in the general population. On average 50 of the 1000 people would still be color blind, but now you know that on average 45 of them are men and 5 are women.[153,154]

These probabilities could change if you knew more about the parents of the subjects. Men have an X and Y chromosome (in the 23rd pair). Women have two X chromosomes. This color blindness trait is carried by only by the X chromosome. Men with this "color blind" X chromosome are color blind. Women with two "normal" X chromosomes are not color blind, those with one color blind and one "normal" X chromosome are not color blind, but are carriers, and those with two color blind X chromosomes are color blind (for this recessive trait). If 9.0% of all X chromosomes are of the color blind variety, the probability that men would get this type of chromosome from their mothers would be 9.0%, so 9.0% of men would be expected to be color blind. The probability that a woman would have two of these chromosomes would be $9\% \times 9\% = 0.09 \times 0.09 = 0.0081$ so ~1% of them would be color blind, as would be all of their sons. The probability they would have no color-

---

[153] However, if this group were 40% men/60% women, 36 men and 6 women or 42 people on average would be expected to be color blind.

[154] Moreover, even finer binning by using the gene pool would lead to better analysis. For example, a much smaller fraction of Eskimos are color blind, so if we knew more about the type of the population we would be able to use more appropriate probabilities.

blind chromosomes would be 0.91 × 0.91 = 0.8281 (given it is 91% for each one) and their sons would not be color blind. The probability that exactly one chromosome is of the color blind variety is 2 × 0.09 × 0.091 = 0.1638, where the 2 appears because it makes no difference which chromosome has it. These women would not be color blind, but would still be a carrier and 50% of their sons would be color blind.

Now, if all of the mothers of the 100 men were carriers but not color blind, half of them or 50 of the sons would be color blind (instead of 9). Essentially all of their mothers would have had fathers who were color blind. So, the probability that a man would be color blind jumps to ~50% if it is known that his maternal grandfather is color blind.

15.6 This Probability Depends on That Probability: Conditional Probability

There are different uncertainties in making probability predictions when some probabilities depend on other probabilities.

Let's say that 85% of the cabs in a town are operated by the Green Co. and 15% by the Blue Co. If they have accidents at the same rate, 85% of the accidents involve a Green cab and 15% a Blue cab. If a witness to an accident says it involved a Blue cab, what is the likelihood the witness is correct? If the witness gives the correct answer 100% of the time, the witness account is correct. If the witness is known to be correct 80% of the time and incorrect 20% of the time, for 100 accidents randomly of Green and Blue cabs, how many times on average would the witness identify a Blue cab? 85 of the cabs in accidents would be Green and with a 20% misidentification rate, the witness would identify 85 × 20% = 17 of them as being Blue. 15 of the cabs in accidents would be Blue and with an 80% correct identification rate, the witness would also identify 15 × 80% = 12 of them as being Blue cabs. So, 17 + 12 = 29 of the 100 cabs would be identified as being Blue, 17 of them incorrectly and 12 of them correctly. So, the witness would be expected to be wrong 17/29 ≈ 0.59 = 59% of the time, and so, in this example, more often wrong than right.[ccclxxvi,ccclxxvii]

Similarly, medical tests are not 100% perfect for a variety of biochemical reasons and this significantly affects the evaluation of testing. Let's say 1000 people take a medical test to see if they have a given disease. Only 2% of them are known to have this disease, so on average 20 of the people have it and 980 do not. With all real tests, sometimes people who have the disease test negative when they should test positive, the *false negatives*; they receive either no or perhaps delayed proper treatment. Those who correctly test positive are called *true positives*. Some who do not have the disease test positive for it when they should test negative, the *false positives*; they worry until follow-up tests clear them. Those who correctly test negative are called *true negatives*. The false rates need to be small for the test to have value, but just how small? Say 5% of those who are sick and should test positive actually test negative, so this is the *false negative rate*. Of the 20 who have the disease, 19 (95% of them) correctly test positive and 1 (5% of them) falsely tests negative. Also, say 10% of those who are healthy and should test negative instead falsely test positive, so this is the *false positive rate*. Of the 980 who do not have the disease, 882 (90% of them) correctly test negative (which is "too few" compared to a perfect test) and 98 (10% of them) falsely test positive (which is far "too many").

A total of 883 people (88.3%) test negative, 1 of them (1/883 or ≈ 0.11%) tests falsely. That one person has no idea of having the disease and will go untreated (and that is "too many"), 117 (= 19 + 98) of them test positive, with 98 of them testing positive falsely (which is "too many") and so 98/117 ≈ 84% are wrongly told that they may have the disease. Only 19/117 ≈ 16% of the people testing positive actually have the disease, but all 117 are advised to take a follow-up test. This follow-up, second test is often a different test, either one with a lower error rate that would be too expensive to administer the first time or one targeted to root out the false positives of the first test.

One example of this testing protocol is that used by physicians to test for syphilis, which is caused by the treponema pallidum bacterium. If it is suspected that a patient has contracted syphilis and the tell-tale lesions are not present, the patient is first given the relatively inexpensive "nontreponemal" test. If the result is negative, the test may be re-administered. If it is positive, the more specific and more expensive "treponemal" test is administered, which has "only" a 1% false positive rate for the general public.[ccclxxviii]

This type of "math thinking" in conditional probability of medical testing, accident identification, and so on is formalized in *Bayes' Theorem*. It was presented to the Royal Society in 1763, two years after the death of the Englishman who formulated it, Thomas Bayes. If there are two events A and B, there is a probability that A and B both occur (Figure 7.2d) (or that both are true in a logical sense, Section 7.1), which means the realm of events where A and B "intersect." Symbolically this is called $P(A \cap B)$. This is equal to the probability of A occurring given that B occurs (denoted by $P(A|B)$) times the probability that B occurs independently of A, $P(B)$. It also equals the probability of B occurring if A occurs ($P(B|A)$) times the probability that A occurs independently of B, $P(A)$. That these two products are equal is the essence of Bayes' Theorem. So, $P(A|B)\,P(B) = P(B|A)\,P(A)$ and therefore $P(A|B) = P(B|A)\,P(A)/P(B)$. Using symbols and equations to express this is less clumsy than expressing it in words.[ccclxxix]

This is exactly the same mathematical analysis we used in the analyses of witness identification and medical test reliability. If the sick people are called S and healthy people H in our medical testing example, the probabilities of being sick and healthy are $P(S) = 0.02$ and $P(H) = 0.98$. The probability that someone known to be sick falsely gets a negative result is the *false negative rate* $P(-|S) = 0.05$, so the probability of correctly getting a positive result is the *true positive rate* $P(+|S) = 0.95$, which is also called the test *sensitivity*. For someone known to be healthy, the probability of getting a positive test is the *false positive rate* $P(+|H) = 0.10$, so the probability of negative test is the *true negative rate* $P(-|H) = 0.90$, which is also called the test *specificity*. Of course, Bayes' Theorem leads to the same numerical conclusions that we obtained in our analysis with "words."[155]

---

[155] The probability that somebody gets back a negative test result independent of whether the person is sick or healthy is the sum of two products: (1) the probability of a negative test for a healthy person times the probability of a person being healthy ($P(-|H)\,P(H) = 90\% \times 98\% \approx 88.2\%$) and (2) the probability of a negative test for a sick person times the probability of a person being sick ($P(-|S)\,P(S) = 5\% \times 2\% = 0.1\%$,), so $P(-) = 88.2\% + 0.1\% = 88.3\%$ as before. The probability that a person has a negative test telling them they are healthy when they are actually sick is called $P(S|-)$ which, using Bayes' Theorem equals $P(-|S)\,P(S)/P(-) = (5\% \times 2\%)/88.3\% = 1/883 \approx 0.11\%$, as before.

A good test for a disease would have a high true positive rate or sensitivity (P(+|S)) and a high true negative rate or specificity (P(–|H)).[156] A *complete* set of information about a test is provided by this pair of rates, by specific ratios of them called the *positive and negative likelihood ratios*, or by other pairs of such rates (but not the pairs of corresponding true and false rates, which always sum to 1.0). To assess the outcomes for a population base, you also need a third number that gives *prior* information on the likelihood of the disease in the entire population, which are the probabilities of being sick or healthy in our example, P(S) = 0.02 or P(H) = 0.98.[ccclxxx]

Test nonideality is handled in various ways. Whipple's disease has symptoms similar to Alzheimer's disease when it spreads to the brain, but its symptoms progress much faster; it is caused by the bacterium Tropheryma whipplei. 3% of the time the test of the spinal fluid for Whipple's indicates the patient with Whipple's does not have it, i.e. the results are negative. Such false negatives can have dire consequences unless the physician strongly suspects the patient has Whipple's anyway and follows up with an invasive, yet necessary brain biopsy to look for the offending and treatable bacteria.[ccclxxxi]

Some real-life false positive rates are not that small. One study found that for a sequence of 14 tests for a range of cancers, the probability of getting at least one false positive finding is 60% for men and 49% for women. So, the average false positive probability was no less than these values divided by 14, or 3-4%. (It would be more than this if a significant number actually had more than one false positive result.) The step after these positive tests is often an invasive diagnostic.[ccclxxxii] PSA (Prostate-

---

[156] In terms of the often-used likelihood ratios, then it would have a very high *positive likelihood ratio*, (LR+), which is the ratio of the probabilities of obtaining a positive test result for a sick person to obtaining the same positive result for a healthy person, = [P(+|S)/P(+|H)]. This is also the ratio of the sensitivity and 1 minus the specificity, = sensitivity/(1 – specificity) = [P(+|S)/{1 – P(-|H)}]. In our example, it would be 0.95/0.1 = 9.5. A good test would also have a very small *negative likelihood ratio*, (LR-), which is the ratio of the probabilities that a sick person would receive a negative result to that of a healthy person getting the same negative result, = [P(-|S)/P(-|H)]. This is also the ratio of 1 minus the sensitivity and the specificity, = (1 – sensitivity)/specificity = [{1 – P(+|S)}/ P(-|H)]. In our example, it would be 0.05/0.90 ≈ 0.055.

Specific Antigen) testing for prostate cancer has similar uncertainties. Approximately 80% of positive PSA test results are false positive.[ccclxxxiii] In one study, men had 12.9% cumulative risk of receiving at least 1 false positive result after a series of 4 PSA tests, which led to a 5.5% risk of having at least 1 biopsy due to a false positive result. The medical community is currently wondering whether the good done by treatment for the true PSA positives exceeds the harm done by having a larger false positive rate and the follow-up treatment, both in terms of medical help and cost.[ccclxxxiv] This is a tradeoff between risk and benefit (Chapter 20).

Such "binary" judgment of a test being positive or negative is made by comparing lab tests with a threshold standard, and is clearest when the results come back with either very high or low values. However, lab tests can come back with a continuous range of values, so changing the threshold changes the declaration of health or illness. A PSA result is considered positive if it is above a threshold value, conventionally 4.0 micrograms per liter, and negative if below it. Lowering this value would decrease the rate of false negative results, which is good, but at the same time, it would increase the rate of false positive results, which would be bad.[157,ccclxxxv,ccclxxxvi] Genetic testing is increasingly used to assess risk vs. benefit of treatment in medicine. When PSA levels are high, several genes in the patient are tested to assess how aggressive the cancer might be and help decide whether to pursue "active surveillance" or immediate treatment.[ccclxxxvii]

Genetic testing has also used to match suspected criminals to evidence from the scene of the crime or determine whether people belong to the same birth family, as in tests of paternity and identifying missing or dead persons.[ccclxxxviii] In addition, it has become very commonly used to trace one's family lineage. The DNA of each of us is different, except for identical twins. Such DNA testing does not test all of our DNA in our 23 pairs of chromosomes and mitochondria; it samples a large but finite set of variants or alleles. Such tests cannot

---

[157] In fact, decreasing the PSA cutoff to 2.5 micrograms per liter doubles the rate of false positive results. Raising the cutoff would increase the rate of false negative results, which is bad, but decrease the rate of false positive results, which would be good. (This is a mathematical tradeoff that becomes a medical judgment tradeoff.)

prove that no one else on Earth could match a given sample, but only that it is very unlikely. Y chromosomes are passed along only from father to son and DNA from a cell's mitochondria (not from chromosomes) is passed on almost always only from mother to child. DNA testing currently includes over 500,000 (autosomal, nonsex chromosome) genetic markers, which are essentially randomly sorted and mixed from those of both parents at conception (aside from the "sex" related chromosome markers). Consequently, siblings share close to 1/2 of their DNA segments, half siblings 1/4 of them, first cousins 1/8, second cousins 1/32, third cousins 1/128, etc. However, these are averages. For example, the ~1/32 or 3.125% of traits shared by a set of second cousins, may range from ~1% to ~6%.[ccclxxxix]

DNA matching test results are well presented in terms of likelihood ratios,[cccxc] defined as the ratio of the probability that the DNA from the crime scene matches that from the suspect vs. that from a random person, as the ratio that the DNA from a dead body matches that from a suspected family member vs. that it matches the DNA from someone not suspected to be a family member, and so on. In a paternity test, the paternity index is this likelihood ratio.

DNA data are best analyzed applying Bayes' statistics to the objective raw DNA results and the sometimes–subjective prior knowledge. Knowing there is a 90% *prior probability* of a suspect being the perpetrator increases the strength of the interpretation of positive DNA test evidence. If there is no prior knowledge, the prior probability is presumed to be 50%. For paternity tests, this means that brothers would be equally likely identified as a father. The failure to distinguish between these raw DNA results and the partly subjective overall results is the *Prosecutor's Fallacy* or *Defense Attorney's Fallacy* because of the common fallacies in presenting such statistical results in court.

A small probability of a false positive in a DNA test does not always translate into a very large probability of the match or identification being positive.[cccxci] For 50% prior probability, likelihood ratios (LRs) of 1, 10, 100, 10,000, and 1,000,000, correspond to probabilities of a match of 50%, 90.9%, 99.0%, 99.99%, and 99.9999%, respectively. The match probability of 99.0% with LR = 100 means you expect the match to be correct in 99 out of 100 trials and wrong in the other one. This probability increases to 99.89% for a 90% prior probability and decreases to 91.7% for a 10% prior probability. (With LR = 1, the final probability is always equal to the prior probability.)

Simulations of genetic tests with 15 genetic markers (as had once been common) by forensic mathematician Charles H. Brenner used random variables to test match select possibilities and 50% prior probability. This showed that good matches can usually be expected between a parent and a child and between siblings, but these tests may not be definitive due to the distributions in the LRs.[158,cccxcii]

The statistical validity of other tests also affects the quality of criminal justice. Many people have been arrested and charged as felons on the sole basis of false positive tests from inexpensive and inaccurate roadside tests. Sometimes they have ended up incarcerated before better tests were belatedly, if ever, performed on samples of the suspected substance. One such test relies on a solution of cobalt thiocyanate turning from pink to blue when exposed to cocaine. Unfortunately, it also turns blue when exposed to more than 80 other substances, including several acne medications, household cleaners, and over-the-counter pain relievers, such as BC powder, which contains aspirin and caffeine.[cccxciii]

15.7 Probability Chains

When you toss a coin 1,000 times, you may want to track if it is heads or tails on the first toss, the second, the third, …, the thousandth. The outcome after 1,000 tosses will be one point in the final binomial distribution (approximated by the normal distribution), say 508 heads, and then when you do it again, it could lead to 482 heads, and then 499, and so on. For each set of 1,000 tosses, there is a *trajectory* to each final

---

[158] The median LR for parents/child is 54,000, but with a small fraction of LRs below 100. The median LR corresponds to a probability of 99.998% with 50% prior probability and the averaged LR is closer to 99.98%. Moreover, the median LR for two siblings (same birth parents) is 16,000, but with many below 100 and some even below 1. This median LR corresponds to a probability of 99.994%, and the average one is closer to 97.8%. The test is less definitive when comparing two half-siblings (one common birth parent), with a median LR of only 14 and 93.3% probability and average probability only 81%, which is not much better than between two random people (LR = 1; 50%). With more advanced testing, these values and ranges may be different.

outcome that is statistical, with different relative numbers of heads and tails at each step, and different possible trajectories for the same outcome. Also, for a particular trajectory and outcome, there is a final toss for which the numbers of heads and tails are equal.

Such random sequences are called *stochastic processes*.[cccxciv] When there is a series of steps in a random process, each not depending on history, it is called a *random walk*, such as when you toss coins or take random steps of equal length in a random direction. The term *Brownian motion* describes the physical random motion of particles. The next step in a stochastic process may or may not depend on its current value and, these random processes and "walks" are often more generally called *Markov processes* and *Markov chains*. (Some, but not all, limit the term random walks to those Markov chains in which the next step does not depend on its current state.) Stochastic processes can be simulated on computers using *Monte Carlo* methods to sample a sequence of random trajectories chosen by pseudorandom numbers; it is so-named because the randomness of gambling and the famous gambling casino in Monte Carlo. They were developed in the middle of the 20th century by mathematicians Stanislaw Ulam and John von Neumann, physicists Nicholas Metropolis, Marshall Rosenbluth, and Edward Teller, and others.[cccxcv]

15.7.1 Random walks

Random walks characterize of the progression of a sequence of gambling bets, the physical trajectories of colliding molecules, the price of options in the financial market (as we will see in the discussion of risk in Chapter 20), and other random (or seemingly random) events. The distribution of final outcomes of a series of random walks is given by the binomial or normal distribution, but how these walks progress is often of great interest.

When there are limiting states (called *the exit distribution*), you could be interested in how the chain progresses after a certain number of events or bets and in how many events are usually needed to reach these limiting states (*the exit time*). In a random walk analysis, the first time a given target is reached is called *the first passage*,[cccxcvi] such as the first time you cumulatively have 5 more heads than tails. The last time it is reached, perhaps because of the limited number of time the coin is flipped, say a total of 100 times, is called the *last passage.*

The classic *Gambler's Ruin* example is a multi-step chain that explains why gamblers usually lose all of their starting money. When the odds are not in their favor, gamblers lose a little on average each bet, so they usually eventually lose. But, they also lose for even odds (50% win probability per event), even when the winning payoff is the bet itself, though this type of loss occurs after more bets on average. Bad gamblers usually do not quit when they are a little ahead; they continue to play and they will eventually lose the money they began with.

If a gambler starts with $50, bets $1 with an equal probability of winning or losing $1 on each bet, and bets over and over again, the gambler's cash eventually will fall to the exit limit of $0, the *first passage* to this limit, and the game will be over. This random walk is analogous to going up a rung in a ladder after a winning bet and down a rung after a losing bet.

We can help this gambler by setting a second, upper limit, so if the gambler reaches this limit (which is above $50) before hitting $0, and quits at the first passage to this limit, the game will be over and the gambler leaves full of cash and not broke. If the upper limit is $100 and the gambler continues until either the $0 or $100 limit is reached, 50% of the time the gambler leaves broke and 50% of the time with $100. One case is illustrated in Figure 15.5, for a simpler example with $10 bets, where the bettor can leave with the upper limit of $100 after 13 bets or instead ignore it (or have no upper limit) and lose all after 31 bets.

The gambler is more likely to leave broke than rich if the winning probability for each bet is below 50%, so the even odds payout is less than $1, or the upper limit is more than twice the starting point (>$100 here). You can determine the probability the bettor will eventually reach the upper or the lower limit[159] and the number of bets these take, as well

[159] If you start with $\$S$ and bet each time $1 that you win or lose, with a ratio of probability of losing a bet to winning it of x, and end when you reach either $0 of $\$W$ (where $W > S$), the probability that you will eventually win and exit with the W dollars is $(x^S - 1)/(x^W - 1)$.

as the distribution of final amounts the gambler has after a given number of bets.

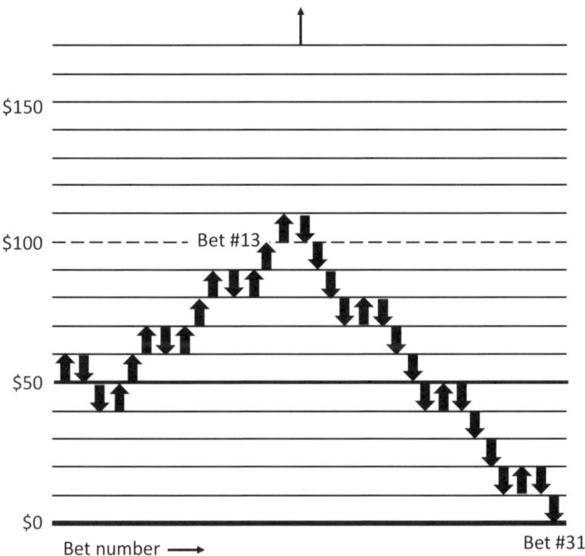

Figure 15.5: Example of a random walk in betting, with $10 bets.

For American Roulette, the ratio of losing a bet to winning is 20/18 = 1.11111111…. If you start with $50 and bet $1 each time until you reach either $100 (rich) or $0 (broke), the probability of winning in this manner is the American Roulette losing/winning ratio to the 50th power minus 1, divided by this ratio to the 100th power, minus 1, (or $(1.111\ldots^{50} - 1)/(1.111\ldots^{100} - 1)$). This is ≈0.51%, which means you will win only 1 out of ≈195 times. For European Roulette, this ratio is 19/18 = 1.05555… and the probability of leaving with $100 is better but still not very good, ≈6.28% or 1 out of ≈16 times. If you had the even higher aspiration of leaving with $150, you replace the 100 by 150 in this result, and find the much lower probability of leaving rich of ≈0.0026% or 1 out of ≈38,000 times when playing American Roulette and ≈0.42% or 1 out of ≈240 when playing European Roulette. Setting higher first passage exit limits also hurts your chances of winning with even odds. Starting with \$50, for a \$100 upper limit this probability is $1/2 = 50\%$ and for \$150 it decreases to $1/3 \approx 33.3\%$.[160] Gamblers lose almost all the time, and they lose even more frequently when their initial aspirations are higher.

How many times can you place bets before the game is over? The classic *Drunkard's Walk* or random walk problem describes a sequence of steps in a random direction, say in 2 dimensions, from a lamppost. The largest distance traveled is the step length times the number of steps taken, but the average location is still at the lamppost. The net displacement is mostly likely in some random direction from the lamppost, with magnitude roughly the step length times the square root of the number of steps away. After 100 steps, this characteristic displacement is ~10 steps away and after 10,000 steps, it is ~100 steps away. This result is also true for the motion or diffusion of molecules in 3 dimensions in a gas[161] and in 1 dimension, as in moving to the left or right or up and down a ladder—for which this distance is very roughly the standard deviation of the binomial distribution. The variance of seeing an event with probability p in n chances is $np(1-p)$. So, for n steps in 1 dimension and equal probability of moving to the left of 0.5 (say p) or the right of 0.5 (so $1-p$), the variance is $n/4$ and the standard deviation is $n^{1/2}/2$. This explains this square root rule (Figure 15.6).

---

[160] If you start with the probability of starting with \$$S$ and leaving with \$$W$ (with $W > S$) of $(x^S - 1)/(x^W - 1)$ (and bet each time \$1 that you can win or lose, with the ratio of the probability of losing the bet to winning it of x), and try to use it for even odds, $x = 1$, you get $(1^{50} - 1)/(1^{150} - 1)$ for the upper exit of $W = 150$ here, which looks like $(1-1)/(1-1) = 0/0$, which is indeterminate. However, you can use a trick, by examining this fraction for $x = 1.1$, and then for 1.01, for 1.001, for 1.0001, for 1.00001, and so on, so as x *approaches* 1. The values are ≈0.0000719476, 0.18693525744, 0.31682058828, 0.33166814567, 0.33316668139, and so on seem to approach 1/3 (just as they approach 1/2 for $W = 100$).

[161] This explains why when an odorous material is released, perhaps you can smell it 3 feet away in 5 minutes, which corresponds to a specific number of steps the offending molecules travel before they collide with one of the molecules in the air. You can smell it twice as far, 6 feet, away after it has made 4 times as many steps, which occurs in 4 times the time or 20 minutes. For a characteristic time t, the characteristic diffusion distance varies as $t^{1/2}$ due to its random walk nature.

In the Gambler's Ruin problem, if a gambler starts with \$50, and bets \$1 each time with 50-50 odds, after $50^2 = 2{,}500$ bets, the gambler will have moved an average of the square root of 2,500 or 50 steps up or down the ladder of riches. So, the game will typically last ~2,500 bets, with equal likelihood of ending up with \$0 or \$100; this estimate ignores the results of passages at these two limits.

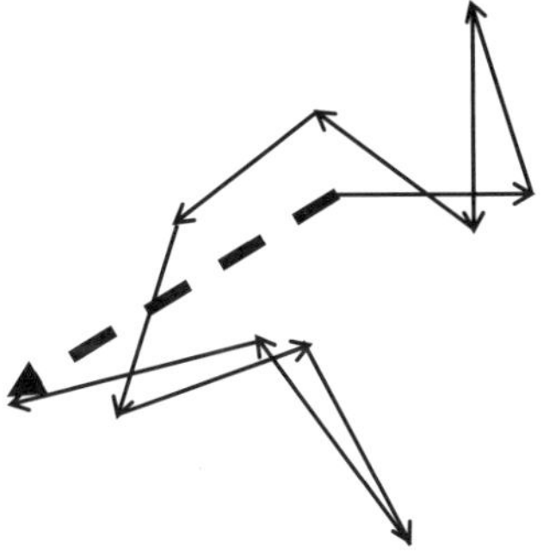

Figure 15.6: Illustration of a random walk of 10 steps of equal lengths in 2 dimensions, and the net result as a dashed arrow from the beginning of the first step to the end of the last one.

Moreover, when the gambler has winning probability < 50%, there is an average downward trend in riches in addition to the randomness caused by winning or losing. This shortens the games ever more as the odds get worse. If you bet \$1 each time in American Roulette, you get back \$2 only if you win (your original bet plus your winnings), and, as seen earlier this chapter, you lose an average of \$0.0526 per bet. Ignoring the random walk, you will lose your initial \$50 amount in $\approx$ \$50/\$0.0526 = 950 bets. For European Roulette, you lose on average $\approx$\$0.0270 per bet and your initial amount in $\approx$ \$50/\$0.0526 = 1,850 bets. In both cases, you are likely to lose after a fewer number of bets than these estimates when you include the random nature.[162]

[162] The standard deviation of the binomial distribution changes very little when the winning probability decreases from 0.5 a bit, because the variance, $np(1-p)$ changes

You are interested in the bettor's first passage to a given value, because the betting may be over then.[cccxcvii] A related concept is the *survival probability*. If there is no upper stopping limit, what is the probability that the bettor will survive and so not fall to \$0 after a given number of bets? With an upper limit, what is the probability that the bettor will survive with continued betting, and will reach \$100 before reaching \$0? A bettor may initially win more than losing and then, say, have \$80, then start losing more than winning, passing through \$50 (a first passage), and then start winning more frequently again and then pass through \$50 (a second passage), and so on. After a given number of bets, the bettor will have passed through the initial \$50 amount a last time, which is the *last passage*. The statistics of when these passages occur are based on the reasoning that leads to the binomial distribution, but have different probability distributions.

In a "game" of many trials such as coin tosses, n, the number of heads is very nearly $0.5n$. You might expect there would be many *passages* through cumulatively equal heads and tails during the game. Also, for a given game, there is a trial, k, that happens to be the last *lead change* or last passage,[163] that you might expect would usually occur near the end, so k would be just a little smaller than n. However, your (and my) intuition would be wrong on both counts.[cccxcviii] For $n = 1000$, it is most probable for the last passage k to occur very near either $k = 1000$, the end, <u>or</u> $k = 0$, the beginning.[164] This last passage probability distribution is known as the *arcsine distribution*, as in Figure 15.7 (which is plotted as a continuous distribution with k/n as the "x" coordinate).[165] It is surprising that the probability that heads and tails will never be equal (and so there is no passage) in the second half of the game is 1/2, independent of how long the game is. So, the winner (or the *more*

---

little. Here, the variance changes from $0.25n$ to $(0.4737)(0.5263)n = 0.2493n$ for American Roulette and to $(0.4865)(0.5135)n = 0.2498n$ for European Roulette.

163 or within 1 if the number of tosses is odd

164 The probability for this last passage symmetrically decreases as k increases from 0 and decreases from 1000, and is symmetric about 500, where it is still significant (i.e., above zero).

165 The arcsine probability distribution function is $1/(\pi(x(1-x))^{1/2})$ in the continuous limit, with x ($=k/n$) ranging from 0 to 1.

fortunate or luckier player, or the luckier side of the coin), will be winning practically the whole time during the game. More heads than tails initially, usually means there will be more heads than tails throughout the game. Fewer heads than tails initially, usually means fewer heads than tails throughout the game. In his classic book on probability, William Feller noted that even trained statisticians expect there to be many more such passages than there actually are,[cccxcix] and so the intuition of even those well trained in math is sometimes wrong!

In a related vein, it might not surprise you that a random walk always eventually returns to the starting point when motion is confined to either 1 or 2 dimensions. But, it may be surprising that the corresponding random walk in 3 dimensions eventually returns to the starting point in only 34% of the walks.[cd]

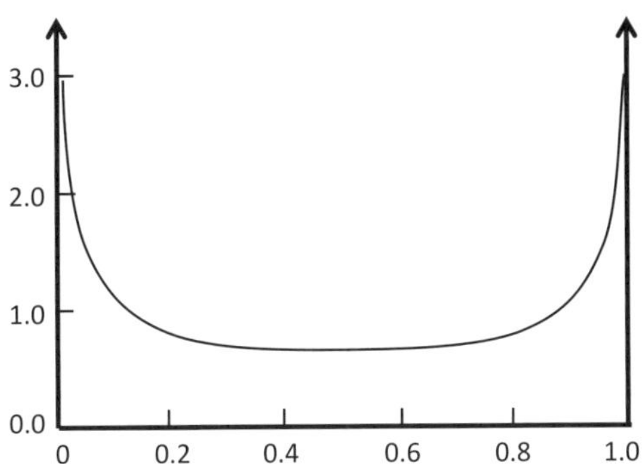

Figure 15.7: The arcsine distribution function.

15.7.2 Are sporting events random events and random walks?

Intuition tells us that the scoring sequences in sporting events should not be random walks. For example, teams seem to have "hot" and "cold" scoring streaks within individual games and winning and losing streaks across a series of games. However, statistical analysis has shown that team and player streaks are merely fluctuations in a series of probabilistically random events, after accounting for the relative strengths of teams and players.

Can the distributions of wins and losses in a league be understood by assuming random events? Yes, they can. The randomness of the outcomes of individual games can be used to compare the level of competition in sports leagues of the top baseball (Major League Baseball, MLB), ice hockey (National Hockey League, NHL), basketball (National Basketball Association, NBA), and football (National Football League, NFL) leagues in the U.S. and Canada and a soccer league in England (English Football Association, FA). Using Monte Carlo simulations (see below), the probability that the weaker team wins in a given game (called an upset) is seen to be smallest for basketball and football (36%, so the stronger team wins 64% of the games on average), and larger for hockey (41%), baseball (44%), and soccer (45%) (counting ties as half victories for both teams). The fraction of games teams win over a season clusters closer to 50% the higher the upset frequency and the more games the teams play each season.[cdi]

Are team hot and cold streaks merely expected statistical anomalies in random events? Yes, they are. Roger Vergin (and several before and after him) showed that in American major-league professional baseball (MLB) and basketball (NBA), the outcome of a given game for a team is independent of the outcome of the team's previous game (on average), so winning and losing streaks do not occur because a team is "hot" or "cold". Of course, teams that are good on average can have longer winning streaks and those that are bad on average can have longer losing streaks.[cdii]

The *Bradley-Terry Model* is used to weight the relative strengths of teams in an otherwise random process.[cdiii] It is part of a more general model of ranking a series of events or objects by comparing pairs of competitors each time. When major league baseball records over decades are analyzed, it is seen that the standings each year (who is first, second, and so on) can be determined using this model, as can the statistics of the lengths of winning and losing streaks. Long winning and losing streaks are merely statistical fluctuations within a series of random events. In recent decades, the weakest team has won an average of 36% of its games and the best team has won 63% of its games.[cdiv]

Does scoring within a game behave as would be expected from a random walk? Yes, it does. This is easiest seen within basketball games, because there is so much scoring. Of course, when team A scores, team B gets the ball, it is more likely to be the next team to score if the teams are competitive. This is called *anti-persistence*. But, aside from this and the relative strengths of the teams, scoring in the NBA is random except at the beginning and end of quarters and at the end of blowouts. NBA games from the 2006-2007 to the 2009-2010 seasons evolved in time as random walks with an average scoring rate is 0.033 points per second, the average being slightly smaller in the second half than the first half. The scoring rate was smaller just as each quarter began because teams usually do not score then and larger at the very end of each quarter because teams try to score with little time left. After a team scores, the probability that the other team scored next was 65.2%, and so the probability that it scored twice in succession is 1 minus this or 34.8%.[166,cdv]

Does the random walk nature of scoring ensure that there are few lead changes and that leads are often safe in a game? Yes, it does. The final score difference in high-level basketball, football, and ice hockey games is approximately equally influenced by these random walk fluctuations and the differences in team strengths. This is similar to the random walk in betting when the odds are not even. The fraction of games with a certain number of lead changes is that expected with a Poisson process (a random process), with anti-persistence. Also, the distributions of times during a game that one team is leading, the last lead change (or passage), and the maximum lead follow the arcsine distribution after accounting for the difference in team strengths and the changes in the rates of scoring at the beginning and ends of quarters or periods.[cdvi] In an NBA basketball game, a 10 point lead with 8 minutes left in the game is "90% safe" when the teams are evenly matched, which means that in 90% of the cases they never lose the lead. More generally,

---

[166] Scoring streaks within a game are consistent with the width of the Poisson distribution. Though which team scores next is largely random, the probability that a team scores does correlate linearly with the lead (even without a large lead near the end of a game), with the team ahead scoring at a slower rate and the team behind scoring at a faster rate than for a tied game.

leads that are at least equal to 0.4602 times the square root of the remaining time expressed in seconds are 90% safe.

15.7.3 More complex Markov chains

Say you know that if it is sunny on a given day, the probability that the next day will be sunny is 80% and rainy 20%, and if it is rainy the probability that the next day will be sunny is 30% and rainy 70%. This gives the probability of sun or rain for the next period based on the current weather, but not on history, and so it is an example of a more complex random walk or Markov chain. If a given day, Day 1, is sunny, on the next day, Day 2, there is a 0.8 chance of sun and 0.2 chance of rain. On Day 3 the chance of sun is 0.8×0.8 (for Day 2 sun) + 0.2×0.3 (for Day 2 rain) = 0.70 and of rain is 0.2×0.8 (for Day 2 sun) + 0.2×0.7 (for Day 2 rain) = 0.30 (which is $1.0 - 0.70$), and so on. If you continued this Markov chain for many more days you would reach the *steady-state* or *stationary solution*, here that the probability of an average day being sunny is 60% and rainy is 40%. The steady-state answer would be the same if you had assumed that Day 1 was rainy. (Problems like this are conveniently phrased in terms of mathematical arrays called matrices and then straightforwardly solved using ordinary and matrix algebra.)

Such models can be refined to make them more realistic, as with more states in the three-state Weather chain (rainy, cloudy, sunny), and applied to other cases, such as the three-state model of Social Mobility from generation to generation[167] and models tracking performance and needs, as in assessing brand preference.[168]

---

[167] This chain tracks upward and downward mobility of the lower, middle or upper classes in future generations using the fractions of those in one of the classes who remain in that class in the next generation and the fractions who enter one of the other two.

[168] To assess brand preference, a Markov chain can use the probability you would prefer your next purchase to be the same brand you currently have, or instead, another brand. To control inventory, the chain can help determine your restocking policy after inventory changes due to sales. You want to have enough items for sale and so make a profit, but not so many that unsold items incur large storage costs. To operate equipment economically and without interruption, the Markov chain can be used to tell you the minimum needed inventory and usage of several repair parts.

15.7.4 Monte Carlo simulations

Modeling processes and situations is sometimes best done by running simulations that employ pseudorandom numbers to reflect uncertainties in initial and other conditions. Say you are building a highway that has a speed limit of 65 miles per hour (mph) and want to see how traffic will flow in it, so you perform a computer simulation. You will need to choose the initial speed of each car and the initial separation of each car with the car in front of it. It would be unrealistic to start each car with the same speed and separation, so you want to make some faster and some slower initially, say between 60 and 65 mph, and some closer together and some farther apart, say between 30 and 100 feet. The initial conditions of each car are best set by using pseudorandom numbers in a Monte Carlo simulation. For a given car, the initial speed could be a pseudorandom number between 0 and 1 multiplied by 5 mph and then added to 60 mph and the initial separation could be another pseudorandom number between 0 and 1 multiplied by 70 feet and then added to 30 feet. You do this for each car and start the simulation and learn about how traffic would flow for one set of initial conditions, and run the simulations again many times with different pseudorandom numbers. Computer codes predicting weather and long-term climate start with different initial weather conditions. Pseudorandom numbers are routinely used to provide the initial conditions and some aspects of the models themselves. This provides a range of predictions that are more reliable because of real uncertainties in both.[cdvii] Monte Carlo simulations are used to assess the uncertainty in weather predictions, including the paths and intensities of hurricanes.[cdviii]

The uncertainties in predicting the future with math requires us to understand the past better, with statistics, to try to get better control of this randomness in our lives.

Chapter 16

The Math of What Was: Statistics – The Good, The Bad, and The Evil

You need to understand the numbers of the past to prepare for and to try to control the future, and to be comfortable with it. So, the "math of doing" needs the subworld of statistics, which is the "math of the past." Statistical analysis has become an integral component of "math thinking" and responsible for the actions of individuals, organizations, and governments. Statistics software tools such as SPSS are available to do the rote statistical computation, but not the "math thinking."[cdix]

One simple and practical example of statistics in everyday life may be relevant to you. You are supposed to take one pill every morning, and so 30 each month for a 30-day month. But, without a fool-proof system, you may forget to ingest a pill some days and on others inadvertently take a second one. Say you swallowed 1 pill on 27 of the 30 days in a given month, no pills on 2 days, and 2 pills on one day, so in that month you ingested at least 1 pill on $27+0+1=28$ days and a total of $27\times 1 + 2 \times 0 + 1\times 2 = 29$ pills. Over the course of 100 months, you may have swallowed at least 1 pill on 26 of the 30 days in 2 of these months, on 27 days in 12 of them, on 28 days in 35 of them, on 29 days in 32 of them, and on 30 days in 19 of them. This is a distribution with an average of $(26\times 2 + 27\times 12 + 28\times 35 + 29\times 32 + 30\times 19)/(2+12+35+32+19) = 2{,}854/100 = 28.54$ days per month on which you ingested at least 1 pill. Related, though different, you may have swallowed a total of 26 pills in 1 of these 100 months, 27 pills in 10 of them, 28 pills in 34 of them, 29 pills in 33 of them, 30 pills in 16 of them, 31 pills in 4 of them, and 32 pills in 2 of them. This is a distribution with an average of $(26\times 1 + 27\times 10 + 28\times 34 + 29\times 33 + 30\times 16 + 31\times 4 + 32\times 2)/(1+10+34+33+16+4$

+ 2) = 2,873/100 = 28.73 pills ingested per month. This same data set was used in both analyses, but the desired statistical analyses were specified differently and so they produced different statistical distributions, with different averages and other statistical attributes.

There are usually broader and deeper issues in such statistical analysis. In this example, we never specified the source or reliability of the data set, but we should have. Did we get data from the entire population of people taking pills, as in conducting a population census or evaluating the performance of players and teams in sports, or did we only sample them from a very large data set because of cost and time limitations, as in surveys and election polls? If we sampled, what was the quality of the survey collection methods, and notably the size of the sampled population? (See the discussion on polling later in this chapter.) Was the sampling large enough to be representative? Even when data are collected properly, the analysis, interpretation, and presentation can be faulty or biased, rendering statistical results useless. This can be devastating, as when data are used to allocate resources, such as the distribution of government funds on the basis of census results.

Care and caution are generally needed in the steps of data acquisition, analysis, and presentation. If there are 5 million people unemployed at a given time and the civilian workforce of the employed and unemployed is 100 million, the unemployment rate is 5,000,000/100,000,000 = 0.05 = 5.0%. If an average of 5 million people were unemployed over a year, data interpretation could be different. Say, all those unemployed during the year were unemployed for half a year, then 10 million people or ~10% of the workforce was at some time unemployed, which might be significant. Moreover, were all unemployed people included in the statistics? If 3 million without jobs did not look for a job in the prior 4 weeks, the U.S. Department of Labor would not have included them in either the currently unemployed or the workforce. If they were included, in our first example the unemployment rate would really be ~ 8,000,000/103,000,000 ≈ 7.8%.[169]

[169] Also, the number of unemployed does not include those who are underemployed, part-time workers, and those who do not work full-time but want to. Furthermore, stay-at-

How are such statistical data acquired: from the entire workforce population, as is done for the entire population in the U.S. Census, or from a sample of the workforce? The U.S. Department of Labor samples 60,000 households consisting of ~110,000 people each month for the data, which is a large sample, especially compared to those used in election polls, as we will see.[170,cdx]

Generally, statistical analysis entails careful data analysis before any conclusions are drawn and action is taken. However, at times data sets tell a story worthy of immediate action without further analysis, even though the exact numbers may be uncertain. A 2016 study showed that 51% of those who suffer a heart attack (sudden cardiac arrest) exhibited warning signs in the four weeks before thc attack, and in 93% of these cases these symptoms recurred with 24 hours of the attack. Also, only 10% of those who suffer a heart attack outside of a hospital and then brought to one are able to later leave the hospital. Even if the data acquisition and analysis procedures were not perfect in these studies and the sample size was not large enough to obtain reliable numbers, the message is clear: be alert about the recurring signs of a potential heart attack and take appropriate action. This advice would be the same whether the 51% fraction of the study were really instead 30% or 70% or the 10% fraction were instead 5% or 20%.[cdxi]

16.1 Statistics: Getting the Data for Them

The goal is to acquire the data in a meaningful way, analyze them well, present the results in a proper manner and a manner that is suitable for the intended audience, and then to use and implement them.[cdxii] Sometimes all collected data are analyzed and sometimes the large set is sampled and then weighted. In many cases, time and cost constraints make it is necessary to obtain statistics by properly *sampling* and

home parents and those who do not look for work because of a disability are not considered part of the labor force statistics.

170 To ensure the meaning of month-to-month and year-to-year changes, a given sampled household is questioned about employment for 4 consecutive months, then not for the next 8 months, and then questioned for 4 more consecutive months and then not again. The statistics are weighted to account for sample variations in age, sex, race and so on.

weighting results from a larger population.[cdxiii] *Bootstrapping* is sampling the same larger set several times to obtain multiple samples, each time after *replacing* the sampled set in the larger set, so a fresh, random sample can be chosen from the complete set.[cdxiv]

Recognizing variables that should define your data acquisition can be the most important task in designing a useful statistical study. They can be tied to what you want to learn, but they also need to include all variables that may be significant in understanding the point of the study. In demographics, age, gender, and race may all play important roles in the actual analysis of the data, even if the dependence of any one of them is not the goal.

You need to obtain a large "enough" sample size to be able to obtain, analyze and interpret a data set well, to ensure the results are *statistically significant*, as we will see later in this chapter. Also, in medical and other studies, getting data from the group you want to study, say composed of people undergoing a certain medical treatment, must be coupled with getting data from a control group, such as a set of people who are not undergoing treatment. Both groups must have shared or common characteristics. Careful selection of control groups and comparison with such groups are equally important in other types of data collection.[cdxv] Survey questions need to be posed in a way that is unbiased.

Statistical studies with carefully controlled data collection have greatly improved our well-being, as in learning the causes of diseases such as heart disease and lung cancer. Medical advice based on such pioneering studies has become so ingrained in our lives that we take it for granted; new studies will likely improve our health even more. Such studies need to involve a sufficiently large survey and observation lists, comparisons with control groups, and proper and random samples. Moreover, data can be gathered from and compared for a specific set of people over time, called a *longitudinal study* or instead from a range of people over a very short span of time, called a *cross-sectional study*. For example, a longitudinal study may follow 40 year olds until they turn 60. The corresponding cross-sectional study would examine a range of people from 40 to 60 years of age, and, for instance, presume that there is a one-to-one correspondence between a 50-year old person in this study and the 50-year old person who had aged 10 years in the corresponding longitudinal study. Longitudinal studies are relatively easy to interpret but harder to accomplish because people need to be tracked for long periods. Cross-sectional studies are easier to conduct and do well for ranges of several variables (such as types and ages of people), but are harder to interpret well to ascertain cause and effect.

The groundbreaking Framingham longitudinal study correlated the incidence of heart disease with lifestyle features. It started in 1948 and is still continuing, with its third generation of participants.[cdxvi] Even early on, it provided insight into the causes of heart disease that had not been recognized at all at the time, but now seem obvious to us: that cigarette smoking, high cholesterol and blood pressure, and obesity increase the risk of heart diseases and exercise decreases the risk.

Cross-sectional studies have many built-in assumptions and a larger chance of error, and are more likely to miss small, but important effects. Still, they have also had outstanding successes and require relatively little time. Smokers have a risk of contracting lung cancer of ~8% (in a lifetime), which is 10× the risk for nonsmokers, ~0.8%. This very large difference is why even cross-sectional studies were able to convincingly isolate smoking as a major risk factor.[cdxvii,cdxviii]

Not all people remain in a longitudinal study. Do the survivors correspond to a random set, without any correction, even if the original set was selected randomly? If not, this is one form of selection bias. More generally, there can be selection bias in how participants are chosen or contacted in all statistical studies, perhaps related to whether or not they are home during the day or have the latest gadget that makes it easier to (selectively) contact them? (At one time this may have meant that they had a land-line phone, and then it meant a mobile phone, and so on.)[cdxix] There can also be bias because we recall things selectively, sometimes wanting to please the data collector.

Selection bias can pop up in less innocent ways as well, as in publication bias. Say a pharmaceutical company funds 50 studies retaining the right to control the publication of results. It allows only the 5 studies supportive of its position to be published, so 100% of all published studies agree with what they want to be supported, instead of only 10% of them. (Such instances have occurred.) Crime statistics can be suspect. Some wonder if the decrease in crime in NYC schools reported in 2017 was due to those trying to improve statistics by not reporting some incidents.[cdxx] Specifically choosing what data to analyze and present and what to throw out is called *cherry picking*. This and other forms of data manipulation can distort any presentation of data. In experimental science, choosing only those data that confirm your theory is one type of scientific fraud.

Data collection can be biased in other ways as well. Airlines are supposed to provide the public with the rates they lose or delay your luggage. The U.S. Department of Transportation has required them to report *only* the frequency that paperwork is filed concerning lost or delayed luggage. Several airlines have improved their statistics, but not their performance, by pre-emptively letting passengers know when their luggage does not make their flight and asking them for delivery instructions. So, passengers do not need to wait without purpose for their luggage, which is good, and don't need to file any paperwork.[cdxxi] This may well be responsible for the ~30% "improvement" in luggage handling reported by one airline. (This regulation loophole from 1987 is expected to be closed eventually.)

Online reviews are one type of collected data that can be biased. The numerical values of these reviews are easily averaged or binned into numerical categories; the written comments take longer for us to digest. Most of us, some say ~66%, who research potential purchases online read online reviews on websites, but a much smaller fraction post reviews of purchased items, so the sample size may be small and not statistically significant. Only about 4% who purchase an item write a review if it is requested. Unfortunately, these samples can easily be biased. Some vendors request reviews only if they think they will be positive and, for unsolicited reviews, some put up roadblocks to posting reviews that are not very positive. Dissatisfied customers may be more likely to post unsolicited reviews. Reviews by those who have not bought the reviewed item, but have an agenda or score to settle, is another source of bias.[cdxxii] Ratings and reviews can be self-selective in other ways as well. Books later in a series tend to have higher ratings than earlier ones. This could indicate that only those who really liked the early books read and ranked the later ones and liked them too, so it is self-selective and biased, or simply that the later books were "better".

Some celebrated statistics are not suspicious in this way but may be very uncertain for more innocent reasons. The 2018 campaign to rid the planet of plastics straws often cited the statistic that Americans use 500 million plastic straws every day. This number had been circulated and treated as hard data for nearly a decade, but was merely estimated by nine-year-old Milo Cress after he called straw manufacturers.[cdxxiii] Rather than a statistic, it was a widely-circulated estimate that became a factoid.

16.1.1 Data have scatter

You start your jog from the same place each day and run the same route. Your GPS tracker tells you that you have jogged a mile when you arrive near a landmark, but sometimes this point is 20 feet past it, sometimes 10 feet before it, and so on. If each time you plot where relative to the landmark you reach a mile, you will see a *statistical distribution* about or near the landmark. These *random variations* are common in measurements. The variations can be due to slightly different routes you took (crossing a street differently), details of how the GPS tracks positions and gives distances, slightly different places where you started your tracker, and so on. If instead, one day you notice that you reach the mile mark 30 feet before the landmark that day, then 20 feet early the next day, 40 feet early the next, and so on, you still see scattered data, but now centered ~30 feet before the landmark. This is a *systematic* change, perhaps due to a new algorithm used to monitor GPS distances.

The spread in collected or sampled data is a real and unavoidable part of the math of life; this is *statistical* uncertainty. In contrast, when you ask a bank to convert dollars (d) into quarters (q), you always get 4× as many quarters ($q = 4d$), which is *deterministic* and not statistical or stochastic, aside from possible errors made by the bank. If a thermometer has scales in degrees Fahrenheit (°F) and in degrees Celsius (°C), when one scale says 68.0 °F, the other definitely reads 20.0 °C, using the relation in Section 4.4. When one says 98.6 °F, the other definitely reads 37.0 °C. When you plot the results in °F on one axis and in °C on the other, all of the points fall perfectly on a straight line, with no statistical scatter. When you plot the temperatures read by two separate thermometers, you may get similar readings if they are close to each other, but there would be some scatter about a straight line fit, whatever the units.

Data variations over different time scales are common in data acquisition. For a particular November, the temperature variations over several days may look like the curve in Figure 4.5b; this shows the daily variation of temperature and the general trend toward lower temperatures as this month progresses. A thermometer placed outside in Central Park in New York City (Belvedere Castle Transverse Road (near 79th & 81st St.)), well sheltered from rain and sun, monitors, among other things, the low temperature each day in November over the course of many years. As noted, the temperature generally decreases during this month each year, but some days are colder or hotter than "expected." After many years you will have a data set of different temperatures at each day with scatter in the data. Statistical analysis shows that over 1981-2010 the average low temperature decreases there from 45.6 °F on November 1 to 37.3 °F on November 30. (For this month the change is fairly linear with the day in the month; quadratic terms are needed for most other months.) There is a dispersion or width of the year-to-year scatter about each daily average value that is characteristic of the year-to-year variability and of *statistical* measurements or *stochastic* processes in general.[cdxxiv] Understanding such variations are essential in managing electrical power grids, which entails planning based on the average daily and seasonal changes in power consumption and renewable energy sources such as solar and wind power, and their stochastic changes as well.

16.1.2 Caution in using statistics when the problem is not well defined

The results of some studies may be inconsistent with each other. Sometimes the statistical distributions of analyzed data sets are excessively broad. At other times they may have two distinct peaks, i.e. they are bimodal (Figure 16.1). Why? This is not necessarily due to faulty data acquisition or analysis, but could occur because there is a key variable that had not been noticed. For example, there is a wide variation in how "hyper" people feel after drinking caffeinated coffee. There is also a wide split in the risk of hypertension and heart attacks among coffee drinkers; the distribution is bimodal. If the reason for this were known, those who have a high risk would be advised not to drink much caffeinated coffee, while the others who derive some health benefit from other chemicals in coffee would be advised to continue drinking caffeinated coffee. The bimodal statistical distribution suggests there are two cohorts of people. In fact, there are, with different set of genes responsible for the CYP1A2 enzyme that metabolizes caffeine. These different enzymes respectively lead to slow metabolism of caffeine (for those with at least one of the two CYP1A2 genes responsible for that enzyme, with the *1F allele (or form of the gene)) and to fast metabolism of caffeine (for those who have *1A/*1A, or two copies of the *1A allele). When the data for these two groups are collected and analyzed separately, the results are distinct and the advice given to the two cohorts is clear: The slow metabolizers should avoid drinking caffeinated coffee, while the fast metabolizers can drink it without concern.[cdxxv] Complicating this, different populations have different frequencies of these alleles and sometimes also have other alleles (that do not seem to behave like the *1F allele).[171,cdxxvi]

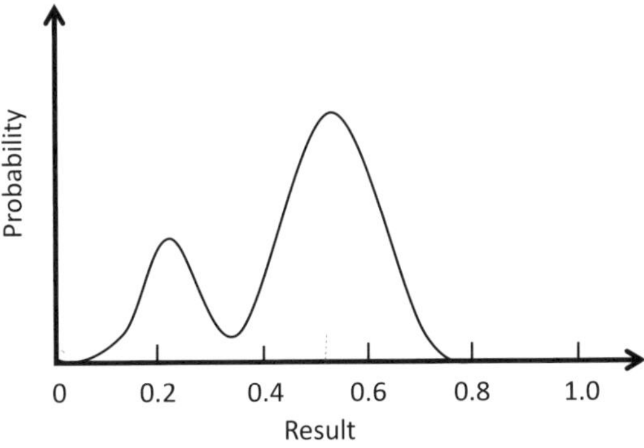

Figure 16.1: A bimodal distribution.

171 The *1A allele is very common in those with several geographical ancestries (≈31% of Northern Europeans and ≈39% of Ethiopians and Sub-Saharan African), but is very rare in others (<0.1% of Japanese), and the *1F is common in many (≈42% of Northern Europeans, ≈50% of Ethiopians and Sub-Saharan African, and ≈63% of Japanese).

16.2 Statistics: Analyzing Them

One statistical measure of a series of observations is its mean or average value, sometimes denoted as μ. To obtain the mean you sum the value of each point, x, and then divide by the number of data points, N.[172] Data can be analyzed to find the mean or average, median (middle value), or mode (most common value). In our earlier example, the number of days per 30-day month you took pills has an average of 28.54, a median of 29, and mode of 28.[173] The distribution of pills per month ingested had an average of 28.73, and, for this particular case, the same median of 29 and mode of 28. Averages μ of an entire population of N points differ from those of a subset of that population, m, that *sample* only n points of the entire distribution. In our pill example, we assumed that we analyzed the entire set of points.

Weighted averages count some numbers more than others in the average for a variety of necessary reasons, as in correcting for the different sized sampled groups in censuses. Each data value is multiplied by the weighting factor and then summed, and this sum is then divided by the sum of the weighting factors to provide the needed *normalization*. In an unweighted sum, the weighting factors are all 1, so the denominator is just the number of samples.[174]

There is typically a spread or dispersion about the (unweighted) mean of a distribution, which is characterized by the *variance* and *standard deviation* of the distribution.[cdxxvii] The variance of the distribution, σ^2, (for now the entire population) is a mathematical measure of how far data points (x) are from their mean, μ. To determine it, you take the difference between the value of each point and the mean and then square it, add these squares for each point, and then divide the

---

[172] The mean of the N data points in the entire population is $\mu = (x_1 + x_2 + x_3 + \ldots + x_N)/N$ and that of the n data points in the sampled set is $m = (x_1 + x_2 + x_3 + \ldots + x_n)/n$.

[173] A middle value only makes sense with an odd number of data points; we had an even number, 100, but the middle two values were both 29.

[174] The weighted sum is $(p_1x_1 + p_2x_2 + p_3x_3 + \ldots + p_nx_n)/(p_1 + p_2 + p_3 + \ldots + p_n)$ for n data points. We described each term above as the x value times the p weighting factor, but have used the more common, and equivalent, order of the p weighting factor times the x value here.

sum by the number of points, N.[175] This gives the average of the squares of these deviations from the mean. The standard deviation, σ, is the square root of the variance.[176] The variance and standard deviations can be obtained for each distribution type presented in the previous chapter, and more generally for all discrete and continuous distributions.[177]

Determining the variance and standard deviation of the distribution of the number of days you took at least one pill uses the mean, 28.54 days. For each of the two months that you took pills on 26 days, the difference from the mean squared is $(26 - 28.54)^2 = (-2.54)^2 \approx 6.45$. Summing this for these two months contributes $2 \times 6.45 = 12.90$ to the variance sum. The sum of squares of such differences for the 100 months of data is 98.84,[178] and the variance is this divided by the number of points, 100, or $98.84/100 = 0.9884$. The standard deviation is the square root of this, $\approx 0.9942 \approx 0.99$.[179] Sometimes the mean of a distribution is presented as the mean plus or minus the standard deviation, $\mu \pm \sigma$, or 28.54 ± 0.99 days per month here.

How data vary with one or more parameters (or variables) are often analyzed using a procedure called *regression analysis*.[cdxxviii] One or more variables that you can control or that are considered input are called the independent, predictor or explanatory variables, such as the day of the month in the above example. The other variables that are thought to depend on them, which you measure or obtain, are called the dependent, response or outcome variables. A trial function relating all of these

---

175 The variance of the entire population of N data points is $\sigma^2 = ((x_1 - \mu)^2 + (x_2 - \mu)^2 + (x_3 - \mu)^2 + \ldots + (x_N - \mu)^2)/N$.

176 The standard deviation of the entire population of N data points is $\sigma = (((x_1 - \mu)^2 + (x_2 - \mu)^2 + (x_3 - \mu)^2 + \ldots + (x_N - \mu)^2)/N)^{1/2}$.

177 The variance in the probability of the binomial distribution of observing an event with probability p in n chances is $np(1-p)$, while the mean is np. For a Poisson distribution of r successful events, it is r, the same as the mean. For a normal distribution, it is σ^2, so the standard deviation is σ, while the mean is μ.

178 The sum of squares of the differences for all the 100 months of data is $2\times(26-28.54)^2 + 12\times(27-28.54)^2 + 35\times(28-28.54)^2 + 32\times(29-28.54)^2 + 19\times(30-28.54)^2 = 2\times(-2.54)^2 + 12\times(-1.54)^2 + 35\times(-0.54)^2 + 32\times(0.46)^2 + 19\times(1.46)^2 \approx 2\times6.4516 + 12\times2.3716 + 35\times0.2916 + 32\times0.2116 + 19\times2.1316 = 98.84$.

179 The standard deviation and variance here are surprisingly (and perhaps suspiciously) close to each other, only because of the (arbitrary) choice of the data set.

variables is devised, sometimes on the basis of theory or expected results. These parameters are adjusted to optimize the fit, commonly by using the *Method of Least Squares* developed by Legendre in 1805 and Gauss in 1809. In this method, the difference of the measured value and that expected from the fit (and so the error) is squared and then these are summed for each data point to arrive at the sum of these squared residuals (called *SSE* (sum of squared errors of prediction) or *RSS* (residual sum of squares)). This sum is minimized by using standard differential calculus methods (differentiating with respect to one of the parameters and then equating it to zero) and this is repeated for each parameter. The resulting equations are solved to give the parameters of a fit that has been optimized to minimize the "error" (dispersion), and then evaluated to assess its value or *significance*, as we will see later. The sum of the squares of the differences of each data point of this optimized fit divided by the number of data points, and so the average of the squares of differences, is called the variance of the fit and the square root of the variance is the standard deviation of the fit.[cdxxix]

In *linear regression analysis*, the parameters appear in linear form, which might be the first step in more careful analysis.[cdxxx] For a straight line fit this means obtaining two constants: the slope (which could be called b) and the "y" intercept (a).[180,181,182] Figure 16.2a shows hypothetical data, with the best straight-line fit shown. In Figure 16.2b there is a second set of data with the same curve fit, but with a larger dispersion or error. In each, the point with the largest x is shown with an

[180] With this fit, $y_i = a + bx_i$ for a given x_i. These parameters for the slope and y intercept are used here because they are easily extended for more complex fits. However, in math classes, the names of these parameters are often reversed and this is written as $y = ax + b$.

[181] In the vertical offset method, the fit is obtained for each measured Y_i at x_i by summing $(Y_i - y_i)^2 = (Y_i - a - bx_i)^2$ for each point i and differentiating the sum with respect to a and then setting it equal to 0, and then repeating this for b. After this is done, the residual error at each point is $e_i = (Y_i - a - bx_i)$, using the optimized values of a and b. The error, or estimator for the variance in the error s^2, of such a fit of N points is obtained by summing the squares of each of these N values of e_i, and dividing it by $N - 2$. The square root of this (s) is a measure of "typical" error in the fit.

[182] And, for a quadratic fit with a, b, and c in $y_i = a + bx_i + cx_i^2$ for each measurement $i = 1, 2, 3, 4, \ldots$, assuming one dependent variable and one independent variable.

error bar (which might indicate ± a standard deviation), and this could be representative for the other points as well.

16.2.1 Covariance and correlation

The variance and standard deviation assess the spread of a given variable about the mean, independent of other variables. How two variables correlate with each other can also be determined through the statistical measure called the *covariance*. The correlation of two variables, x and y, is given by the *Pearson's correlation coefficient*, which is obtained using the *covariance* of x and y and the individual standard deviations of x and of y.[cdxxxi]

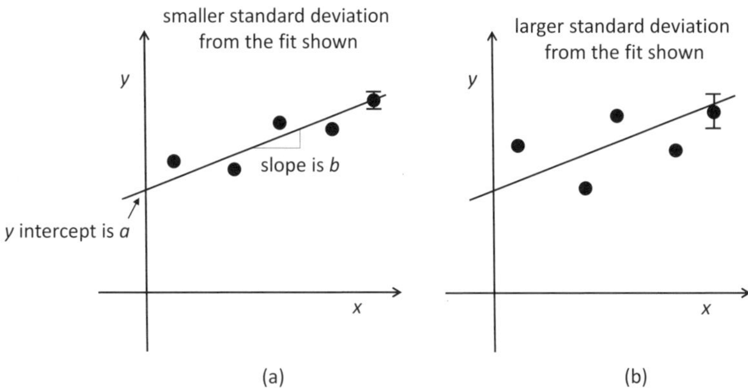

Figure 16.2: Linear regression analysis.

As noted before, to get the variance of a variable, x, you find the difference of x for a data point and the mean of the variable, square it, sum the squares over all data points, and then divide by the number of data points N. You find the variance of a second variable y in the data set in the same way. To find the covariance of these two variables, you find the difference of a variable and the mean of the variable for one data point, x, do the same for that y associated with the same data point (because they are paired), multiply the two differences, sum the products over all N data points for the entire population, and then divide by

N.[183,cdxxxii] The population correlation coefficient or the population Pearson correlation coefficient, called ρ, assesses how statistically correlated x and y are. It equals this covariance divided by the product of the standard deviations of x and that of y, for the entire population (which is the square root of the product of their variances).[184] This coefficient ranges from 1.0 for total positive correlation to -1.0 for total negative correlation (anticorrelation), with 0 indicating no correlation at all (uncorrelated).

To illustrate this, we find the correlation coefficient of three x variables that are respectively associated with three y variables. This example has far too few data points to be of any statistical value, but it enables you to see the numbers in action. The means of x and y are both assumed to be 0 to make the calculation simpler (Figure 16.3). In Case 1, the three x variables -1, 0 and 1 are respectively associated with three y variables, -4, 0, and 4, respectively, so -1 and -4 are one x and y pair, 0 and 0 are a second pair and 1 and 4 are a third pair. The numerator of the variance of x is $(-1-0)^2+(0-0)^2+(1-0)^2=1+0+1=2$, and that of y is $(-4-0)^2+(0-0)^2+(4-0)^2=16+0+16=32$. The numerator of the covariance is $(-1-0)\times(-4-0)+(0-0)\times(0-0)+(1-0)\times(4-0)=4+0+4=8$. The covariance and standard deviation denominators cancel, so the correlation coefficient is $8/(2\times 32)^{1/2}=8/64^{1/2}=8/8=1$. x and y are not merely positively correlated here but totally correlated here, which makes sense because in each pair the value of y is the same positive factor, 4, × that of x. In Case 2 we keep the three x variables the same and change the three y variables to 4, 0, and -4, respectively, so the three pairs are -1 and 4, 0 and 0, and 1 and -4, and $\rho=-8/(2\times 32)^{1/2}=-8/64^{1/2}=-8/8=-1$, so x and y are not merely negatively correlated but totally anticorrelated, which again makes sense because in each pair the value of y is the same negative factor, -4, × that of x. In Case 3, the y variables are

---

[183] The population covariance for N data points is $\mathrm{cov}(x,y) = [(x_1-\mu_x)(y_1-\mu_y)+(x_2-\mu_x)(y_2-\mu_y)+(x_3-\mu_x)(y_3-\mu_y)+\ldots+(x_N-\mu_x)(y_N-\mu_y)]/N$, using the population averages of x and y: $\mu_x=(x_1+x_2+x_3+\ldots+x_N)/N$ and $\mu_y=(y_1+y_2+y_3+\ldots+y_N)/N$. The population variances for x and y are $\sigma_x^2=((x_1-\mu_x)^2+(x_2-\mu_x)^2+(x_3-\mu_x)^2+\ldots+(x_N-\mu_x)^2)/N$ and $\sigma_y^2=((y_1-\mu_y)^2+(y_2-\mu_y)^2+(y_3-\mu_y)^2+\ldots+(y_N-\mu_y)^2)/N$.

[184] The population correlation coefficient is $\rho_{x,y}=\mathrm{cov}(x,y)/(\sigma_x\sigma_y)$.

2, -4, and 2, respectively, so $\rho = 0/(2 \times 24)^{1/2} = 0/(48)^{1/2} = 0$ and so x and y are uncorrelated. In Case 4, the y variables are 1, -4, and 3, respectively, so $\rho = 2/(2 \times 26)^{1/2} = 2/52^{1/2} \approx 0.277$ and x and y are positively correlated, but only partially. In Case 5, the y variables are -1, 4, and -3, so $\rho = -2/(2 \times 26)^{1/2} = -2/52^{1/2} \approx -0.277$ and x and y are negatively correlated, but only partially anticorrelated.

We will see later in this chapter that mere correlation does not imply causation or common causation. In Chapter 20 we will use the math of correlation to show that diversifying your financial assets can lower your risk.

	y, Case 1	*y*, Case 2	*y*, Case 3	*y*, Case 4	*y*, Case 5
x = -1	-4	4	2	1	-1
x = 0	0	0	-4	-4	4
x = 1	4	- 4	2	3	-3
Case correlation coefficient	1	-1	0	0.277	-0.277
Interpretation	Totally correlated	Totally anti-correlated	Uncorrelated	Partially correlated	Partially anti-correlated

Figure 16.3: Correlation analysis example.

16.2.2 Sampling and polling

So far, we have determined the mean μ and standard deviation σ for a full data set, such as a complete population or a complete set of test takers or voters. If there are too numerous data points to do this, you may be only able to *sample* it, say n random points of the total of N points, and then you would need to assess how representative your sample is of the entire distribution by choosing this limited number of samples.

One central concept in this field is that when you make many measurements of the same thing many times, the average value you find should be close to the expected value, and it becomes even closer with even more trials.

The sample mean, now called m (instead of μ) is calculated the same way, but with only n points sampled instead of the points of the entire population, N.[185] The standard deviation, or really the *estimator* of it, is calculated the same way for the n points, but with two differences. The difference of the sampled variable and the measured sample average is now determined for each data point, squared and summed. This sum is divided by n and its square root is determined to give the *uncorrected sample standard deviation*, s_n. However, this tends to underestimate the standard distribution of the entire distribution, increasingly so for increasingly small sample sets; the "biases" are typically >1% for 75 or fewer samples.[cdxxxiii] This bias from the actual standard deviation is much less for sample sizes down to about 10 when you divide the sum of the squares by $n - 1$ instead of n; this gives the *corrected sample standard deviation* or, more simply, the *sample standard deviation*, s (instead of σ).[186] For normal distributions, the sampling error is even smaller for very small samples, even down to ~3, when you divide by $n - 1.5$ instead of $n - 1$.[cdxxxiv]

If the sum of the squares of these differences for the quite small set of 5 samples happens to be 500.0, the uncorrected sample standard deviation would be $(500.0/5)^{½} = (100.0)^{½} = 10.0$. The sample standard deviation would be $[500.0/(5 - 1.5)]^{½} = [500.0/3.5]^{½} \approx 11.95$, which is much closer to the actual standard deviation of the distribution (perhaps ~12.0) than is 10.0. If the sum of these squares from a larger set of 100 samples of a particular (and different) distribution happens to be 10,000.0, the uncorrected sample standard deviation would be $(10{,}000.0/100)^{½} = (100.0)^{½} = 10.0$. For this larger sampling, this is now close to the sample standard deviation $[10{,}000.0/(100 - 1.5)]^{½} = [10{,}000.0/98.5]^{½} \approx 10.076$, which itself is still a bit closer to the actual standard deviation (perhaps ~10.1 in this case).

Analysis of the correlation in a sample of n pairs of points in the population is similar to that for the entire population, and gives the *sample correlation coefficient* or the *sample Pearson correlation coefficient*, called r (instead of ρ). As in determining the standard deviation, the sampling number n often needs to be replaced by a

---

[185] The sample mean of n sampled points is $m = (x_1 + x_2 + x_3 + \ldots + x_n)/n$.

[186] The sample standard deviation of n sampled points is $s = (((x_1 - m)^2 + (x_2 - m)^2 + (x_3 - m)^2 + \ldots + (x_n - m)^2)/(n - 1))^{1/2}$.

different factor, such as $n - 1$ for small samples in determining the variance, and such population factors again cancel in determining the correlation coefficient from the samples.[187]

The crucial issue is how close the measured sample mean you measure, m, is to that of the actual full distribution, μ. The standard deviation of this measured mean relative to the mean of the full distribution is the *standard error of the mean* or σ_{mean}.[cdxxxv] This is the standard deviation of the full distribution, divided by the square root of the number of samples, n.[188] So, for 25 samples from a distribution with an average of 80 and a standard deviation of 10, the standard error of the mean is $10/25^{1/2} = 10/5 = 2$. This means that the average of the sample will be within the range from 78 to 82 (or within 80 ± 2) $\approx 68.3\%$ of the time for 1 standard deviation (σ) and from 76 to 84 (or within 80 ± 4) $\approx 95.4\%$ of the time for 2 standard deviations (2σ). With more samples, say 100, the standard error of the mean decreases, here to $10/100^{1/2} = 10/10 = 1$, and these two ranges narrow, to 79 to 81 and 78 to 82, respectively.

One important yet often maligned example of sampling is the polling of a population prior to an election. Let's say we poll n potential voters, say 1000, whether they intend to vote for candidate A or B. The goal is to determine the fraction of the general public that will vote for A, which we will call p, or for B, which is then $1 - p$. The average number of those polled who say they are voting for A would be np and for B it would be $n(1 - p)$. If 50% will vote for A and B, we would expect an average of 500 of the polled would say they would be voting for A and 500 for B, but the actual numbers for each poll would be near but not exactly the same as these. This distribution depends on the standard deviation of the poll, which determines the *margin of error* of the poll.

The distribution of potential outcomes of this poll is a binomial distribution and for this specific example it is equivalent to that of

---

187 With n of the n sampled data points replaced by $n - 1$, the sample covariance is $ss_{xy} = ((x_1 - m_x)(y_1 - m_y) + (x_2 - m_x)(y_2 - m_y) + (x_3 - m_x)(y_3 - m_y) + \ldots + (x_n - m_x)(y_n - m_y))/(n - 1)$, and the x and y sample variances $ss_{xx} = s_x^2 = ((x_1 - m_x)^2 + (x_2 - m_x)^2 + (x_3 - m_x)^2 + \ldots + (x_n - m_x)^2)/(n - 1)$ and $ss_{yy} = s_y^2 = ((y_1 - m_y)^2 + (y_2 - m_y)^2 + (y_3 - m_y)^2 + \ldots + (y_n - m_y)^2)/(n - 1)$, with the x and y sample averages $m_x = (x_1 + x_2 + x_3 + \ldots + x_n)/n$ and $m_y = (y_1 + y_2 + y_3 + \ldots + y_n)/n$. These can be used to obtain the sample correlation coefficient $r = ss_{xy}/(ss_{xx}\, ss_{yy})^{1/2}$.

188 $\sigma_{\text{mean}} = \sigma/n^{1/2}$

tossing an unbiased coin 1000 times. The variance in a binomial distribution is (as above) known to be the number of events (such voters or coin tosses) times the probability the event outcome occurs times 1 minus this probability (which is the probability of the other outcome), or $np(1-p)$. (So, the ratio of the standard deviation and the number of events decreases as this number increases, as $n^{1/2}$, and there is less relative uncertainty with large samples.[189,cdxxxvi]) For our specific example, the variance is $1000 \times 0.5 \times (1-0.5) = 1000 \times (0.5)(0.5) = 250$, and so the standard deviation of the poll is $(250)^{1/2}$ or ≈ 15.8.[190] If you divide the standard deviation 15.8 by the number of votes, 1000, you get the fraction ≈1.58%, and for 2 standard deviations you get ≈3.2%. More precisely, 2σ corresponds to ≈95.45% confidence and $\approx 1.96\sigma$ to 95% confidence, which means that 95% of the time the results for a candidate will fall between the expected value and plus or minus 1.96 standard deviations of the poll. For a poll of 1000 potential voters this fractional poll "sampling margin of error" is $\approx 1.96 \times (\pm 1.58\%) = \pm 3.1\%$. With 300 polled, the fractional margin of error is larger, ±≈5.8%, and with 2000 polled, it is smaller, ±≈2.2%.

So, what is the potential error in the oft-noted poll of 1000 people of two candidates? From a purely statistical basis, polling 1000 people whether they will vote for candidate A or B has a statistical uncertainty ± ≈3% of being correct for either candidate 95% of the time.[cdxxxvii] So, a candidate polling at 52%, would be expected to have the support of 49% to 55% at the election approximately 95% of the time. This is not a "statistical tie" or "statistical dead heat"[cdxxxviii] because it is more likely than not the candidate would win the election; still, it would not be a shock if the candidate lost. If the candidate polled at 55%, the prediction would be more definitive. (Because a statistical error that predicts too few votes for candidate A is a statistical error that predicts too many votes for candidate B, the sampling margin of error of the difference of

---

[189] As before, this is $\sqrt{np(1-p)}/n = \sqrt{p(1-p)}/\sqrt{n}$.

[190] We can get this variance another way. If we assign a value of 1 for a vote for A and of 0 for B, the average would be $500 \times 1 + 500 \times 0$, divided by 1000, or $0.5 = 50\%$, and the variance would be $500(1-0.5)^2 + 500(0-0.5)^2 = 250$.

the votes for A and B is twice the margin of error for either one, or ± ≈6% here.)[cdxxxix]

This 6% range (from -3% to +3%) represents an imperfect selection of the eventual voters for a candidate due only to the statistical nature of an ideal polling process. However, analysis of past elections shows that such polls taken close enough to the election that voters have pretty much decided on a candidate, have a range that is closer to 12-14%, because of additional uncertainties that bias the statistics.[191,cdxl,cdxli,cdxlii,cdxliii]

Voter turnout affects the predictions of polls, and in turn is affected by the polls. In the 2016 U.S. presidential election, turnout was 53% in states where the winner was not in question, and so the Electoral College voters were clearly going to one of the candidates, and 62% where the results in the state were in question.[cdxliv] Linked to this, fewer people appear to be participating in surveys in general[192,cdxlv] and sometimes the phrasing of the poll question biases a poll.[193,cdxlvi] Moreover, pollsters with the same raw data sometimes end up with differing polls results, and this can also account for part of this increased error.[cdxlvii]

---

[191] So, 95% of the time the candidate who received 52% of the votes of the 1000 polled would have the support of perhaps 45% to 59% at the election, so victory is likely for this candidate but less likely than if statistics provided the only uncertainty. What causes this much larger margin of error? The way people are being targeting for polling may be biased toward one candidate or the other—as could occur when contacting them only by phone or by new communication technologies. The supporters of one candidate may be more or less likely to vote than those of the other. The supporters of a candidate doing poorly in the polls seem to be less likely to respond to a poll request than supporters of the leading candidate. In the 2008 U.S. general election, 55% of those who claimed they would not vote, and so were not included in polls, actually voted, while "only" 87% of those whose who claimed they would vote, and so were polled, actually voted. The poll would be biased if the voting preferences of these groups differed.

[192] As noted in the participation rates in Pew Research Centers telephone surveys, which dropped from 43% in 1997 to 2012 to 14%

[193] A poll concerning a politician A and potential corruption can be phrased more neutrally, as in "Do you have concerns about corruption in the administration of politician A, if elected?" or less neutrally, "Given the well-known history of the candidate, do you have concerns about corruption in the administration of politician A, if elected?".

The polls and results of the 2016 U.S. presidential election illustrate several features of the math of U.S. elections, which should be understood independent of your political persuasion. Hillary Clinton won the national popular vote by 2.1%, though she lost the deciding Electoral College vote. However, on the day after the election, with all of the mail ballots still not counted, the popular vote seemed to be closer, with her ahead by only 0.2%. The polls immediately prior to the election generally gave Clinton a 3-4% lead over Donald Trump. (Of 10 major polls, five gave her a 4% lead, two gave 3% and one gave 2%, while Trump was ahead in one by 2% and in another by 3%.)[cdxlviii] Since these polls usually had a stated 2σ statistical error $\pm \approx 3\%$, the early popular vote outcome appeared to be statistically unlikely for any single poll, and even more unlikely when considering them in aggregate, and many wondered what had gone "wrong." But, with the final count in, the results were seen to be well within the expected statistical error of the polls.

In any case, the election outcome depended on the Electoral College vote tally. The number of Electoral College votes a state has is the number of representatives it has, which is proportional to population (to the nearest integer), plus the number of senators, which is 2, aside from Washington, D.C., which has three Electoral College votes. Eighteen of Trump's victory margin of 74 Electoral College votes[194] came from his winning nine more states (counting Washington D.C. as a state). However, the decisive mathematical difference in the election was not his winning more states, but the "binning" nature of winner-take-all rule in virtually each state and his narrowly winning several states.

16.2.3 How good is the fit to the data?

Though objectively assessing how good a fit is to the data set is essential, making decisions on how to assess the mathematical analysis is surprisingly still controversial. The *chi-squared test* [cdxlix] (or chi-square test or the χ^2 test) uses the chi-square distribution[cdl] to help determine if a fit is "good enough." This means that the data sampling set and fit are *statistically significant*, and this test quantifies this. It is also used to

---

[194] These are those actually won by him—though some were not cast for him.

check if data are indeed random and independent, when they supposedly are. You can use this test to assess the suitability and quality of a linear fit to the data obtained by regression analysis or a *null hypothesis*, which could be that the outcome in the tossing of coins or dice is unbiased. For coins, you expect to get heads and tails 1/2 the time and for dice to get any number from 1 to 6 exactly 1/6 of the time, but the results are never exactly this because the number of tosses is limited.

The *number of degrees of freedom* is a parameter in this test. For coin tossing, there are 2 outcomes, but this corresponds to 1 degree of freedom or uncertain outcome because if 48 out of 100 tosses are heads, you know for certain that 52 are tails. For dice, there are 6 possible outcomes and 5 degrees of freedom, because if you know the number of outcomes of getting a 1, 2, 3, 5, or 6 you know the number of times you get a 4.

The most-used chi-squared test is the *Pearson's Chi-squared Test,* first investigated by Karl Pearson in 1900.[cdli] For each case considered there are an expected number of times for each outcome and a measured number of times for each (for the example of dice, the number of tosses divided by 6). For each possible outcome (say of getting 1, 2, 3, 4, 5, or 6), the difference of the expected and measured numbers is squared and then divided by the expected number, and then these are summed for each outcome (from 1-6 for dice). This gives a χ^2, which is compared to the expectations from the chi-squared distribution. We will return to this when we address using the results of statistics.

16.3 Statistics: Presenting Them

Statistics need to be presented in a manner that is suitable for the intended audience, by targeting its needs and background, and by doing so in an ethical way.

They are often presented numerically or graphically without bias, so you can visualize the statistical data and draw accurate conclusions. Standard methods include scatter plots showing the data points of one variable vs. another, line graphs showing a fit to the data (sometimes with the standard deviations, "errors" or error bars indicated), and plots showing both the data and the fits, as in Figure 16.2. The data can also be shown with bar graphs (Figure 15.1a,b), pie charts, and other graphical tools. Bar graphs can show results at specific values of variables or classifications, such as car sales for each model color. Results can be grouped in some manner, for example, summed over a range of values or bins, such as from 0 to 1, from 1 to 2, and so on or results in the first, second, third and fourth quarters of a year; these are called *histograms*. (The bins need not be of equal widths.) Sometimes data are presented as the actual numbers that were obtained, and sometimes they are normalized to be fractions of a total result or they are weighted in some manner. The nature of the data set and what you want to show dictates which method is most useful.

The information presented in reports, such as Table 21 in the U.S. Department of Health and Human Services study called *Health, United States, 2017*, illustrates the care needed in presenting statistics. This table presents the overall death rate in the U.S. in 2016 as being 849.3 per 100,000 and also as 728.8 per 100,000.[195] Both death rates count deaths the same way and over the same period, but mean different things, as is clearly noted in the table. The rate of 849.3 is the crude rate, which is the total number of people who died that year divided by 100,000. The rate of 728.8 has been age-adjusted to the age distribution of the reference year of 2000. This was done for a very good reason, to improve year-to-year comparisons by reflecting changes in the death rates and age distributions. To do this, the number of deaths per age group in 2016 was multiplied by the fraction of the total population it represented in 2000 and these products were summed to give this age-adjusted, weighted rate. (If you followed the same procedure but instead used the population fractions in 2016, you would end up with the crude rate again.) These differences are clearly apparent in the data for males and females: the age-adjusted death rates were 861.0 for males and 617.5 for females and the crude death rates were 880.2 for males and 819.3 for females.[cdlii] Similarly, the rates of contracting a specific disease can be expressed as a

---

[195] In Section 13.4 we saw that lifetimes in exponential decay equal the reciprocal of the decay rate, but beware—you cannot use this relation in other cases. Here, the reciprocal of 849.3 deaths per year per 100,000 people, is 117.7 years per person. This is <u>not</u> the average lifespan of the people in the study because the defined rate is a sum of the rates over many ages, and so it does not represent simple exponential decay.

rate for a given gender, age group, and so on. The rate of contracting many diseases has been declining in many populations. The reasons for this are not always clear and they are the subjects of intense statistical scrutiny.

Actuarial tables include period or static life tables, which give the current probabilities of death for people of a given age and gender over the coming year, and cohort life tables, which present the probabilities of death for a cohort of people, such as those of the same birth year and gender, over their lifetime. These cohort tables are obtained by necessity by using the cross-sectional data from period life tables and not from separate, longitudinal studies tracing a group over many years.[cdliii] For example, a cohort life table shows that of 100,000 U.S. people born alive who would have become 70 in 2016, 72,843 of them survived if they were men and 82,573 survived if they were women.[cdliv] The probability that an exactly 70-year-old would die before reaching exactly 71 was found to be 2.3122% for men and 1.5413% for women, and their life expectancies were 14.40 and 16.57, respectively.[196] This was, for instance, used to find the number of men surviving to 71, which is the number of 100,000 live births who survived to 70 times 1 minus the 1-year death probability at 70, or $\approx 72{,}843 \times (1 - 0.023122) = 72{,}843 \times 0.976878 = 71{,}159$. So, each yearly death rate helps determine the cohort life table for the next age group. Though such statistics are frequencies and not predictions of future mortality rates, they are used as the starting point for evaluating mortality probability. These can change, and usually decrease, due to improvements in medicine, public health, and general safety.

How you round off (Section 8.1.1) statistics affects how they are presented and then interpreted. Inflation rate data are officially given in percentages to one decimal place. A reported increase from 0.2% to 0.3%

[196] The probability that an exactly 50-year-old in 2016 would die before reaching exactly 51 was 0.5007% for men and 0.3193% for women, and their life expectancies were 29.69 and 33.26 years, respectively. The probability that an exactly 60-year-old would die before reaching exactly 61 was 1.1533% for men and 0.6848% for women, and their life expectancies were 21.61 and 24.60 years, respectively. The probability that an exactly 80-year-old would die before reaching exactly 81 was 5.7712% for men and 4.2539% for women, and their life expectancies were 8.34 and 9.74 years, respectively.

from one month to the next might be associated with an increase that this much smaller than 0.1% and closer to approximately 0.0% (as when the 0.2% had been rounded down from 0.248% and the 0.3% rounded up from 0.252%) or the larger increase of almost 0.2% (as when the 0.2% had been rounded up from 0.152% and the 0.3% rounded down from 0.348%).[cdlv]

Imaginative graphical and pictorial representations and abstract models of statistical data can help you better understand the essence of the results.[cdlvi] Unfortunately, they can also be devised to trick you into thinking that the data trends are either exceptionally large or small, when they are not.[cdlvii] Plotting the relatively small decrease in annual salary from $50,100 to $50,000 over several years with a salary axis ranging from $50,000 to $50,100 makes the decrease seem very large (Figure 16.4a), while plotting an increase in salary from $50,000 to $60,000 with a salary axis ranging from $0 to $200,000 makes this large increase seem very small (Figure 16.4b). Some have humorously shown how to draw different plots through the same data set to send whatever message that is desired.[cdlviii]

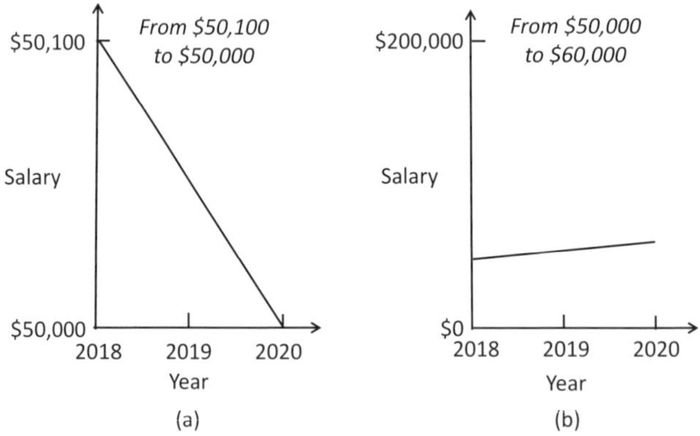

Figure 16.4: Deceptive plotting of statistics.

Presentations with partial data presented can be deceptive, even when the data are accurate. For example, reporting that opioid prescriptions in one community changed from 200 to 180 prescriptions per 100 people per year from one year to the next, a 10% decrease, and from 2 to 4 prescriptions per 100 people per year in another, a 100% increase, tells a clear, unbiased story. By reporting only the 10% decrease in the first community and the 100% increase in the other, implies that the problem is not severe in the former and is terrible in the latter, while the opposite is true. Such actions are designed to control your response to the data.

Other forms of selective or carefully controlled presentation of data can be misleading as well. Falsely comparing different sets of data is deceptive, such as representing changes in salaries as if they were changes in total compensation with benefits or vice versa.[cdlix] If football announcers note that a team had a record of 3-1 (3 wins and 1 loss) after trailing by 10 points or more in a game, you are being led to think that the team did a very fine job in coming back when behind by a large score. If in fact, that same team had a record of 3-22 after trailing by 9 or more (so they are 0-21 when behind by 9 points), you are being misled. Television commercials in 2018 promoting the sale of gold as an investment, extolled the increase in gold prices from ~$300 per ounce in 2001 to $1600-1800 per ounce in 2011-2012. They neglected to say that prices then dropped to ~$1200 per ounce and remained at that level until the time of the ad.

Averages can be deceptive, even without sinister intent. The average predicted path of simulations of Hurricane Joaquin in 2015 showed it hitting the American mainland at New York City, which led to great local concern. However, no simulation predicted that the City would be hit. The most probable trajectories were clustered about it traveling either more westward and hitting land between New Jersey and South Carolina or veering eastward into the Atlantic Ocean (as it eventually did).[cdlx] A well-known "statistical" joke concerning averages illustrates this: "Did you hear about the statistician who drowned in a lake averaging only 2 inches in depth?"[cdlxi] (It was much deeper than this in some places.)

A statistic may be accurate and nondeceptive as presented, but may still warrant further scrutiny, perhaps because it represents an average over different categories; proper binning may give needed insight. From 1996-2016, the annual U.S. inflation rate was 2.8%, a weighted average of many items with quite different annual changes in costs. Annual changes for textbooks and college tuition were 10%, for childcare and healthcare 6%, food and beverages and housing 3%, new cars, household furnishings and clothing 0%, cellphone service -2%, software and toys -3%, and for televisions -5%.[cdlxii] So, how these affected you depended on how much of each type of expense you incurred. An article addressing the reaction to the escalating cost of higher education showed that in the U.S. from ~1990 to ~2017, consumer prices increased by almost 100%, medical care by ~200% and college tuition by almost 400%.[cdlxiii]

Sometimes presented statistics may seem confusing only because the data sets are not explained well enough. In 2010, the average pace for running a mile in U.S. 5k races was 11:47 minutes (11 minutes and 47 seconds). This was surprisingly longer than the 9:31 in the longer 10k (6.2 mile) races, which in turn was, as expected, shorter than the 10:33 for the longer half-marathon (13.1 miles), and which in turn was unexpectedly longer than the 10:00 for the marathon (26.2 miles).[cdlxiv] For an individual, the time per mile would always be expected to be less when the overall distance is shorter, but this was for self-selected groups of athletes with widely different levels of training and accomplishment in races with varying entry qualifications.

At times, statistics can evolve into self-fulfilling prophecies, becoming the cause for an effect. The fraction of students accepted to a school is a metric called the selectivity and the fraction of accepted students that decide to attend the school is another metric called the yield. The most "highly desired" undergraduate schools in the U.S. have low selectivity fractions and large yield fractions. The desirability of most of these schools may not have changed much over the years, but their selectivity fractions have decreased greatly, so they appear to have become even more desirable. One reason for this may have been the increasing pool of students, but another was the increasing number of applications filed by each student (out of concern of being admitted anywhere and due to the increased ease of the common application process). As students started applying to more schools, school selectivity numbers decreased and they needed to apply to even more schools just to have a good chance of being admitted somewhere. Say 1,000 students want to attend 1 of 10 equally "desirable" schools, each hoping to enroll 100 of them. Before the selectivity upswing, each student applied to 4 schools, expecting to be admitted to 2 of them. There were a total of 4,000 applications, on average 400 to each school. Each school admitted 200 students, expecting 100 to accept their offer, and so had a selectivity of 200/400 = 50% and a yield of 100/200 = 50%. Now the schools become "hot" and the next year students fear that with the decreasing selectivity fractions they won't be accepted anywhere, so each applies to even more schools, say 8 schools, now expecting to get into 2 of them. There are a total of 8,000 applications, on average 800 to each school. Each school still admits 200 students, expecting 100 to accept their offer, but now each school has a better selectivity of 200/800 = 25% and the same yield of 100/200 = 50%. Nothing much has changed, but students worry more and need to pay application fees to apply to more schools. The schools need to have larger or more overworked admission staffs, but can proclaim that their lower selectivity fractions have made them even more desirable.

16.4 Statistics: Taking Action Based on Them

Decisions are often made and policy is developed after applying criteria to statistics.[cdlxv] They can be faulty when based on statistical analysis using inaccurate data, nonsense when based on distorted, fabricated, or guessed data, or foolish or misguided when based on data interpreted or presented improperly or in an unclear manner.

Mortality and disease rates, as noted above, are constantly employed for many purposes. Notably, actuarial (or life or mortality) tables from the Social Security Administration are calculated using the statistics of U.S. mortality[cdlxvi] and are routinely used to determine life and medical insurance rates, individual and Social Security retirement plans (including annuities), demographics, and so on.

Such statistics educate, mystify, and haunt those with cancer. After being diagnosed with abdominal mesothelioma in July 1982, Stephen Jay Gould wrote the now famous essay "The Median Isn't the Message."[cdlxvii] He noted that though the median lifespan after diagnosis of his form of cancer was eight months for then current treatments, the mortality distribution had very long tails for shorter and longer lifespans, in part due to variations in treatments and genes. He noted that because of these very long tails toward longer lifespans he might be one of the lucky ones to live a long life. He lived for another 20 very productive years, perhaps because of improved treatments, good genes, and good luck, and actually died from a different, unrelated cancer.

You make everyday decisions on the basis of means and variances of statistical distributions. Elevators are usually faster than escalators on average, but variances in escalator times are much smaller, and this entices some to ride escalators.[cdlxviii] You are using the statistics of variances and covariances when you decide to diversify your investments, as we will see in Chapter 20.

16.4.1 Is the fit statistically significant?

Is the fit to the data good enough to take action? The chi-squared and related test statistics are commonly used to analyze data and to help you make decisions on the basis of the goodness of the fit. Often you want to decide whether or not you can reject a baseline, *null hypothesis* that says the results would be expected only by chance and you may want to do so within a stated statistical significance. Or, you may want to learn whether there is insufficient evidence to reject the null hypothesis at that level of significance, and so an *alternative hypothesis* is possible and the results may not be due to chance.[cdlxix,cdlxx] The term "significance" is a mathematical assessment of the level of statistical certainty and does not mean the findings are or are not significant *per se*. For the Pearson's test, this amounts to comparing the value of chi-squared of the measurements determined using this null hypothesis (and which you calculate, as we will soon see) to the "p-value" for that chi-squared in the chi-squared distribution (which have been calculated already and are obtainable from published graphs and tables). If it is more than this value, the null hypothesis can be rejected to "1 – test significance" at that level of significance, and if it is less than it, there is insufficient evidence to reject it. In essence, you are comparing observed variances with expectations from a normal distribution.

To obtain the value of chi-square, *Pearson's cumulative test statistic*, you find the difference between the number of observations of a given type (O) and its expected value (E) assuming the null hypothesis is correct, then square it and divide it by this expected value, and then sum it for the different types of observations.[197] The chi-square distribution is different for different numbers of *degrees of freedom*. As noted earlier, in tossing a coin there is 1 degree of freedom and in tossing a die there are 5 degrees of freedom. By similar reasoning, in choosing a ball that you expect to randomly be either green, blue, red, or yellow and with a diameter of 2, 4, or 8 inches, there are $(4-1) \times (3-1) = 3 \times 2 = 6$ degrees of freedom.

An example should make this clearer. Consider the null hypothesis that a die is unbiased, as we might expect it to be. If you roll a die 60 times, you would expect that it would be a 1, 2, 3, 4, 5, or 6 on the average 10 times each if it were unbiased, but because of statistical variations you definitely do not expect to find 10 times for each. For respective total rolls of 5, 6, 9, 10, 10, and 20 times, chi-squared = $(5-10)^2/10 + (6-10)^2/10 + (9-10)^2/10 + (10-10)^2/10 + (10-10)^2/10 + (20-10)^2/10 = 25/10 + 16/10 + 1/10 + 0/10 + 0/10 + 100/10 = 142/10 = 14.2$. For these 5 degrees of freedom, tables show that *the upper tail critical* values, $1-p$, are 11.071 at 0.95 probability and 15.086 at 0.99 probability. Because 14.2 exceeds 11.071, <u>the null hypothesis of an unbiased die can be rejected and you can conclude that the die is biased, to a 95% significance</u> (so the *p*-value is less than 5%, which is often called a statistically significant level). However, because 14.2 is less than 15.086, <u>there is insufficient evidence to reject the null hypothesis of an unbiased die and conclude that the die is biased, to a 99% significance</u> (so the *p*-value is less than 1%, which is often called a highly statistically significant level). In biological studies, the commonly accepted threshold for significance is 95%.[198,cdlxxi] Though they are based on specific calculations, these assessments may seem a bit arbitrary and subjective, as we will see.

---

[197] This is $(O_i - E_i)^2/E_i$ summed over each type *i*.

[198] If instead in 60 tosses, you found a 1 and 3 a total of 9 times each, 2 and 4 a total of 8 times each, a 5 for 14 times, and a 6 for 12 times, chi-squared would be: $(9-10)^2/10 + (8-10)^2/10 + (9-10)^2/10 + (8-10)^2/10 + (14-10)^2/10 + (12-10)^2/10 = 1/10 + 4/10 + 1/10 + 4/10 + 16/10 + 4/10 = 30/10 = 3.0$. In this case, there is now insufficient evidence to reject the null hypothesis at 95% significance because 3.0 is less than 11.070 (and so this null hypothesis of an unbiased coin may be appropriate in this case).

Such tests are phrased in terms of rejecting the test fit (the null hypothesis). This means that the fit is so good that the probability of it not being true is so small that it lies in the far wings of the normal distribution (Figure 15.2). The basing of decisions on the statistical significance or the lack thereof is the classic interpretation by Ronald A. Fisher, which has usually prevailed in the past century. The math by Pearson and others utilized is this approach uses is clear and correct. Still, the *p*-values obtained when different sets of samples are chosen can be very different from each other, unless the sample sizes are very large, and so the reliability and the reproducibility of this test is less than is often claimed.[cdlxxii] Moreover, some have noted that this all-or-nothing or true-or-not-true approach is not a good recipe for making decisions and taking action, as we will now see.[cdlxxiii,cdlxxiv]

In the course of developing ways to improve the brewing of Guinness beer, in 1908 chemist William Sealy Gosset began to develop different statistical analysis methods, building on Pearson's work. This included the *"Student's" t-distribution* and the associated *t*-test,[cdlxxv] which can be used to see if two sets of data are significantly different. (He used "Student" as a pseudonym in his scientific paper with his results because Guinness did not allow publication by employees.)[cdlxxvi] Gosset utilized it to assess the substantive or economic significance of different variables, such as storage temperature, cask type, and yeast chemistry in brewing beer—even when the sample size of experiments was relatively small, and to take action based on this. This approach of using statistics to design a small number of experiments for maximum impact is called the *statistical design of experiments*, and it strongly influenced the subsequent work of Fisher. However, Gosset's approach focused less on whether or not the test is "statistically significant" and more by how much an item or process is improved by changing the size of something by a certain amount.

In *The Cult of Statistical Significance: How the Standard Error Costs Us Jobs, Justice, and Lives (Economics, Cognition, and Society),* Stephen Ziliak and Deidre McCloskey argue for the decision making based on Gosset's approach and against the classic testing of hypothesis approach of Fisher. To illustrate this they pose the existence of two, and only two, diet pills: "Oomph," which on the average reduces weight by

20 pounds, but with a large uncertainty of weight loss of roughly plus or minus 10 pounds, and "Precision," which reduces weight by only 5 pounds on the average, but with a much smaller uncertainty in this loss, plus or minimum one-half pound.[cdlxxvii] Which would you recommend to someone who needs to lose a lot of weight? Common sense says it would be better to use Oomph and lose 20 ± 10 or a range from 10 to 30 pounds rather take Precision and lose only 5 ± 1 or a range from 4 to 6 pounds, because the size of the loss will very likely be much larger, and this is the actual goal. However, strict statistical metrics and hypothesis testing would favor Precision because its statistics are cleaner.[199]

16.4.2 Metrics

Increasingly, statistics are being condensed into a few pithy numerical summaries termed metrics and these metrics are often all that you see. They are meant to help you get to the core of the issue and make decisions. There are also some common metrics based on non-statistical numerical evaluation. Some metrics are fair and provide deep insight. Others are assigned undue significance, may be confusing or off target, become the goals themselves, or are deceptive and you can make poor decisions if you use them.[cdlxxviii]

Some metrics are quite new, while others are longstanding, such as the use of degree days to track the range of temperatures about the mean. This metric is used by energy concerns to assess how much heating or cooling may be needed over a period. The magnitude of the differences between a reference, such as 65 °F, and those daily averages that are higher are added into the sum of cooling degree days for that period; those that are lower are added into the sum of heating degree days.[cdlxxix]

Metrics are frequently used in the business and investment worlds, some in plain view and others that are hidden from the public. The Dow

---

[199] The "signal-to-noise ratio" in the results can be expressed as the magnitude of the difference between the observed weight loss effect and the hypothesized null effect (of taking no pill at all), divided by the variation of the observed weight loss effect; the square of this is closely related to the chi-squared statistical measure. This ratio is small for Oomph $(20 - 0)/10 = 2$ and large for Precision $(5 - 0)/0.5 = 10$, so the data are cleaner, clearer, and less uncertain for Precision, though the size of the desired outcome is much larger for Oomph.

Jones Industrial Average is a statistics-based, public metric you may encounter daily. It is the sum of the prices of 30 currently specified stocks, divided by a divisor, which was 0.14748071991788 on June 26, 2018, and which is adjusted for stock splits and other weighting factors.[cdlxxx] It is more heavily weighted toward changes in higher-priced stocks by a given percentage than lower-priced stocks and includes too few stocks to be truly representative, so it may not be a very good metric.[cdlxxxi] Another metric, the Standard and Poor's 500 is weighted by market capitalization.[cdlxxxii]

Investors sometimes use metrics, such as price-to-earnings ratio (the ratio of the stock price to earnings per share) to assess the valuation of a stock.[cdlxxxiii] Historically, it has had an average value of ~15 (from 1870-2016 using the Standard & Poor's 500), but has more recently averaged ~20-25. It is typically different for different types of industries, commonly being lower for telecom stocks (in the low teens) and higher for high-tech stocks (even as high as ~40).[cdlxxxiv] A different business metric is the ratio of the compensations of the company CEO to that of the median earning employee, which is commonly several hundred. It is used by some to assess the company income distribution and social consciousness.[cdlxxxv] In a similar vein, the ("Karl") Marx ratio is the ratio of profit per worker to median work pay. It has a median of 0.82.[200,cdlxxxvi] Companies with a high Marx ratio reward their shareholders relatively well, while those with a low ratio reward their employees better. Similarly, the Organization for Economic Cooperation and Development (OECD) assesses a country's income inequality by its "Gini Coefficient," which is the fraction of its income that would need to be redistributed to achieve income equality among its people. Of course, this can be calculated using the statistical distribution of incomes that either excludes or includes possible government transfers to low-income households, and so is one of the many metrics that needs to be used with caution.[cdlxxxvii] The ratios of salaries in the 90th and 50th percentiles and of the 50th and 10th percentiles are also used to assess income distributions.[cdlxxxviii]

---

[200] This was the median on May 3, 2018, of the 394 companies in the Standard & Poor's 500 that had reported their median compensation.

Also within the business world, metrics are used to assess the data on advertisement viewership to help set advertisement rates. These depend on the measured number or frequency of impressions, exposures or opportunities for the ad. A metric assessing the cumulative effectiveness of an ad might be proportional to the frequency of exposures or only those above a given frequency threshold, or it may increase with frequency and then *saturate*, tailing off to a constant value (as in Figure 4.3).[cdlxxxix] New metrics have been developed for the digital world.[cdxc] In 2016, Facebook noted that the metrics for ads they post were flawed due to data miscalculations, including *underestimating* the number of advertisement videos that had been fully viewed.[cdxci]

Metrics also impact the government world. The extent of poverty can be determined with a metric obtained using income and resource statistics. This metric can then be compared to either an absolute standard or a relative goal and this can lead to government action of some sort. The poverty line standard in the U.S., as developed by Mollie Orshansky of the Social Security Administration in 1963, is the cost of the cheapest sustainable diet per person multiplied by the number of people in the household, then multiplied by three to account for other expenditures. This is the U.S. official metric. It is an absolute standard in that it does not depend on how much others have or spend. However, it does not include benefits received that are not directly in cash, such as Food Stamps and tax credits, and does not account for the changing cost of housing, health care, clothing and utilities relative to food or for regional differences in costs. The Supplemental Poverty Measure adopted by the Census Bureau in 2011 was largely developed by the National Academy of Science (NAS) in 1995. This metric sets the poverty line at the spending level on essential items that is exceeded by two-thirds of households, plus 20% for miscellaneous needs, so it is partly absolute and partly relative,[201] and has been adjusted and used to

[201] It includes all types of income, minus taxes, medical and childcare costs, and accounts for geographical differences in housing costs. It yields a rate of poverty that is a little higher than the first metric.

account for the higher-than-average cost of living in New York City.[202,cdxcii]

Metrics are used to evaluate performance in some occupations and some of these might induce less-than-professional behavior. Surgeons and hospitals are often judged by the mortality rate within 30 days of an operation, the lower the rate the better. But some surgeons may avoid necessary, though risky surgery because their 30-day postoperative death rates may rise, and this would reflect badly on their careers. Moreover, surgeons and hospitals have been known to do whatever they can to keep clearly, terminal patients alive for 30 days and no longer, to keep their mortality rates low. On the positive side, medical professionals use the Apgar score, a non-statistical metric based on observations, to rapidly evaluate the health of newborns to see if immediate medical action is needed. Still, within the area of health, but more mundane, is the often-used "rule-of-thumb" or metric that is it is safe to eat food that has been on the floor for less than 5 seconds, the so-called "5-Second Rule." Unfortunately, it is now known that fewer bacteria are transferred from the ground to the food in 5 seconds than in 50 or 500 seconds, but enough are transferred that this metric or rule is not really valid.[cdxciii]

The world of metrics has become dominant in scholarly studies occupations and the university evaluation of such activities. The impact of a scholar's career-to-date work is sometimes assessed by the scholar's *h-index* (Hirsch index),[cdxciv] a metric that increases as the number of that person's publications that are widely cited (referenced) in other papers in scholarly journals increases. Someone with an h-index of 40 has published 40 articles with 40 or more citations, but not 41 articles with 41 or more citations. Statistical physicist Sidney Redner has used statistical analysis to show that most people have an h-index very nearly equal to half of the square root of the total number of citations for all of their publications.[cdxcv] The quality of a scholarly journal is sometimes similarly judged by its *impact factor* for a given year, which is the number of times articles published by that journal in the previous two

---

[202] This New York City Center for Economic Opportunity (CEO) metric was introduced in 2008 using the NAS metric, leads to a poverty rate in NYC that is a little higher than that determined by the first two gauges.

years were cited, divided by the number of articles in that journal the previous two years. Higher h-indices and impact factors are thought to confer more prestige, though this is questionable.

Metrics are widely used to rate and rank your leisure activities and entertainment. Users sometimes rate tweets by "the ratio," i.e. the ratio of the number of replies to the sum of the numbers of "likes" and retweets. Large ratios (>>2:1) suggest an awful tweet, while small ratios mean the tweet may be on its way to becoming viral.[cdxcvi] Zagat has long rated restaurants, starting in 1979 in New York City, using the integers 1 to 30 in each of its three categories, restaurant food, décor, and service. In late July 2016, Google acquired Zagat and changed these (nonstatistical) metrics to a 1 to 5 star rating to make them more in line with other types of ratings.[203,cdxcvii]

Competing metrics may emphasize different features. Still within the realm of entertainment, the Rotten Tomatoes and Metacritic websites both "average" many film reviews to provide a single score metric from 1 to 100% and 1 to 100, respectively. However, these statistical metrics analyze data differently.[cdxcviii] The Rotten Tomatoes metric presents the percent of reviews that they judge to be positive, without distinguishing between strongly and mildly positive and strongly and mildly negative reviews.[204] Metacritic is a weighted metric. It assesses the degree of favorability of each review on a scale from 1-100 and then averages them, giving more weight to the reviews of well-known, established critics.[205,cdxcix] The same set of reviews can lead to very different metric outcomes for the same film. For example, the 2016 movie *Lights Out* was very mildly liked by virtually all critics. Consequently, it received a 95% from Rotten Tomatoes, but only a 60 on Metacritic. The problem with such one-number metrics is that they do not reflect the range and

---

[203] They were then more finely-grained with 50 grades because ratings then had decimal points, such as 3.8.

[204] Scores of 75-100% are "Certified Fresh" indicating good movies, 60-75% are "Fresh" and so good but not as good movies, and 0-59% are "Rotten" for movies that are deemed to be rotten.

[205] Scores of 81-100, 61-80, 40-60, 20-39, and 0-19, respectively mean "universal acclaim," "generally favorable reviews," "mixed or average reviews," "generally unfavorable reviews," and "overwhelming dislike."

intensity of the reviews. A 95 score from Rotten Tomatoes could have also meant that virtually every reviewer thought the movie was tremendous. A 60 score from Metacritic could have also meant that ~60% of all the reviewers loved the movie and ~40% hated it, or several well-established reviewers were not thrilled about the movie, but that many others really liked it. A two-number metric would seem to be better, one reflecting the mean or the median of the reviews and the other their spread or dispersion (i.e., the standard deviation).

Perception is very important in defining metrics and then in utilizing them to evaluate success. Applied mathematician Joseph B. Keller once offered an example of using the theory of lines or queuing (Chapter 18) that shows how stark changes in perceived success can occur with very small actual changes in performance. It interrelates statistical metrics, optimization protocols of traffic, and perception. Say that planes land once a minute and take off once a minute at an airport, and for most of the day there are 10 airplanes circling waiting to land. Each spends 10 minutes waiting, and if they arrived at the airports on time, all arrive 10 minutes late. If the threshold for calling a plane officially late were 15 minutes then the 60 planes that land every hour would be officially on time and none would arrive officially late. If now five extra planes suddenly arrive and then the arrival rate reverts to the former rate, there are now 15 planes circling in steady state, each waiting 15 minutes to land. Now 60 planes still land every hour but now each is 15 minutes late, so no planes land officially on time every hour and 60 arrive officially late every hour. The choice of this specific metric has changed perceived performance and statistical interpretation from superb to terrible, for a very modest 5-minute change in performance.[d] Currently, the U.S. Department of Transportation defines an on-time arrival as one when the plane reaches the gate within 14 minutes of the scheduled time. Some airlines have padded their arrival times more than others, to "improve" their reported on-time performance absolutely and relative to the competition, so choose your airlines wisely.[di]

16.4.3 Metrics in sports

Statistical metrics are being used increasingly in professional and amateur sports. More data are being collected during athletic games and they are being analyzed using increasingly sophisticated statistical methods. Some say such sports analytics are now being used too much and are negatively affecting how games are played.

In baseball, the "traditional" metrics, such as batting average (fraction of official at-bats resulting in hits) and runs-batted-in (RBIs), have been supplemented and sometimes replaced by the "advanced" empirical statistics or advanced "analytics" that adherents think correlate to game outcome better. They were pioneered in baseball by Bill James in 1980 and dubbed by some as *sabermetrics*, from the Society for American Baseball Research (SABR), and are now common in many sports.[dii] For example, in baseball the net output of a hitter is to create runs, which Bill James thought to be equal to the statistic "runs created by a player." In its most basic form, it is the sum of hits and walks (or bases on balls), multiplied by total bases (= singles + walks + 2 × doubles + 3 × triples + 4 × home runs), divided by the sum of at bats (official times at bat) and walks.

The all-in-one metrics Wins Above Replacement (WAR), such as the version by Sean Smith,[diii] and Win Shares (WS),[div] by Bill James, evaluate the number of wins a baseball player contributes to the team above that that would have been provided by a replacement player of marginal performance over the course of a season or career. These two metrics for evaluated players track each other very nearly linearly.[206,dv]

One team metric builds on the reasonable expectation that teams scoring more runs than they allow win most of their games on average and that the bigger the difference the larger fraction of their games they win. This led to the metric developed by Bill James, the so-called *Pythagorean Expectation*, which says that the expected fraction of wins by a team over a season equals (Run scored)^2/[(Run scored)^2 + (Run allowed)^2], named after the Pythagorean theorem, which relates the

---

[206] The number of team victories a player is directly responsible for in a 162-game major league season equals one third of his win shares. A pitcher's win shares typically turn out to be roughly equal to his number of victories in the season. For position players (non-pitchers), All-Star players typically have 20 win shares and most valuable player (MVP)-level players typically have more than 30 win shares.

lengths of the sides of a right triangle. "Predictions" later improved when the exponent 2 was replaced by 1.83.[dvi]

Using such statistics to build a winning baseball team with minimal cost was highlighted in the book *Moneyball: The Art of Winning an Unfair Game*.[dvii] In building their 2017 World Series Champion team, the Houston Astros relied on baseball analytics more than most other teams (and perhaps on other things as well). Still, they learned that they needed to improve the human interaction element to build the culture needed for team success.[207,dviii] Some say such extensive data acquisition and analysis leads to longer games and the time between balls in play (those that are being fielded), making games boring.[dix]

Basketball is also quite amenable to statistical analysis and metrics. Sports analyst John Hollinger has developed the player efficiency rating (PER) that boils down a National Basketball Association (NBA) player's statistics into one number. The league average PER every season is pinned at 15.00. After the 2018-2019 season, the highest career PER was that of Michael Jordan, 27.91.[dx] Though clearly better players usually have higher PERs than clearly poorer players, there is much subjectivity in this and other all-in-one metrics and in ranking players by using them.

In basketball, a "triple double" is a metric that is often-cited as being highly desirable. The most common type indicates a player has scored 10 or more points, collected 10 or more rebounds, and dished out 10 or more assists in a game. But is that performance really better than one with 20 points, 11 rebounds, and 9 assists, which is not a triple double?

Sports metrics continually evolve to improve their effectiveness. For example, the RPI metric, used to rank college basketball teams and predict outcomes in head-to-head competitions since 1981, was modified to overcome its shortcomings in analyzing results and making predictions.[dxi] Moreover, in 2019 it was replaced by another metric to evaluate teams for the NCAA end-of-season tournament, the newer NCAA Evaluation Tool (NET).[dxii]

---

[207] They also found they needed to explain to fielders the reasons for the shifts in the defensive positions recommended by analytics and to acquire seasoned players, sometimes past their most productive years, to provide sage insight to younger players.

Though less amenable to sports analytics, some statistical metrics are also used in football. The National Football League (NFL) passer ratings for quarterbacks converts the passing statistics of completion, touchdown, and interception fractions and yards per passing attempt into another often-cited metric. It is the sum of four components, which are then multiplied by 100 and divided by 6. These are the: (1) % of passes that are completed minus 30, then multiplied by 0.05,[208] (b) passing yards per attempt minus 3, then multiplied by 0.25,[209] (c) % of attempts that are touchdown passes, multiplied by 0.2,[210] and (d) 2.375 minus 0.25 times the % of attempts that are interceptions.[211,dxiii] This analysis might seem complex, but is an ideal example of "just needing to go through the numbers." The highest rating possible is 158.3 (= (2.375+2.375+2.375+ 2.375) × 100/6), meaning the quarterback has reached the top cutoff in each category. Passer ratings of 100 or more clearly denote extremely good performance, but this metric is a bit arbitrary. Do higher values for each component signify better performance and should each component be assessed equally? Should quarterbacks be penalized for receivers dropping passes and tipping them to the defense for interceptions? Should quarterbacks be rewarded for the receivers making great catches from poorly thrown balls or for short passes that become touchdowns due to great running by the receivers?[212,dxiv]

Sports metrics not only affect professional athletics, but the general public of amateur athletes. How you rate as a runner for longer distance events uses as a metric your age-graded performance level, defined as your average speed in that event divided by the world-record speed for your age group and gender. (This equals the world-record time for the event divided by your time.) If your age-graded performance level is

---

[208] but with a lower limit of 0 for percentages <30% and an upper limit of 2.375 for percentages >77.5

[209] but with a lower limit of 0 for <3.0 yards per attempt and an upper limit of 2.375 for >12.5 yards per attempt

[210] but with an upper limit of 2.375 for percentages >11.875

[211] but with a lower limit of 0 for percentages >9.5

[212] Also, should rushing yards by quarterbacks be included and not just passing yardage? They are in the new Total Quarterback Rating (QBR) metric, but does this necessarily make it a better metric of quarterback performance?

over 60% you are at least at a local class level, over 70% at least regional class, over 80% at least national class, over 90% at least world class, and near 100% you are at the world-record level.[dxv]

16.4.4 Correlation is not causation

Taking action on the basis of a statistically established correlation can be foolish. Economist Thomas Sowell noted in *The Vision of the Anointed: Self-Congratulation as a Basis for Social Policy*: "One of the first things taught in introductory statistics textbooks is that correlation is not causation. It is also one of the first things forgotten."[dxvi]

When two statistical trends are correlated or associated with each other, there is no reason to presume that one caused the other or that both are caused by the same thing. Serious studies in the 1960s showed that coffee drinkers had a higher incidence of lung cancer than the rest of the population. This correlation was true but misleading. It is now known that coffee drinking does not cause lung cancer, but smoking does and there was a correlation between those who smoked and drank coffee. In studies trying to link lung cancer to behavior, the drinking of coffee confounded researchers and so was a *confounding factor* in the analysis.[dxvii] Similarly, if disorders or traits A and B are prevalent in a given segment of the population, you could say they are correlated, but you cannot say that A causes B, B causes A, or that A and B are caused by the same other property of these people, C.

Sadly, the media routinely presents serious reports of correlations in terms of cause and effect. Of note, some forays into such correlations are intended to be satirical: There was a $\geq$ 95% correlation (correlation coefficient $\geq$ 0.95) between total revenues generated by arcades and the number of computer science doctorates awarded in the U.S. from 2000 to 2009; U.S. spending on science, space, and technology, and the number of U.S. suicides by hanging, strangulation, and suffocation from 1999 to 2009; the divorce rate in Maine and the per capita consumption of margarine from 2000 to 2009; U.S. crude oil imports from Norway and the number drivers killed in collision with railway trains from 1999 to 2009; and the number of math doctorates awarded and the amount of uranium stored at U.S. nuclear power plants from 1996 to 2008, but these correlations are nonsense and not causal.[dxviii] There is also a very strong correlation between the number of Nobel Prizes awarded to a country (through October 10, 2011) per 10 million persons and its chocolate consumption per capita, for the 23 countries with Nobel laureates. The correlation coefficient was an astounding 0.791, and it increased to 0.862 when Sweden was excluded from the analysis.[dxix]

There are indeed serious and earnest efforts in using correlations, such as in finding medical or biomarkers that can be used for predicting, diagnosing, monitoring, and treating diseases and finding indicators of the outcomes of treatment to determine a "surrogate" endpoint of disease. Such endpoints could come from blood and urine tests or symptoms and would then substitute for actual clinical endpoints of disease as needed.[dxx]

Correlations can have subtle legal implications. Pauline Kim has noted that the U.S. "Age Discrimination in Employment Act" allows employers to take into account "reasonable factors" that may be highly correlated with a protected characteristic, such as how cost may be affected by age, as long as they don't rely on the characteristic explicitly." This type of action is not allowable in the federal anti-discrimination statute that covers race and gender. [dxxi,dxxii]

16.4.5 Using statistics to detect when numbers are less random than expected: Benford's Law

Some series or distributions of numbers you encounter in the statistics of business ventures are expected to follow some rules, and if they do not, it might indicate fraud worthy of government prosecution.

For a truly random set of numbers, say of all three-digit integers, the probability that the number will begin with a given digit is the same for digits 1 to 9, which is $1/9 \approx 11.1\%$. In more restricted sets of random numbers, say from 1 up to a given number, the smaller the first digit the more likely it is. For example, for randomly chosen numbers from 1 to 220, the most common first digit would be expected to be 1. The digits 1 through 9 are each the first number for 11 of the 99 numbers from 1 to 99, 1 is the first digit for each number of the 100 numbers from 100 to 199, and 2 is the first digit for the 21 numbers from 200 to 220. So, the probability of the number starting with a 1 is $(11 + 100)/220 = 111/220 \approx 50.45\%$, with a 2 it is $(11 + 21)/220 = 32/220 \approx 14.55\%$, and with any integer from 3 to 9 it is 11/220 = 5.0%. (See that 50.45% + 14.55% + (7 × 5.0%) = 100%.) For more general sets of maximum numbers, these probabilities are ≈ 30.10%, 17.61%, 12.49%, 9.69%, 7.92%, 6.69%, 5.80%, 5.12%, and 4.58% respectively for first digits 1 through 9, and this is known as *Benford's Law* (or the *First-Digit Phenomenon*). This probability distribution happens to equal the log of the digit plus 1, minus the log of the digit.[213] In 1881, astronomer Simon Newcomb noted that the first digit in the page numbers in a series of books he used was more likely the smaller it was, and then discovered this distribution. In 1958 physicist Frank Benford confirmed it for a series of datasets. It was formally derived by Theodore Hill in 1995.[dxxiii]

Benford's Law is used by auditors to detect anomalies in accounting records, because proper records generally follow this distribution. For these purposes, 0 is not considered a digit, so the first digit would be 1 for 128 and for 0.00128. Enron's 2000 financial data and Bernie Madoff's monthly fund statements deviated from the Benford pattern. When no receipts are needed for a series of expense accounts, say, below $50, an excess of the first 4s could indicate fraudulent claims were being filed, as has been seen. Such anomalies can also be due to innocent and quite plausible reasons.[214,dxxiv] For example, if those filing expense accounts are allowed to combine expenses to lessen paperwork, up to a maximum of say $100, many would quite rationally combine expenses up to this limit, so you would see more 9s as the first digit than otherwise expected.

Our lives are being increasingly affected by the rapid expansion of data collection and analysis by the tools of statistics and other math methods. This trend may make you comfortable or uncomfortable, but, in any case. you must recognize it and purposefully deal with it.

---

[213] For digit d, this is $\log_{10}(d+1) - \log_{10}(d) = \log_{10}((d+1)/d) = \log_{10}(1+1/d)$.

[214] Some other sets of positive numbers do not follow this rule, such as square roots and reciprocals of unbiased sets of numbers, and numbers with digit restrictions such as telephone numbers.

Chapter 17

The Math of Big Data

Data science affects you directly and indirectly, and in ways that may seem to be either full of promise or quite ominous. It entails acquiring data into databases, exploring and identifying patterns in this information, using the databases to make predictions of performance, and assessing how certain the predictions are.[dxxv] The subworld of big data entails many largely "private" and many largely public data streams. It involves the analysis of numerical data, our focus here, as well as other forms of information, such as text. It encompasses and is intertwined with statistical analysis, optimization methods, and risk vs. reward assessment. Decisions are now routinely made on the basis of data to make businesses more profitable, examine life patterns for health care and insurance, establish government policy and resource allocation, and manage traffic conditions in real time. Data analysis has become essential in medical testing and DNA analysis (Section 15.6), and in assessing the distinctiveness of and matching in fingerprint testing[dxxvi] and facial recognition. Moreover, as noted by cognitive psychologist Steven Pinker, data can promote objectivity and insight when evaluating the "big picture": "My own view of the world was radically altered when I looked at data instead of headlines. If history is about all the wars, all the disasters, you're missing all this incremental improvement that can only be ascertained through data."[dxxvii]

There have been miraculous successes of data collection and analysis. In 1854, parts of London were hit particularly hard by one of the recurring onslaughts of cholera. A physician, John Snow, suspected that water contaminated by sewage was the cause. By indicating on a map the living quarters of those who had died from cholera in one region in London, he showed that the area served by the Southwark and Vauxhall (S&V) water company had a much higher death rate from cholera (315 deaths per 10,000 houses) than that in a largely adjacent area, with similar demographics, served by the Lambeth company (37 per 10,000) and in the rest of London (59 per 10,000). He accounted for the possible importance of other factors in the statistical analysis and, then correctly correlated deaths from cholera with the source of a family's water supply. Moreover, he knew that the Lambeth water company drew its water from the Thames River upriver from where sewage was discharged and so the water it supplied was relatively clean, while the S&V drew its water below the sewage discharge and so its water supply was contaminated. He correctly concluded that water contaminated by sewage was the cause of this and other cholera outbreaks. Snow's use of data collection and analysis to solve one of the public health issues of the day (and of any other day as well) was innovative and brilliant and proved to be a giant step forward in public health and human progress. It ushered in the field of epidemiology.[dxxviii,dxxix]

Use of big data wisely failed miserably on April 10, 2017, when security people forcibly ejected 69-year old Dr. David Dao from United Air flight 3411 in Chicago for refusing to leave the seat he had paid for and was occupying. The plane was overbooked and the standard financial inducements United offered to booked passengers did not free up the seats needed to transport four airline employees. Four already-boarded passengers were randomly selected for involuntary bumping; three complied and one, Dr. Dao, did not, and this led to one of the more-unseemly events in airline history. What led to this, aside from United's bad judgment in treating its customers with disrespect? Decades of data collection and simulation algorithms had taught airlines how to optimize operations and profits by running flights that were nearly booked and fewer flights overall.[dxxx] From 2000 to 2015 the "average" plane increased from being ~72% full to 84% full (weighted by travel miles in national and international flights by U.S. airlines).[dxxxi] But, sometimes fluctuations in flight occupancy expectations and the needs for the airlines to transport their own personnel are larger than expected (in statistics, these occasions are said to be many sigma (σ) from the mean—

or rare events), aside from understandable ones due to bad weather. In 2016, United convinced 62,895 of its more than 86 million booked passengers to give up their seats by using financial inducements. They bumped 3,765 without their acquiescence; however, apparently none of them had been forcibly dragged from their seats.[dxxxii] Airlines appear to be bumping fewer passengers nowadays, by using their data more wisely.[dxxxiii]

Data have been publically collected and then analyzed for millennia. The Roman Republic used census counts to keep track of adult men for military service.[dxxxiv] The U.S. has conducted an extensive census every decade since 1790. Phil Sparks, a director of the Census Project has noted "The census is the gold standard of data collection, not just in the United States but in the world …."[dxxxv] As with all data acquisition, census data collection can be suspect. Not everyone is counted in a census, and some groups may be undercounted more than others. Self-reporting leads to errors. Older people used to claim they were older than they actually were to make themselves more respected, but now because of changes in societal values, people report that they are younger than they are.[dxxxvi] Collection of weather conditions across the country and world have rapidly progressed this past century, and this has helped improve weather forecasts. Publically gathered data are sometimes used later for other open purposes. For instance, the 2016 University of Washington analysis of the rates of death from different causes in each U.S. county from 1980 to 2014 used publically available death records from the National Center for Health Statistics.[dxxxvii] Past wind data can be used in mathematical models to show where to place wind turbines to optimize the use of and revenue from wind energy.[dxxxviii]

Moreover, data acquired in public places can be used to monitor, control, and improve public facilities. They also can be used to monitor people and their belongings for collective analysis for the public good and, less desirably, to monitor specific people. In cities, data can stream from sensors and cameras in public places to examine infrastructure, the environment, people, buildings, and elements related to safety and security. Automatic paying of tolls for roads and bridges, such as EZPass in the New York and other areas, tracks your car (and you, despite government protestations to the contrary). As Michael Malone has noted:

"The Big-Data Future Has Arrived."[dxxxix] The interplay of big data, privacy, commercial interests, and the public good continues to evolve.[dxl]

Data have been collected and deposited into massive private and public databases. You have direct or indirect access to some of them, either for free or for a fee. One database you likely frequent online is the *Internet Movie Database* or *IMDB*, which you can access directly[dxli] or through summaries provided by sites such as *Box Office Mojo*[dxlii] and *The Numbers*.[dxliii,dxliv]

The massive collection of data and statistical metrics has become an expanding business in sports to optimize business plans and to amuse the public. I have frequently looked at the free major league baseball database *Baseball Reference*.[dxlv] Some data are from the recording of outcomes, such as hits in baseball. Some come from "tapes" and data processing, as in baseball to obtain detailed defense statistics beyond fielding attempts, errors and so on,[dxlvi] and to analyze the exit speeds of hit balls, their launch angles,[dxlvii] and where they go.

Many supposedly private interactions you have online and in communications produce data about you that can be stored and used forever. Each medical interaction becomes part of your history. Data streaming from body sensors is becoming a fact of life for the population, and not only for people closely monitored in hospitals. Each financial interaction (aside from those using cash) becomes part of your permanent record. Every time you take a survey, you have added to the data stream. This occurs whether or not you want it to be monitored and saved.

Data about you are tracked by private and public organizations through your Social Security number, cellphone numbers, credit card numbers, and so on. Assigning each person in the U.S. a Social Security number started in 1936 to track the Social Security accounts and this number began to be used as the *de facto* general-purpose identification number in the 1960s, as the use of computers for data files burgeoned. It has now morphed into largely personal information not to be used for tracking. You should be concerned about the decisions being made about you on the basis of data streams from social media and your other online interactions (including the permission boxes you routinely check online).

So, where does math enter in this? **Everywhere!** Data analysis and tracking and the resulting decisions are being made on the basis of statistics and of algorithms that are not explicitly mathematical functions, but that still entail mathematical correlations of data. For example, it has been reported that those evaluating loan applications consider applicants with more than 40% of their cellphone personal contract entries listed with first and last names to be 16 times as reliable as those who have very few entries with both names.[dxlviii] (In this case this information was obtained with the loan applicant's permission.)

One thing is certain. The age of big data has arrived and it has done so because the math of data analysis is just so good in doing what it needs to do! In 2009, Google data scientists Alon Halevy, Peter Norvig, and Fernando Pereira noted, as had physicist Eugene Wigner before, that much of physics can be expressed by math equations, which Wigner had dubbed "The Unreasonable Effectiveness of Mathematics in the Natural Science."[dxlix] However, they added, it is difficult to use math formulas to neatly characterize sciences involving humans, such as economics, and it is then better to make use of what they dubbed "the unreasonable effectiveness of data."[dl]

The statistical methods we have examined here have largely been available for some time and are used routinely to analyze data. However, collecting and analyzing the rapidly expanding volume of data requires even more sophisticated and diverse mathematical methods (algorithms) that have become to be increasingly fast and efficient.[dli] Sometimes the need is to target very specific goals, rather than a very thorough analysis. Developing all of these methods is now a major area in computer science and applied mathematics. The anticipated future needs of data analysis were well outlined over a half century ago in the classic 1962 work by John Tukey.[dlii] He stated that new development should be guided by new needs and evaluated by how well the new methods work. The wisdom of his roadmap was prescient, and not as obvious as it may seem now, because it has not always been followed. Some data analysis has focused more on mathematical elegance than practicality.[dliii]

Early on, statistician Leo Breiman noted there are two approaches in using statistical modeling, also called data analytics, to reach conclusions from sets of data.[dliv] The goal in both is to be able to make *predictions*, meaning to take input, independent variables, plug them into the model, and arrive at response variables, and to obtain *information* concerning the nature of the process that converts the input to the output. The approach of the "Data Modeling Culture" is to determine and then optimize a stochastic (or statistical) model (one including the randomness of the processes) that relates known input to the measured output. The model is shown to be valid, and so it is *validated*, by determining the "goodness" of the fit. Most statisticians historically used this computational statistics type of approach (~98% in 2001), using methods such as those discussed in the previous chapter. The approach of the "Algorithmic Modeling Culture" is to relate the input to the output using algorithms (approaches). The dependences may be too complex to be described by functions, but even without an explicit mathematical form, inputs can be related to outputs and this is all that is needed. Models are validated by how well they predict outcomes, as by using decision trees (as we will see in our discussion of fault tree analysis of risk in Chapter 20), neural networks (or neural nets, which are inspired by the network of neurons in our brains), and Bayesian networks (graphical models that utilize the conditional probabilities using Bayes' Theorem, which we introduced in our discussion of probability in Section 15.6).[dlv] Historically, most data analyzers did not typically use this latter approach (as little as only ~2% in 2001), but its use has been expanding exceedingly rapidly, especially for massive data streams.

This latter method without explicit programming is *machine learning*.[dlvi] In machine learning, machines, meaning computers, "learn" by using input data, and as such it is an application of the broader area of *artificial intelligence* (AI), in which machines conduct smart tasks.[dlvii] The four goals of AI are to perceive and categorize items, put them into context for understanding and learning, make predictions using past and current information, and to make plans. Advances are occurring in each, with the most success currently in the first goal.[dlviii] Computer scientist Arthur Samuel, co-winner of the 1987 Computer Pioneer Award did seminal work in machine learning starting in 1949.[dlix]

Different levels of direction can be provided in machine learning. In *supervised learning* the goal is for the computer to optimize an algorithm, i.e. to learn a rule that maps the given examples of inputs to outputs, such as between input and output data sets. Such direction is not provided in *unsupervised learning*, in which the algorithm is directed to learn the structure, correlation, and meaning of the inputs. Finding interesting trends and correlations in data sets is the goal of the exploratory mode of operation called *data mining*. Algorithms are devised to search for trends that are not being targeted specifically, but which can still be "deduced" from the data. *Reinforcement learning* entails developing and using algorithms in a dynamic environment to achieve a goal, perhaps with limited feedback.

Data analysis by statistical modeling and algorithm machine learning methods are examples of optimization problems (as in the next chapter). Either may prove better for a given set of data and one specific method in that given approach may prove best. Different models in the two approaches could have the same small fitting error, though their interpretations of the process relating inputs to outputs can be very different. Their predictive capability can differ, and this concern is central in optimizing the model, as is analyzing how "robust" the model is. Robustness characterizes how a model is relatively insensitive to small, seemingly insignificant changes to the data set; assessing it is known as *sensitivity analysis*. The concepts of interpolation and extrapolation, discussed in Section 8.2.4, are simple examples of the predictive capability of a model. In machine learning, typically 2/3 of the set of input and output data is used to develop the model for learning or training and 1/3 is used to test to see how good the model is.[dlx,dlxi] Such use of the database to devise a model and then test it is standard in much of mathematical modeling.

These methods are being used increasingly to allocate resources and, of direct interest to you, to dynamically price gasoline, airplane tickets, and other fast-moving of products.[dlxii] Real-time and historical data are rapidly processed to assess how customers and competitors will respond to price changes, to best optimize revenues. First, a database is built using historical data about the market and competitors to help "train" algorithms. New data are collected and compared to this database, and then used in the algorithms along with the desired business goals and constraints to implement new prices. Then these new prices and sales are fed into the database to update prices again later. In conjunction with this, the structure of the algorithm is being optimized constantly.

Machine learning is being applied increasingly in Wall Street-type trading. In hedge funds, the "traditional" quant develops hypotheses about patterns in data and, if they are confirmed, writes an algorithm to test them against historical data. Then the quant uses the algorithm in live trading to see if these tests prove to be successful. A "machine learning" quant writes machine learning code and trains it on historical data, then sees how if it performs in tests with live data, and if it is successful puts it into live trading (possibly after modification).[dlxiii]

Deep neural nets (DNN), multiple layered networks, are being trained by using radiologist diagnoses from sample images of previous patients, to diagnose diseases by using medical images of new patients. This is being done to relieve radiologists from the work burden associated with making more "mundane" diagnoses and also to have such diagnoses be made more reliably.[dlxiv] They are also being applied to diagnose other conditions,[dlxv] such as childhood diseases,[dlxvi] and to "objectively" assess whether people are candidates for end-of-life palliative care. This is done by determining whether the patients likely have less than one year to live.[dlxvii] Allotting resources this way may become even more controversial. Other uses of machine learning have upsides, as in voice recognition and image analysis (for self-driving cars). However, they have potential controversial downsides, such as inaccurate characterization (due to insufficient or improper AI training for image recognition[dlxviii]) and the manipulation and faking of images and statements by people. Data-based algorithms can also be used to either mitigate or promote discrimination in job interviewing, hiring and promotion decisions.[dlxix]

Biased data will train an unbiased algorithm to produce a biased model that will yield biased predictions.[dlxx,dlxxi] Training DNN algorithms with biased FICO (for Fair Isaac Corporation) credit scores may result in biased loan recommendations.[dlxxii] Some data sets may be biased because the information was obtained in relatively easy rather than accurate ways. This is known as the *Streetlight Effect* (because you tend to look for things outside at night more near streetlights). There is a recurring concern about the completeness of the U.S. Census, particularly when the method of data collection changes. This is a concern now with the onset of online census counting.[dlxxiii] Questionable data are thought to have caused overestimating the risk of the spreading of the Ebola virus in West Africa in 2013-2016 and underestimating the ultimately successful local initiatives used to control it. They may have contributed to inaccurately assessing the impact of Hurricane Sandy in 2012. Decisions may have been made by relying on the many more tweets concerning events that emanated from Manhattan rather than on those from the much more devastated New Jersey shore.[dlxxiv]

Even with valid data collection, the quality, purpose, and impact of modeling and analysis with these data may be up to debate. In *Weapons of Math Destruction*, Cathy O'Neil argued that the black-box, largely hidden models of big data control the evaluation of our loan applications, health, workplace performance (such as the quality of teachers and policing), parole applications by prisoners (as in Chapter 20), and so on. She noted that such largely arbitrary models have been designed with little social responsibility in mind.[dlxxv] Her book's subtitle summarizes her foreboding thesis: "*How Big Data Increases Inequality and Threatens Democracy*." She did not question the math itself, just how it is often used to control our lives, often with tools that are unproven or riddled with faulty reasoning. Bad algorithms become self-fulfilling prophesies that control performance in unintended and negative ways, with feedback loops that spiral with little control. Clearly, in our rapidly evolving world, the responsible use of data is intertwined with developing and applying the math of data analysis, as is highlighted by the acronym FATES, which summarizes the goals of Fairness + Accountability + Transparency + Ethics + Safety (doing no harm) and security (guarding against malicious behavior).[dlxxvi]

When you are told to believe what someone is selling or telling you on the basis of massive data collection and the use of the latest algorithm, simulation or model, you don't have to believe them. It is best to question the math assumptions in each step because they often rule your life.

We have seen that this second world of math can be used to assess probabilities and help in collecting and analyzing data. You can use it as input to prognosticate and make the decisions that control your lives. We now need to learn about the math of the decision-making process itself and see if you are comfortable with it and its outcomes.

Part IV

The Third World of Math: The Math of Making Decisions and Winning

Sequences, probability, statistics and data analysis, and machine learning as discussed in earlier chapters are components of the second world of math, the "math of doing." This can provide the input for you, companies, and governments to make decisions. This includes exploring the probabilities in gambling, compounding interest and annuities, medical testing, infectious disease control, computing, encryption, data analysis, dynamic pricing, sports statistics, social science, and so on. Armed with this, you can use the third world of math or the "math of making decisions and winning" to make decisions to control your life. It is widely used in the worlds of economics, finance, and areas involving interpersonal relationships. Much of it has been developed more recently than the math we have already encountered. It can be challenging to cleanly separate these second and third worlds because they sometimes interlock.

Chapter 18

The Math of Optimization, Ranking, Voting, and Allocation

Making decisions entails optimization. A classic example of optimization is gerrymandering, which has long been used to devise election districts more favorable to the party in power by spatially "binning" the voters to optimize their chances of staying in power.[dlxxvii] In one recent case, the shape of a U.S. congressional election district was characterized as "Goofy Kicking Donald Duck"[dlxxviii] because of how it looked; it deviated greatly from the traditional criteria that include compactness and continuity. Imagine a state of 2,000,000 voters with 60 state representatives from Party A and 40 from Party B. The election districts have been carefully constructed by Party A so the districts they control each has 12,000 Party A and 8,000 Party B voters, while each one of the 40 districts represented by Part B representatives has 16,000 Party B and 4,000 Party A voters. Party A has 60% of the representatives but only (60 × 12,000 + 40 × 4,000)/2,000,000 = 44% of the overall voters. Party B has only 40% of the representatives but (60 × 8,000 + 40 × 16,000)/2,000,000 = 56% of the voters. Data science and other mathematical tools have contributed to more effective and robust ways to gerrymander election districts and to make them less subject to fluctuations in voting outcomes. But, they have also improved ways of detecting such "outlier" outcomes of voting district simulations that indicate that gerrymandering has indeed occurred.[dlxxix]

The subworld of the math of optimization explored in this chapter determines what is the cheapest, most reliable, fastest, or fairest method to achieve a goal. In the context of efficient, optimized programming, mathematician Donald E. Knuth once said "Premature optimization is the root of all evil," meaning that you should spend most time optimizing the factors that have the greatest impact in improving the computation speed.[dlxxx] More generally, optimization should focus on the factors most key in improving the outcome.

Optimization produces outcomes that are supposedly certain for a set of initial conditions. But of course, when you are told something has been optimized you should challenge the statement: For whom is it being optimized? and Has the optimization procedure been truly validated and tested? Dynamic pricing, noted in the previous chapter, does indeed optimize the market: by optimizing profits for the vendor. We take for granted that the electric power grid is run reliably. This entails carefully optimized performance based on known seasonal and daily cyclic changes in renewable energy sources,[dlxxxi] with feedback from stochastic variations of usage and renewable power sources. Such optimization benefits both the power companies and the public.

A range of factors enter into optimizing decisions. Statistics are used in determining on-time performance and queuing (waiting on lines), as we will see later in this chapter. Data science tools are also used in making decisions. The Dutch have developed a successful algorithm that instructs coaches how to deploy their speed skaters in Olympic events to maximize their chances of winning gold medals, even given uncertainty.[dlxxxii] Perception is important in assessing the success of optimization, as is the choice of statistical metrics (Section 16.4.2).

In the next chapter, we will see that the subworld of the math of game theory poses ways of choosing the best strategies and decisions for actions on the basis of what opponents may do to optimize their own choices. In the final chapter, we will assess the subworld of math of optimizing reward relative to risk on the basis of the uncertainties of risk and in the face of known rewards.

18.1 Finding the Biggest and the Smallest

Optimization often means you want to maximize or minimize one or more items, which can sometimes be represented by functions.

When you climb up and then down a mountain, with every step your elevation is higher until you reach the peak and then it becomes smaller with every step after the peak. With every step very near the peak the increase in your elevation is very small and positive before the peak, zero at the peak or very small, and negative after the peak. If your optimization goal were to maximize your elevation, you would need to learn how to find where this maximum is.

One way to do this is with a function that gives the mountain elevation vs. lateral distance or step number. For example, you might model this height as 100 minus the square of the lateral distance called x (and so it is $100 - x^2$, with the height and x in feet). As x increases from -3, -2, -1, 0, 1, 2 to 3, the height changes from 91, 96, 99, 100, 99, 96, to 91. It makes sense to examine the differences in such a function as the variable decreases by a small fixed amount, here by 1, from 3 to 2, 2 to 1, 1 to 0, 0 to -1, -1 to -2, and from -2 to -3. These differences change from $(91-96=)$ -5, $(96-99=)$ -3, -1, 1, 3, to 5. This difference becomes zero somewhere between the range from 1 to 0 and that from 0 to -1, and it seems to occur at $x = 0$. Decreasing the step interval to 0.1, as x increases from -0.3, -0.2, -0.1, 0, 0.1, 0.2, 0.3, the function changes from 99.91, 99.96, 99.99, 100.00, 99.99, 99.96, to 99.91 and the respective differences (from 0.3 to 0.2 and so on) to -0.05, -0.03, -0.01, 0.01, 0.03, to 0.05. Repeating this with successively smaller intervals, confirms that the value of the function is a maximum at the variable value where the differences are 0, which is $x = 0$ here (Figure 18.1a). We have just maximized this function by using differential calculus, (without the calculus notation) by "setting the first derivative," also called the slope, to 0.[215]

You can perform the same procedures with the function 100 plus the square of the lateral distance $(100 + x^2)$ to find the location and height of

---

[215] The relevant rules of differential calculus are: The derivative of a constant is 0, of ax is a, of bx^2 is $2bx$, of cx^3 is $3x^2$, and so on. If a function is called $f(x)$, its derivative would be called df/dx (and pronounced "d" "f" "d" "x", as in Section 4.7). The derivative of a sum of terms is the sum of the derivatives of each term. In the parlance of calculus, we took the derivative of the function $f(x) = 100 - x^2$ and obtained $-2x$. We set it equal to 0 to find what we will see is an extremum, and in fact a maximum here, and found the variable that maximizes the function occurs with $x = 0$, and so the value of the function is 100 minus the square of 0, or 100. The second derivative of the function, d^2f/dx^2 (pronounced "d" "squared" "f" "d" "x" "squared") is the derivative of df/dx. It is -2 here, indicating this extremum is a maximum, as we will see.

the bottom of a bowl. Now, the function changes from 109, 104, 101, 100, 101, 104, to 109 for x changing from -3 to 3 as the previous case using integral values of x, and the differences change from 5, 3, 1, -1, -3, to -5, and so on. But now where the differences are zero, again when x is 0, there is not a maximum but a minimum (Figure 18.1b).

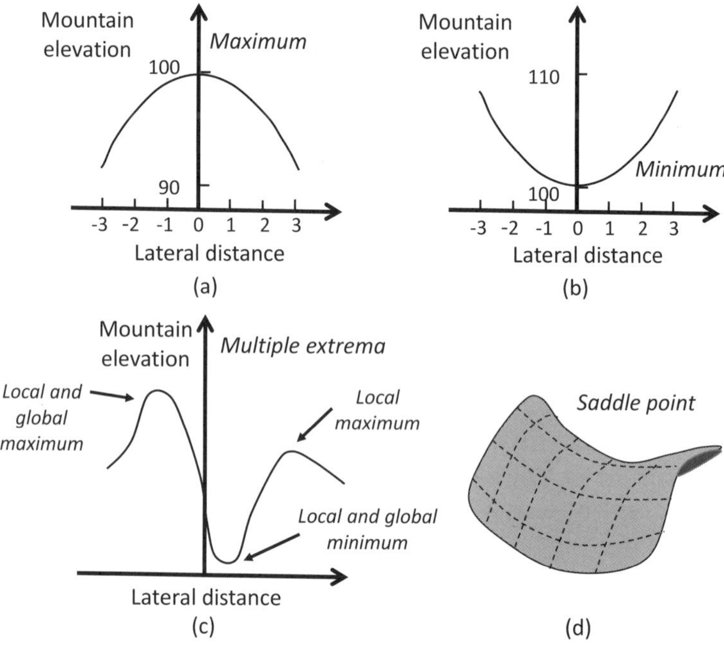

Figure 18.1: Maxima, minima, and saddle points.

This procedure actually finds an *extremum*, which means it could be either a maximum or a minimum. You need to examine the differences of the differences (which in calculus is called the second derivative) to see which it is. For our first case, with $100 - x^2$, the differences of -5, -3, -1, 1, 3, to 5 are (first -5 – (-3) and then so on) -2, -2, -2, and -2.[216] The value associated with x at the extremum (here the middle two -2s)

[216] They all have the same value because the function is quadratic.

is negative, so the function is at a maximum.[217] For the second case, with $100 + x^2$, the differences of 5, 3, 1, -1, -3, to -5 are 2, 2, 2, and 2, which are positive indicating a minimum.

For some functions (as those describing many mountain ranges) there is more than one extremum, so each maximum or minimum may be that only *locally*. Of all of these local maxima and maxima, there will be one *global* maximum and one *global* minimum (Figure 18.1c).

Some functions depend on more than one parameter, so optimization needs to account for variations in each parameter. For example, there are really two lateral distances (the perpendicular x and y directions) that need to be considered in analyzing a real mountain or bowl. Imagine cutting a spherical shell in half so it looks like a bowl. It has a point that is a minimum, when you change either x or y. When placed upside-down, this bowl has a point that is a maximum for either change, as would a mountaintop. In contrast, on a saddle, there is a minimum along one direction, say x, but a maximum along the other, say y, which is called a *saddle point* (Figure 18.1d).

The maxima and minima of appropriate model functions can be determined to optimize the situation. Sometimes this is subject to constraints, and there are mathematical methods that can do this.[218]

18.2 What is the Cheapest (and Most Reliable) Outcome?

Optimization math is widely used in industry, technology, and economics to find the least expensive option for a specified needed level of reliability, in part by modeling processes and costs as functions and then applying differential calculus.[dlxxxiii] The optimized problem may be causal (based on specific rules, such as engineering design principles, resulting in deterministic answers) or probabilistic and stochastic. Examples of the former include optimizing aircraft trajectories and auto design, controlling chemical reactions, and making mechanical devices robust. Optimizing return on investment involves a mathematical assessment of risk, and so it entails stochastic optimization. Optimizing

---

[217] In calculus parlance, we have found the second derivative, also called the curvature, and showed it is negative here.

[218] This is called the method of Lagrange multipliers.

factory output entails analysis in the face of deterministic tradeoffs and uncertainties.

"Cost" functions are used in economics to relate variables and then examine the cost of items to help assess tradeoffs.[dlxxxiv] For instance, the Cobb-Douglas production function gives the production in terms of controllable variables such as the value of labor input and capital input.[dlxxxv] In the simplest version, this cost function is a product of these two factors, with each respectively raised to different powers,[219] which show how sensitive output is to these factors (the *elasticities*). Optimization of production is determined by using this cost function and constraints. To maximize crop yield, a cost function can be developed to characterize crop yield as a function of fertilizer use and watering.

The *Law of Diminishing Returns* in economics, finance, and much of life means that at some point additional resources no longer result in significant enough additional benefits to warrant them (Figure 4.3). This defines an optimization problem. If the value of a product initially increases by \$4 for each minute you work on it, the product value will be \$40 with 10-minutes work. In differential calculus, this rate of increase is called the derivative of the value with respect to your time, while in integral calculus this sum or "integrated" added value is determined using these rates. This rate of increase usually begins to decrease after a while and eventually may become so low that the total value increases too little for the additional time you spend on it. If you value your free time as \$1 per minute, the point of diminishing return occurs when the rate of product value increase decreases to this level. If the value increased at a rate per minute that stayed at \$4 during the first 10 minutes and suddenly decreased to \$3 during the 11th minute, to \$2 during the 12th minute, and then to \$1 for the 13th minute, you would likely stop after 12 minutes, with the total product value winding up to be \$40 + \$3 + \$2 = \$45. We have essentially characterized the cost function for this problem.

---

[219] For production, Y, in terms of controllable variables, such as the value of labor input, L, and capital input, K, this is $Y = AL^{\beta}K^{\alpha}$, where A is a constant. If the constants $\alpha + \beta = 1$, production scales with input, so doubling capital and labor also doubles output. For $\alpha + \beta < 1$ and > 1, doubling input changes production by less than or more than double.

When should you stop proofreading a document? This is an example of diminishing returns that is not often explicitly analyzed with numbers. You want to limit errors in documents you have written and to otherwise improve them (or works of art and music and so on), but when have you reached the threshold of merely wasting your time? I admit to spending too much time on proofreading. (Perhaps, you think I should have proofread this book a bit more.) More wisely, Eugenia Cheng has noted, "I proofread closely but accept the fact that something as long as a book is destined to have typos in it." and added "I am not trying to optimize the absolute outcome of my effort but rather the ratio of outcome to effort."[dlxxxvi]

18.3 What is the Fastest (and Cheapest) Outcome?

"Time is money" expresses the principle behind a range of the optimization problems, including setting trip itineraries, waiting on lines, boarding airlines, reducing traffic, and interviewing potential employees.

18.3.1 Optimizing the Traveling Salesman Problem

The *Traveling Salesman (or Salesperson) Problem* (TSP) is a classic optimization challenge in minimizing distance, and therefore cost. You are given a list of cities and the distance between each pair of them. Starting at one city, you visit each other city once and return back to this first city. The challenge of the TSP is to find the shortest possible distance (or tour length) that could be traveled by the salesman.[dlxxxvii] The concept of "cities" is easily extended to that of customers, other points, and so on, and the concept of "distance" to that of time, cost, and so on. Since first posed in 1930, the TSP has been studied and applied many times to a broad range of other applications as well, including planning school bus routes,[dlxxxviii] enabling campaigning politicians to visit as many communities as possible in a state,[dlxxxix] designing connections between components in microelectronic chips, and efficiently looking one-at-a-time at several sources with telescopes in astronomy.

The task of finding the shortest route needed to visit Washington D.C. and one major city in each of the 48 continental U.S. states began circulating in the math community in the 1930s. This challenging 49-city problem was solved exactly by mathematicians at Rand in 1954 by hand by using the then most current math tools, who found it to have a tour length of 12,345 miles.[dxc] In 1962 Proctor & Gamble widely-promoted the lesser, though still nontrivial challenge of minimizing the travel distance between 33 specific cities in the U.S.[220]

The TSP is thought to be a computationally-intensive type of problem (Sections 13.8 and 13.9), with the time needed to solve it by brute force without shortcuts appearing to increase "exponentially" with the number of cities, n, roughly as n^n.[221] Even with some computational "short-cuts,"[222] the TSP has only relatively recently been solved for up to many thousands of cities.[223,dxci] Often the exact solution is not needed, because answers that are within a few percent of the optimal tour length are often sufficient. The *Christofides algorithm*, developed in 1976, is still one of the best in providing a useful "worst-case" analysis answer, and one that provides an answer within 1.5× of the actual answer.[dxcii]

The geometry of the locations of the TSP stops is very important. One simple case is for bus stops equidistant around the circumference of a circle. If the distance between neighboring such stops is 20 miles and there are 10 stops, there are 10 travel distances, from stop 1 to 2, 2 to 3, 3 to 4, … 9 to 10, and then from 10 to 1, for an overall trip distance of $10 \times 20 = 200$ miles (Figure 18.2a). If instead, these 10 equidistant stops were along a straight line, there would be 9 travel distances from stop 1 to 2, 2 to 3, 3 to 4, … 9 to 10, for $9 \times 20 = 180$ miles. The return trip from stop 10 to 1 is the same distance, so the entire trip would take 360 miles

---

[220] For any chosen starting city, there are 32 choices for the first destination, then 31 for the next one, and so one, of $32 \times 31 \times 30 \times \ldots 3 \times 2 \times 1 = 32!$ permutations. Because reversing the order of the cities leads to the same route, there are $32!/2$ or $\sim 1.3 \times 10^{35}$ distinct possible routes. If 1,500 trillion arithmetic operations could be performed in a second (the record in 2009) and if each route takes only one operation (a generous assumption), the "brute-force method" of checking each route would take $(1.3 \times 10^{35})/(1.5 \times 10^{15}) \approx 8.7 \times 10^{19}$ seconds or ~27 trillion years (given there are 31,536,000 seconds in a year).

[221] For $n + 1$ cities there are $n!/2$ routes. Using Stirling's formula, $n!/2 \sim (2\pi n)^{1/2}(n/e)^n/2$.

[222] Such as "dynamical programming methods," which run in times scaling as $n^2 2^n$.

[223] Deciding whether there is a TSP solution that is shorter than a given distance, the *decision* problem, is a nondeterministic-polynomial or "*NP*" class problem (from Section 13.9), but verifying that a given route is the shortest one, the *optimization* problem, may not be.

(Figure 18.2b). In both cases, the calculated distance is minimized for that geometry and this number of stops.

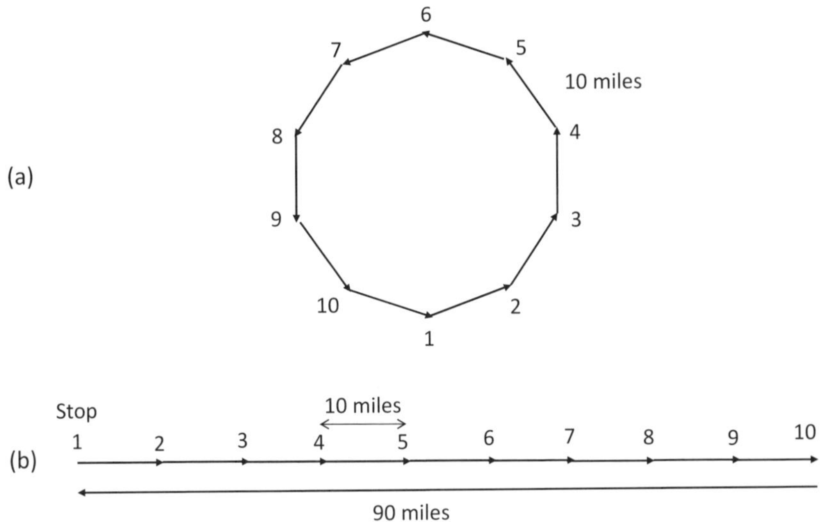

Figure 18.2: Traveling salesman problem, for simple routes.

Choosing the next nearest stop on the route does not always lead to the most efficient route.[dxciii] For example, the third leg of a bus route in Figure 18.3b is longer than that in Figure 18.3a (12 vs. 9 miles), though it is shorter overall (54 vs. 65 miles). This also (correctly) suggests that more generally, routes with crossing or zig-zagging paths are longer (Figure 18.4).[dxciv]

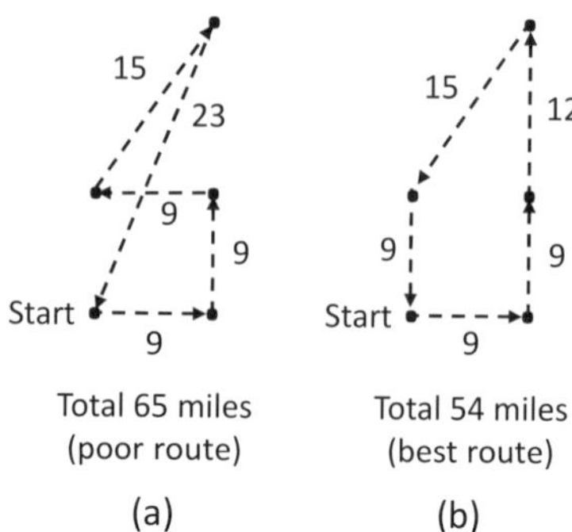

Figure 18.3: Traveling salesman problem, more realistic routes.[dxcv]

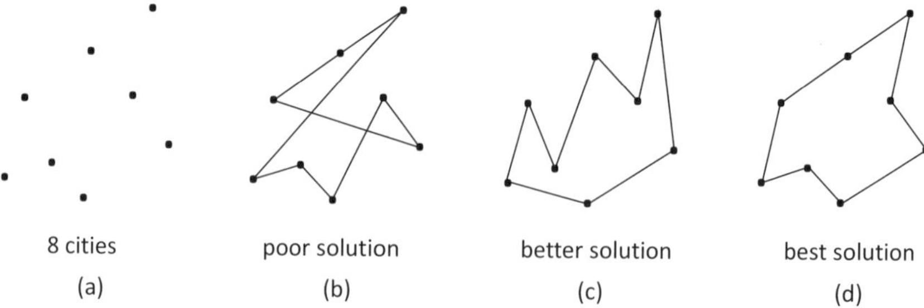

Figure 18.4: Traveling salesman problem, general routes.

18.3.2 Optimization in waiting on line

While I am impatiently waiting on a line or queue and wondering when to leave it, I am not the least bit comforted knowing there are well-developed math optimization procedures that minimize waiting times. Queueing math controls how you wait to pay for purchases and vote, wait for an elevator, and wait when you are put on hold on the phone and told there are five customers ahead of you and that your business is very important to the company. Data scientist Richard Larson claims that some people spend 1-2 years of their lives waiting, mostly in congested traffic.[dxcvi] Some have likened the psychological effects of waiting to (at least some milder forms of) torture.[dxcvii] Institutions have tried to lessen complaints about waiting by diverting us in clever ways, such as by placing mirrors next to elevators (so the wait seems shorter) and directing luggage at airports to more distant carousels (so the actual the waiting time is less, though the total time from deboarding to luggage pickup does not decrease). Society and companies try to optimize the number of vehicles that can be hailed (taxis, Uber, Lyft and so on) to minimize customer waiting time, maximize driver income, and, for society, minimize road congestion.[dxcviii] The goal of math queuing theory is optimization that is fast, fair, efficient, and inexpensive.[dxcix,dc]

Queuing theory was first developed in the early days of telephones to optimize operations because the capacity for having many simultaneous calls on a network was extremely limited. This is hard to imagine in our age of phone service that is always available due to large network capacities and rapid switching. The queuing math developed then is now still routinely used to speed up objects moving in production lines where tasks are performed at different rates; to avoid congestion, including slowdowns and stoppages of cars moving along a highway, particularly near entrances for road traffic, and planes waiting to land and takeoff, when volume is above a threshold and there are "unexpected" variations; to minimize the time spent by patients wishing to enter a hospital that has few available beds; and to control the time you spend waiting on line to vote, get a response from a call center, board a plane, or to pay for goods at a market, clothes store, or coffee shop.

In discussing arithmetic progressions in Section 13.1.1, we showed that the line of patrons waiting at a coffee shop grows at a rate of 3 customers per minute, aside from fluctuations, if it can serve an average of 2 customers each minute, the *service rate*, and 5 customers enter the line each minute, the *arrival rate*. This occurs until frustrated customers leave the line, which is called *reneging* in queuing theory parlance, or see a long line and decide not to join it, which is called *balking*. If instead, the shop can serve an average of 5 customers each minute and only 2 customers enter the line each minute, any existing queue shortens at a rate by 3 customers per minute until there are only 0 or 1 customers usually waiting at any time (with added random fluctuations). On average, there are 2/5 or 0.4 customers in line. Because there can be only an integral number of customers, this means that no customers are waiting 60% of the time on average and one customer is waiting 40% of the time.

The key is that the ratio of average arrival and average service rates in the first case is 5/2 and so is greater than 1 and in the second it is 2/5 and so is less than 1. The overall service rate is the service rate per service person × the number of service people. The average overall service rate is approximately the average service rate per service person × the average number of service people.[224] Lines are shorter and move faster when there are more service people, as at cash registers, but this option is more expensive, and this is the inherent tradeoff between waiting and cost.

In recent years, some clothes stores and supermarkets have changed from operating with people waiting on lines at each counter (*multiple servers, multiple lines*) to waiting on one master line feeding the next available counter (*multiple servers, single line*). When this single queue is very long, it snakes around, and so this is called a *serpentine system.* Models show that the single line approach improves the overall arrival and service rates, and therefore the average wait time, only when people are not allowed to jump between lines in the multiple lines method.[dci] However, seeing one single long line may deter potential customers. In any case, the single line system decreases customer dissatisfaction because it seems fairer. If you arrive on the line first, you will be served first, and not see others flock to a newly-opened counter.[dcii] There are exceptions to queuing protocol. In emergency rooms, most people who are in only mild distress will not be upset if a critically ill or injured person is treated before they are.

Voter balking and reneging is serious because it reduces voter turnout; such lost votes are termed *deterred* votes. Of course, lines move faster and are shorter when there are more voting machines, but machine acquisition, set-up, and operation are expensive, so there is the same tradeoff between waiting and cost. When the total number of voting machines available in a voting administrative unit, which is often a

---

[224] The average of a product of two terms is usually not exactly equal to the product of the averages of the two terms.

county in the U.S., is fixed, administrators need to allocate machines among the several voting precincts in the community optimally by their gut feeling or by validated optimization algorithms. The obvious allocation method of equalizing the ratio number of "active" or registered voters to voting machines in each county voting precinct apparently does not minimize waiting times or costs.[dciii]

Arrival and service rates in stores change with season and also during the day, as they do for voting on Election Day, and queuing models must reflect this. Many people go to coffee shops before work, people servicing counters take breaks, voting machines may suddenly break down, and so on. Such slowly-varying changes can be predicted from experience and so are deterministic variations in the models. Others are more random, and so are stochastic, and so optimizing rates over short time bins is not sufficient and using only average rates in models is faulty. The standard approaches of probability theory of Chapter 15 are used to describe the arrival and service rates and determine the average line lengths as part of an analytic model of queues.

In one popular model of voting, the arrival rate is assumed to follow Poisson statistics (Section 15.1.2) and these random arrivals are called "Poisson arrivals." This model further assumes that voters take different amounts of time to vote, which is described by the exponential probability distribution function (such as Figure 7.3b, but with probability plotted vs. waiting time). For such "exponential service," the number of voters who take longer and longer times to vote decreases exponentially. This model gives the analytic answers for the probability that a voter or other "customer" will need to wait in line, the average waiting time, and the probability that the wait will be shorter than a certain targeted time.[225] These are "steady state" results that do not depend on the number of people waiting in line initially when the polls

[225] In this Erlang C or *M/M/c* model, the probability that a wait is needed is $P_w = ((c\rho)^c/c!)/(((c\rho)^c/c!) + (1-\rho)\ (\Sigma\ (c\rho)^n/n!))$, where Σ indicates here a sum from n equal to 0 to $c-1$. The variable $\rho = \lambda/(\mu c)$, with λ being the arrival rate, μ the service rate per machine, and c the number of voting machines in the precinct or agents at a call center or checkout counter; ρ needs to be < 1. The average waiting time is $T_w = P_w T_s/(c(1-\rho))$ and the probability that the wait is less than a target duration t is $P_t = 1 - P_w\, e^{-c(1-\rho)(t/T_s)}$.

opened, which is the initial condition. It assumes that prospective voters always enter the queue and frustrated waiters never leave the queue.

This Poisson statistics model is known as the *Erlang C model* (or the *M/M/c model*), after the Danish mathematician Agner Krarup Erlang who derived this basic equation of queuing theory in the early 20th century to determine the number of telephone lines (or "trunks") needed to provide a given number of phone users with effective, cost-efficient service.[dciv] The important parameter in this model for any queue is the ratio of the overall arrival rate, here given by the product of the number of calls from all phone subscribers and the average duration of each call, to the overall service rate, given by the number of phone lines or phone operators (whichever causes the limitation). The model gives the probability that a call is delayed because there are not enough lines or operators, the average delay of calls, and the average fraction of calls that are answered in a given time.

For example, if there are 12 calls per minute (the arrival rate) that last an average of 4 minutes, the traffic intensity on average is (12 calls per minute) × (4 minutes) = 48 calls. If there are 55 agents or booths, the service rate is (55 agents)/(4 minutes per call) = 13.75 calls per minute. The agents will be occupied 48/55 ≈ 87.3% of the time on average, the agent occupancy fraction, which must be less than 1. This also equals the ratio of arrival and service rates 12/13.75 ≈ 87.3%. From the average rates it might seem that you will need to wait for only ≈ 100% – 87.3% = 12.7% of the calls; however, the Erlang C formula shows that you will need to wait on ≈23.9% of the calls, because there is a distribution of rates and not always the average rate. The average waiting time is ≈8.2 seconds, but only ≈84.6% of the calls are answered within 15 seconds.[dcv]

In service engineering, this model gives rise to the *square-root staffing rule*. For a steady-state traffic rate (the arrival rate/the service rate per machine, 48 in the above example), the staffing rate should be approximately this rate plus a factor times the square root of this rate. (The factor is determined by the desired service level and is on the order of 1.) This means that if you expect a temporary increase in business traffic and want to maintain the same level of service, the total number of agents or machines you will need will increase proportionately slower than as this new rate, and in some instances it will increase nearer to its square root.[dcvi]

More realistic analytic models provide even better predictions of waiting times. They can also be used to optimize allocation among voting precincts. Waiting on line can be controlled during peak voting times by designing capacity assuming 1.2× the average voting arrival rate, which is a "safety factor."[dcvii]

Though such analytic methods make "cleaner" predictions, results for unusual customer and voting patterns can be tested by using Monte Carlo simulations on computers (Section 15.7.4). Of course, such planning fails when machines break during election day. It also fails if the voters in a precinct are not divided equally among the voting booths. For example, dividing up voters for two booths, one with voters with last names starting with the first half of the alphabet, the 13 letters A-M, and the other with the second half, the 13 letters N-Z, fails because more than half of voters usually fall in the former category. (An average of ~64% are in the first part in the U.S., though in some communities it is much more.) [dcviii]

This tradeoff between customers remaining customers because they are not waiting long times and operators wanting to minimize cost, thereby increasing waiting times, is sometimes termed the *Goldilocks Principle*. This is derived from *Goldilocks and the Three Bears*. The extremes are not satisfactory, such as a bed being too soft or too hard or soup being too hot or too cold, but a "sweet spot" in the middle is fine, as with a bed or soup being just right; these are points of least dissatisfaction [dcix] This sounds simplistic and humorous, but it is truly an optimization principle used to evaluate tradeoffs.

18.3.3 Optimization in waiting in traffic

Sometimes you are stuck on lines or in traffic and cannot remedy the situation.

Are airplane boarding methods optimized to minimize boarding time? Different airlines board different parts of the plane at different times and their boarding protocols change every so often, though sometimes this appears to be done to optimize revenues from those willing to pay for priority boarding and not to optimize boarding speed.

In any case, trying to board all passengers in a short time works worst when the plane is essentially full. The most limiting event in boarding appears to be the time it takes for passengers to load their luggage in the overhead bins. Physicist Jason Steffens ran simulations running different patterns and concluded that some boarding patterns are clearly much faster than others, though there is no unique fastest method.[dcx]

To board a plane with 20 rows with 3 seats on each side of the aisle, you need to explore the $120! \sim 7 \times 10^{198}$ different ways to board a full flight. As we have seen, checking all possibilities would take too long even with the fastest computers, so it is best to start with well-known ordering patterns (back sections first, then sections nearer the front; window seats, then middle seats, then aisle seats; ...) and randomly-chosen ordering patterns (chosen using pseudorandom numbers, as in Section 8.3) to determine the boarding durations for specific assumptions. This includes the time needed to walk to a seat, the minimum separation of standing passengers, the time needed to store luggage, how to handle the stochasticity, and the assumed bordering order. To optimize the time for a given simulation, a random pair of the ordering of passengers is switched and this becomes the new order if the boarding time decreases, and then this processes continues. For reasonable assumptions, in the optimized approaches the first passenger boards in one of the window seats in the last row, the next passenger boards two rows in front of the first, the third boards two rows in front of the second, and so on, until the front of the plane is reached, or with variations on this.[226]

---

[226] The longest boarding time occurs with sequential boarding of individual passengers from the front to back (seats 1, 2, 3, 4, 5, ... 119, 120; where 1-6 are from left to right for row 1, 7-12 for row 2, ... 115-120 for row 20) and small variations of this order. This is closely followed by sequential boarding of individual passengers from the back to front (120, 119, 118, ... 3, 2, 1), which takes 98% of the longest time. Both clog the plane. Then come: ordered blocks from back to front (74%); random boarding (45%); all window, then all middle, then all aisle seats (43%); what has been called the modified optimal approach (42%); and the best ones that are aptly called the optimal boarding approaches (21%). As noted, in the optimized approaches, the first passenger boards in one of the window seats in the last row, the next passenger boards two rows in front of the first, the third boards two rows in front of the second, and so on, until the front of the

Overcrowding is thought to be the main cause of delays in the New York City subway system, accounting for more than one-third of delays, because subway ridership has increased from 4 million riders a day in the 1990s to almost 6 million in 2017.[dcxi] Few subways lines have 70% or greater on-time rates (and so arrive at the final stop within 5 minutes of the scheduled time). The time a train spends unloading and loading passengers at a station, the *dwell time* is supposed to be ~30 seconds, but increases rapidly above a threshold volume, as it takes longer times for passengers to exit and enter subway cars. The cure is to run more cars, an expensive proposition, and mitigate destructive operating protocols.[dcxii] Moreover, passenger congestion in train travel is less when high volumes of passengers exit and enter trains in "through-running" stations, rather than at terminals. At terminals time is lost because trains need to change directions. Penn Station in Manhattan is an exceedingly busy terminal and a major bottleneck New York transportation.[dcxiii]

Driving in traffic is an optimization theory problem that usually fails in practice because those trying to optimize their own positions hurt the overall flow of traffic. Free flowing traffic on highways has a capacity of approximately 2,000 vehicles per lane per hour and, due to the slower speed, it peaks near 2,200 vehicles per lane per hour near 45 mph. Driving fast, changing lanes when not necessary, and tailgating all contribute to slowdowns near peak capacity and ultimately to clogged roads. These problems that might not occur with more enlightened driving, including maintaining good distances between cars.[dcxiv] These bad driving habits also cause accidents, which lead to the ultimate form of traffic. Surprisingly, excessively courteous driving ('Minnesota nice'), such as letting others merge into your lane whenever they want, can cause delays in congested traffic. Forcing zipper merging (one, then the other, …; by signs directing it or local custom) can lessen delays. This

---

plane is reached; this order is 120, 108, 96, 84, … or, starting with the other window seat, 115, 103, 91, 79, … This procedure is essentially repeated. There are several ways of doing this that are essentially equivalent. In essence, in the slowest boarding methods, few store their luggage at the same time, while in the fastest ones, many store their luggage at the same time. The success of optimized boarding presumes that the passengers will be present at the boarding time and will cooperate, and that there is sufficient overhead luggage storage capacity.

was demonstrated in a construction zone in Missouri in 2015; it reduced a 4-mile backup to 1 mile and traffic delays by half.[dcxv] Traffic models that account for traffic lights, such as *Webster's Formula*, include terms that account for the randomness of traffic.[dcxvi] Congestion at traffic lights in urban areas can be lessened with improved control, including real-time control.[dcxvii] One expects that many of these human-caused congestion problems will be lessened with self-driving cars.

18.3.4 Optimization in selecting employees

How do you optimize your choice of potential hires (finding the ideal match) from a pool of candidates, when you know you can "interview" a specific number of candidates, the relative order you meet them is uncorrelated with their relative quality, and you want to interview as few of them as possible?

This tradeoff between choosing the best of a set of candidates and spending the least time interviewing them is called the *Secretary Problem* (or, tongue-in-cheek, the *Marriage Problem*), and entails *Optimal Stopping Theory*.[dcxviii] If there are three candidates, with the best called 1, the second best 2, and the worst 3, there are 3! = 6 orders of meeting them: 1 (first) 2 (second) 3 (third); 1 3 2; 2 1 3; 2 3 1; 3 1 2; and 3 2 1, but you do not know what the order will be. Of course, you would make the correct decision all the time if you could meet all the candidates, but that would be time-consuming with many candidates. If you decided to choose the first candidate you met, here you will have chosen the best one 2/6 = 1/3 = 33.33…% of the time (with the first two orders). Say, instead you decide never to choose the first candidate you meet, but the first candidate who is better than the first and then stop interviewing; if you interview all, you choose the last candidate. For these six orders, you would be selecting 3; 2; 1; 1; 1; 2, or the best one 3/6 = 1/2 = 50% of the time. This is a better outcome than by choosing the first one and usually less time consuming than interviewing them all.

When the number of applicants increases, the probability of finding the best with the first interview plummets and the time needed to interview all of them skyrockets, but the outcome of the extension of this algorithm remains quite good, as can be shown by using probability analysis. For many candidates, you select the best candidate you have interviewed after first meeting and not selecting $1/e \approx 36.8\%$ of them, and then the probability of choosing the best candidate, the ideal match, is $1/e \approx 36.8\%$.

18.4 What is the Fairest Outcome?

18.4.1 Ranking and voting using rankings

We are obsessed with rankings, whether they are used for serious endeavors (Should I apply to this university because it is ranked higher than another?) or for leisure (Who is it the most valuable player in the NBA and who is the best all-around basketball player?). What are the more objective and more subjective elements of ranking? What mathematics is used?[dcxix] Is the ranking perceived to be fair? We have already encountered ranking in several contexts (including in Chapters 15 and 16). It usually entails optimization and ordering.

Ordering integers, say from 1 to 5 in descending order, 5, 4, 3, 2, 1 is objective and cannot be said to be unfair or misleading. How about ranking information from sets of data fairly? The order of ingredients on food labels in the U.S. is by descending mass. Someone who wants to limit the intake of sweeteners may not see them ranked as the first or second most abundant ingredient in a packaged food product, but they may be because the customer is being misled about the many ways to add and to label added high-calorie sweeteners.[dcxx] If there are 8 grams and 6 grams of the first two ingredients, and then 5 grams invert sugar, 4 grams high fructose corn syrup, 3 grams evaporated cane juice, and 2 grams rice syrup, the total of 14 grams of added "sugars" would really be the dominant ingredient. (There are 28 grams in an ounce.) Presenting this total first may be the fairest way to rank the ingredients.

Ranking depends on the comparison group. Ulysses S. Grant graduated West Point in June 1843 ranked number 21 out of 39 graduates, which is very near the middle. However, his ranking looks better when he is compared to all 82 in his entering class (in which he finished higher up: at the bottom of the top quartile).[dcxxi]

Ranking is strongly tied to ratings. We saw earlier that there is increasing resistance to rating employees on numerical scales, which is essentially ranking them, because this practice can hurt employee morale.[dcxxii] Ranking colleges, such as done annually by *U.S. News and World Report* and by the *Wall Street Journal*, entails a range of metrics of performance and outcomes based on data that are weighted in subjective and arbitrary ways. They drive universities to optimize themselves to improve their rankings and become "better, more attractive" schools.[dcxxiii]

We will see how game theory methods are used to allocate scarce resources, such as spots in medical schools, by utilizing the ranking of schools by the applicants and the ranking of applicants by the schools in a way that is supposed to be fair. But first, let's see how the ranking of candidates in elections might lead to "fairer" voting.

18.4.2 Different ways to vote, with and without rankings

Alternative voting procedures can address the problems that occur when a third or a third and fourth candidate siphons off votes from the leading two candidates. This has affected many U.S. presidential elections, and particularly so more recently with the Ross Perot candidacy in 1992 and the Ralph Nader candidacy in 2000. As noted in Section 16.2.2, the Electoral College system has a big impact in deciding the presidency in several ways. Some think it unfair when the winner of the popular vote plurality loses in the Electoral College and consequently the election, as has occurred recently in 2000 and 2016. Others think the system is fair. In any case, it has helped decide elections when several candidates strongly split the vote in the country and so it provided stability at times of crisis. In the 1860 election, Abraham Lincoln won 59.4% of the Electoral College vote, the needed majority, while gaining only 39.8% of the popular vote among of the four major candidates, a plurality though very far from a majority.

What is the fairest way to determine who is the winner of an election when there are more than two candidates? Should the winner be someone other than the one getting the most first-place votes? Can the majority be said to rule, but in different ways with perhaps different outcomes? Usually, the one with the most votes, who has a plurality, wins. However, plurality voting (PV) may not elect a candidate that the majority likes the most. Alternative systems include a series of ways of casting and analyzing ballots that are cast by eligible voters and those in which runoff elections are conducted. Many of these utilize rankings.

Sometimes an outright majority is needed for a victory, as at the nominating conventions for the Republican and Democratic candidates for the U.S. presidency. When there are three or more candidates on the ballot, some may not think it fair to declare the candidate with a mere plurality the victor. Sometimes there is a runoff election between the top two candidates if the candidate with a plurality, but not a majority, does not receive a percentage of votes cast above a stated minimum. In the Democratic primary election for New York City mayor on September 10, 2013, Bill de Blasio received 40.8% of the vote and avoided a primary runoff with Bill Thompson, in second place with 26.1%, because he received more than 40.0% of the primary vote.[dcxxiv]

Runoff elections work best when small groups of voters select one candidate or small groups of candidates. Voting for modern-era players to the NFL Football Hall of Fame entails a sequence of votes by the 48-person Selection Committee, in which the field is winnowed down to successively smaller lists that are then voted on. First, the list is narrowed down to 25 using rankings, then to 15, then to 10, and then to 5 candidates. Then, each of the 5 final candidates is voted on individually and each needs at least 80% of the vote to be selected and enshrined.[dcxxv] The initial steps select the most preferred candidates, to optimize their relative ranking, and the final step assures that the winner is very widely respected, to assure an absolute standard is being met.

However, conducting runoffs can be costly for larger communities and may not truly represent the will of the people when one person is to be selected. If candidates A, B, and C receive 35%, 33% and 32% of the vote in the first round, in a second round, with the lowest vote-getter C no longer on the ballot in the runoff, A may beat or lose to B. However, C might have been able to win a one-on-one competition against A or B, perhaps because C was a centrist candidate who might have been acceptable to a majority in two-way races. One way to avoid a runoff system is with a single-election voting cycle in which the candidates are ranked. The outcome of the election depends on the details of the ranking and on how the ordering is evaluated. This and other alternative voting methods hinge on being mathematically consistent and being perceived to be fair.[dcxxvi] However, different alternative voting methods that are consistent and seem to be fair can produce different outcomes.

For example, if candidate A gets 40 of the 100 votes casts, or 40% of them, candidate B 35 votes (35%), and C 25 votes (25%), then A wins if a simple plurality decides the election, and B comes in second place and C third. But, with slight changes in the rules of voting, all with seemingly fair procedures, the winner could be A, B or C. If ranking is involved, a voter may prefer A to B and B to C, which we will call a vote for ABC. There are then six possible votes ($6 = 3 \times 2 \times 1 = 3!$): ABC, ACB, BAC, BCA, CAB, and CBA. For simplicity let's say that votes are cast for only three of these possibilities: ACB, BCA, and CBA. With respective vote tallies of 40, 35 and 25, each of the three candidates would receive the same number of first-place votes as before and A would still win if only the top choice receives points as before. If instead the top choice is allotted 3 points, the second choice 2 points and third choice 1 point, then A receives $(40 \times 3) + (35 \times 1) + (25 \times 1) = 180$ points, B receives $(40 \times 1) + (35 \times 3) + (25 \times 2) = 195$ points, and C receives $(40 \times 2) + (35 \times 2) + (25 \times 3) = 225$ points. The person with the third most first-place votes, C, now wins, B comes in second, and A in third. If these same three vote tallies were instead for ABC, BCA, and CBA respectively, B would come in first (with 235 points), C in second (185 points), and A in third (180 points).[227] Allotting different numbers of points for each place in this *point or Borda system* is one way to change the outcome; however, merely changing each point allotment by the same number, such as from 3, 2, 1 to 2, 1, 0 or 4, 3, 2, does not alter the rankings, of course.

A point system like this is used in sports league voting for the most valuable player and other awards. The Cy Young Award is awarded annually to the best pitchers in Major League Baseball, one in each of the two leagues, as voted on by 30 chosen members of the Baseball Writers' Association of America. Starting in 2010, each voter has ranked the five best pitchers for that year, with the best receiving 7 points, the second best getting 4 points, the third collecting 3 points, the fourth receiving 2 points, and the fifth best pitcher getting 1 point.[dcxxvii] This system weights

---

[227] A receives $(40 \times 3) + (35 \times 1) + (25 \times 1) = 180$ points, B receives $(40 \times 2) + (35 \times 3) + (25 \times 2) = 235$ points, and C receives $(40 \times 1) + (35 \times 2) + (25 \times 3) = 185$ points.

first-place votes very heavily, which seems fair, but this selection method is not without controversy, especially in years when more than one pitcher stands out. In 2016, the American League Cy Young Award winner was Rick Porcello, who received 8, 18, 2, 1, and 1 votes for first to fifth place, respectively, for a total of 137 points. The second-place finisher was Justin Verlander, who received 14, 2, 5, 4, and 3 votes for first to fifth place, respectively, for a total of 132 points.[dcxxviii] Two voters did not rank Verlander as one of the top five pitchers, which contributed to his not winning the award, even though he had many more first-place votes than Porcello (14 vs. 8).

Bucklin voting uses rankings, but without points assigned to each. If no candidate receives a percentage of first-place votes above 50%, a majority, the percentages of first-place and second-place votes of each candidate are calculated. If only one receives over 50%, that person is the winner; if more than one gets more than 50%, the one with the higher percentage wins. If no candidate is above 50%, in the third round the percentages of first, second, and third-place finishers are added, and this procedure is continued until a winner is found. In this method, the candidate with the highest median ranking wins. In our example with 40% voting for ACB, 35% for BCA, and 25% CBA, in the second round A gets 40%, B receives 35% + 25% = 60%, and C has 40% + 35% + 25% = 100%. Both B and C are over 50%, but C has the higher percentage and therefore wins.[dcxxix]

These rankings and combinations can be used to conduct *instant runoffs*. Using this same example after the first round, A is the top choice of 40% of the ballots, B in 35% of them, and C in 25% of them. The lowest top-vote getter C is eliminated before starting the second round, so ACB becomes AB (40% of the votes prefer A to B), BCA becomes BA (35% prefer B to A), and CBA also becomes BA (25% prefer B to A in these ballots). In the second round B is the leader in 35% + 25% = 60% of the ballots, and now would be declared the winner. The winner of the instant runoff can change with the details of voting algorithm (or method). This method can be extended to more than three candidates.[228,dcxxx]

A related, but distinct way of determining the winner given voter rankings was developed by the 18th century French mathematician The Marquis de Condorcet, as has been noted by Nobel laureates in economic science Eric Maskin and Amartya Sen.[dcxxxi] The winner of the election would be the one who would beat each other candidate in all or most one-on-one matchups in this *round robin or Condorcet system.* Using our first example, A is ranked ahead of B by 40% of the voters (for ACB and only for ACB) and of C by 40% of them (for ACB), neither a majority. B is ranked ahead of A by 60% of the voters (35% for BCA and 25% for CBA), a majority, but is ahead of C by only 35% of them (BCA). C is ranked ahead of A by 60% of the voters (35% for BCA and 25% for CBA) and ahead of B by 65% of them (40% ACB and 25% CBA), both majorities. C beats both A and B in one-on-one matchups and would be declared the victor. As noted earlier, without ranking in three-way races the centrist candidate may come in third and not qualify for a runoff, but would win this competition. In this particular scenario, the outcome would be the same as for the one in which we assigned points. Sometimes the results are less definitive. With small changes in the voting results, this winner-take-all method could lead to A beating B, B beating C, and C beating A.

Another way to avoid a runoff system using a single-election voting cycle, but without ranking, is *approval voting (AV)*, as promoted by political scientist Steven Brams[dcxxxii] and others.[dcxxxiii] It has been used since the 1980s to elect the leaders in several professional societies and it usually selects the Condorcet winner. Each voter says “yes” or “no” to each candidate and the candidate with the most “yes” votes wins. This is simpler than ranking and your choices are made less definitively, but

[228] When there are more than three candidates, the relative rankings can be used to conduct a runoff between the top two candidates, or to conduct a sequence of a runoff with the lowest vote-getter eliminated before the next round, and this is repeated until a candidate receives a majority. These two procedures can produce different winners. Of note, your voting preferences may not be counted in the final round of an instant runoff, which is called *ballot exhaustion*, unless you are allowed to rank all of the candidates or at most one fewer, and so, say, rank 7 or 8 candidates in a field of 8.

assures that you will be voting for candidates you find acceptable. As with some of the other methods, this approach is subject to gaming; for instance, this approach does not arrive at a true resolution if all voters decide to select only one candidate. When multiple candidates vie for more than one available slot, such as two judgeships or three memberships on a committee, this method is modified to the *proportional approval voting (PAV)* method.[229,dcxxxiv]

All of these voting methods use only the votes actually cast. Deciding not to vote can also influence the outcome of an election, as can gaming. Some of these alternative approaches are clearly practical in specific types of elections and are in limited use.

To some, these methods seem too complex for use in large general elections. *How does this affect you?* Because, they may be the future. Ranked choice voting has been gaining traction in some local elections.[dcxxxv] In 2016 Maine approved the instant runoff system for federal races and in statewide primary elections. It was first used there for the U.S. House and Senate races on November 6, 2018, amidst a bit of confusion and cries that it was not fair.[dcxxxvi] In the race for Maine's 2nd Congressional District, Bruce Poliquin led the four candidates in first-choice votes after the first round, but had fewer than 50% of them. After the instant runoffs, the candidate who was second in the first round, Jared Golden, won the election.[dcxxxvii]

Ranking can also be important in the allocation of resources.

---

[229] For example, if three candidates A, B, and C vie for two slots, the three outcomes could be AB (meaning that A and B are elected), AC, or BC. People vote in one of 6 ways: for AB (to select both A and B), AC, BC, A (to select only one person, A), B, and C. Each outcome is assigned points from each vote depending on the quality of the match. The outcome is determined by assigning $1 + 1/2 = 1.5$ "satisfaction" points to one of the outcomes for a perfect match (if the vote has the same two options as that outcome), 1 point to those outcomes for a partial match (if the vote has one option of that outcome), and 0 points for no match (if the vote has neither of two options of that outcome). If 20 people vote, with 8 for AB, 2 for AC, 6 for BC, 4 for C, 0 for A or B, the AB outcome gets $1.5 \times 8 + 1 \times 8 = 20$ points, the AC outcome gets $1.5 \times 2 + 1 \times 18 = 21$ points, and the BC outcome gets $1.5 \times 6 + 1 \times 14 = 23$ points. So, the approval for BC is highest and this combination wins, even though the plurality voted for AB. This method can be extended to three or more positions in the outcome.

18.4.3 Allocating scarce resources fairly without increasing prices

You can optimize the outcome when a scarce resource is being distributed and increasing the price is not an option, by using one approach in the math of game theory. (We will explore game theory more generally in the next chapter.) Classic examples of its use include the selection of students by highly desired schools and the distribution and exchange of human organs, which are examples of the more general matching and optimizing the needs of two sets of populations. In the admission to schools that charge fixed tuition rates, the schools are really the "sellers" and the students the "buyers," but in such competitive interactions with a limited number of items to be sold (limited enrollment slots, whether or not tuition is charged), the schools appear to be the buyers, with the power to make choices.

Economist Alvin E. Roth, working with Atila Abdulkadiroglu and Parag Pathak, developed an algorithm that matches the many applications to the relative few slots available in medical schools, medical residencies, and specialized high schools, as in New York City. It optimizes the outcome for all students and schools. In each case, *cooperation* (a term often used in game theory) among from the schools is permissible. (Such cooperation is forbidden in the usual admissions process for undergraduate and graduate students.) In a common version of this method, students submit lists of schools they would like to attend in their preferred order. The schools first consider all students who ranked them first and decide whether to provisionally accept them or to reject them. Then they look at all still-unaccepted students who ranked them second, and decide whether to provisionally accept or reject them. Sometimes they may see a candidate they prefer to someone they had provisionally accepted earlier but do not have space for them, so they provisionally accept the new student and reject the student they had provisionally accepted earlier. This process continues until all the student school choices have been exhausted; however, it can be continued to assign some of the remaining students to schools. Some students may still remain unmatched at the end of the process. This provisional or tentative acceptance approach always has a solution and is stable. Also, it is optimal for the students (the "proposers") and not the schools (the "receivers" of the proposals), because it allows students to go to the school they want the most—as long as it is to a school that wants them—and so they can aim at higher ranked schools without hurting their chances of being selected by another good school that they had ranked lower.[dcxxxviii]

Say four students A, B, C and D rank two schools, called 1 and 2. A and B rank them as 1, and then 2, and C and D rank them 2, then 1 (Figure 18.5). In the first round the schools decide whether to admit the students who ranked them #1; School 1 selects Student A and rejects Student B, while School 2 accepts Student C and rejects Student D. In the second round the school decides on students who ranked them #2 and who have not been chosen already; School 1 selects Student D, and School 2 accepts Student B and now decides to reject Student C, whom it had provisionally accepted in the first round. If there are no more rounds, Students A and D would be admitted by School 1, Student B would be admitted by School 2, and Student C would remain unmatched.[dcxxxix]

Ranking by students

Student	School ranked first	School ranked second
A	1	2
B	1	2
C	2	1
D	2	1

(a)

After first round

Student	Outcome
A	Accepted by 1
B	Rejected by 1
C	Accepted by 2
D	Rejected by 2

(b)

After second round

Student	Outcome
A	Still accepted by 1
B	Now accepted by 2
C	Now rejected by 2
D	Now accepted by 1

(c)

Figure 18.5: Fair allocation example.

This optimization algorithm is an offshoot of the famous allocation approach to *The Stable Marriage Problem*, by David Gale and Lloyd Shapley in the 1960s, which optimized how people from two equally-sized sets could choose partners by ranking them (if indeed marriage partners were allocated in this manner).[dcxl] (This differs from the algorithm for selecting mates in the Marriage Problem of Section 18.3.4.) Roth and Shapley were awarded the Nobel Memorial Prize in Economic

Science in 2012 "for the theory of stable allocations and the practice of market design"—in good part due to these advances.[dcxli,dcxlii]

Another allocation or matching method is useful when a donor is willing to donate an organ, such as a kidney, to a specific patient, say a family member or friend, but there is a donation mismatch. The donor in a given pair is instead matched to a recipient who can receive the kidney, a stranger, from a different donor-patient pair and the patient in that pair is successfully matched to a donor, also a stranger, but likely from yet another donor-patient pair. In the end, all prospective donors in the paired sets donate an organ and all prospective recipients in this same paired sets receive an organ donation, all relatively quickly. For the simple case of 3 pairs, Donor A/Patient A, Donor B/Patient B, and Donor C/Patient C, a successful match might be: Donor A donating to Patient B, Donor B to Patient C, and Donor C to Patient A. Such transplant matching can be successful only if there are many such pairs in the database.[dcxliii]

Allocating decisions by society can affect the course of your life in other ways as well, as with the fair and effective allocation of vaccines. When vaccine production is limited during an influenza season, the U.S. Centers for Disease Control and Prevention (CDC) allocates vaccine to different geographical areas proportionately to population, which might be thought to maximize equity. However, this means if the flu hits California before the vaccine is ready, it still gets the same proportion of vaccine as does New York, where the flu may not have struck yet. According to data scientists Anna Teytelman and Richard Larson, perhaps ~30% fewer people may contract the flu by allocating vaccine to different areas by instead trying to maximize effectiveness, based on the assessed, data-driven need for a given geographic area.[dcxliv] Locations where the flu has not yet hit would have a greater need than those where it had essentially run its course and would get more vaccine than other areas. In this allocation, vaccine production would be the limiting temporal feature of the distribution process, and as such could be called the *rate-limiting step*.

The social justice of the allocating economic resources has been studied by Kenneth Arrow, who shared the Nobel Prize in Economics in 1972 for other work, and Amartya Sen; they built on earlier work by

Italian engineer and economist Vilfredo Pareto at the beginning of the 20th century.[dcxlv] A *Pareto Improvement* is a change in allocation that helps at least one individual, without leaving anyone else worse off. A distribution is *Pareto Optimal or Efficient* if there is no change that can help someone without hurting somebody else. Say people A, B, and C need to choose how to allocate cake (and each one wants the largest piece possible) among four choices: (1) A and B get half and C nothing, (2) A and C get half and B nothing; (3) B and C get half and A nothing; and (4) A, B, and C each gets a third. This distribution set is Pareto optimal because though changing from one choice to another helps someone, it hurts at least one other, so no Pareto improvement can be made from any of them. Such optimization analysis is related to the alternative voting methods we discussed earlier this chapter and the interdependent choices that we will address using game theory in the next chapter.

We have assumed that optimization largely produces outcomes that are certain for a set of initial conditions, even when we included the stochastic nature of arrivals on a line. You may need to also optimize situations with uncontrollable, interdependent factors, and will need to use the math to do this with the enticing name of game theory.

Chapter 19

The Math of Gaming

You are walking on a narrow sidewalk or down a narrow corridor on a collision course with somebody walking toward you, and neither of you shows signs of changing course. What do you do? More generally, how do you make a decision to optimize your outcome when there is the uncertainty of the decision to be made by your opponents? The math subworld of game theory was developed to pose ways to choose the best strategies and decisions for actions on the basis of what opponents (other individuals or groups of people) may do to optimize their own choices. This can be important to you on several levels.

We encounter applications of game theory in everyday life. In our hypothetical collision course, collisions are sometimes avoided at the last moment by decisions by each *player* to walk straight or swerve to the right or to the left, and sometimes collisions are not avoided. This is an example of the formal *game* often called *Chicken*. Disputes in many households on how to cut a slice of cake in two, when both "players" want the larger (or smaller) slice are implicitly resolved by using game theory. One player cuts the slice into two and the other player has the first choice of slices.[dcxlvi]

Game theory math can help you control your own life and that of your society when there are the largely uncontrollable uncertainties of interdependent decisions, as we will see in this chapter. Game theory was also applied in the voting and resources allocation examples presented in the previous chapter. The balance of risk vs. reward, discussed in the next chapter, is inherent in all forms of game theory, so game theory is also one part of the math of risk.[dcxlvii] Some think the name "game theory" is a misnomer, and this area should really be called "Multiperson Decision Theory" (though this sounds less enticing).[dcxlviii]

The theory of games combines the mathematical concepts in logic, probability, ranking, sequences, and absolute and local maxima to analyze situations, choices, and outcomes in situations that we colloquially call "gaming each other." Two or more rational decision makers or *players* or *combatants* try to make the decisions and choose *strategies* or *courses of action* in their best interests that have *outcomes* with associated *payoffs*, given the types and level of potential conflict and cooperation. It is used explicitly or implicitly in economics, business markets, politics and geopolitics, warfare, and in our everyday life. Game theory usually does not apply to conventional games of chance and chess, for which there are clearly mathematically better and poorer decisions, or to decisions that may be made only in part by what one player thinks the other will do, maybe by bluffing or deception. Generally, there is an optimal way to play games when one wins and the other loses, but it is more difficult when the interests of the players are partially aligned.[dcxlix,dcl,dcli,dclii]

Each decision maker decides which of the strategies in a well-defined model of the game would optimize that decision maker's best interests in a rational way, even though that person may not have all the desired information and controls the outcome only partially.[dcliii] When players know their own rules and payoff functions and those of the other players, the game is called one of *complete information*. When the players move one at a time and not simultaneously, and know all prior moves before they make their next one, it is a game of *perfect information*. Such ideal games can be studied, along with more realistic situations. In real life situations, we often do not know the game as it is viewed by the other players, make moves in controlled sequences, or know the other players' moves before we decide to move. Analysis, information, and timing are key.[dcliv] Some think that President Jimmy Carter failed in handling the Iran hostage crisis because his analysis of the strategies, outcomes, and payoffs as viewed by Iran was incorrect.[dclv]

Even without the use of a mathematical framework, gaming has long been pervasive in history, fiction, and the home in other instances as well. Purportedly, John and Robert Kennedy went through many layers of "if we do this, they will do that, and then we would do that and they that, and so on" as they developed the strategy in handling the United States-side of the Cuban Missile Crisis with the Soviet Union in 1962.[dclvi]

The 1970 movie *Little Big Man* has a fictional account of events leading up to the real Battle of Little Bighorn massacre of General John Armstrong Custer and his U.S. Army 7th Calvary Regiment troops in 1876. Custer and his (fictional) scout Jack Crabb, whom he called Muleskinner, game each other, knowing each hates the other and wants a bad fate for the other. General Custer: "What do you think I should do, Muleskinner? … Should I go down there (to the potential engagement site) or withdraw? … ." Jack Crabb: "General, you go down there … . There are thousands of Indians down there, and when they get done with you, there won't be nothin' left but a greasy spot …. You go down there if you got the nerve." General Custer: "Still trying to outsmart me, aren't you, Muleskinner? You want me to think that you don't want me to go down there, but the subtle truth is, you really don't want me to go down there."[dclvii] Crab tells the truth thinking Custer won't believe him and will think it safe to proceed. Custer responds and the rest is "history."

Many date the beginning of game theory to the 1928 proof of the *Minimax Theorem* by mathematician John von Neumann, which states that there are optimal strategies in zero-sum two-person games (where whatever one wins the other loses). The best a player can be assured of, the *maximin* (by pursuing the maximin strategy), equals the worst that person's opponent can limit that person to, the *minimax* (by pursuing the minimax strategy).[dclviii] Some earlier work in this area had been done by mathematician Emile Borel in 1921. Much of the traditional game theory and applications, without only "I win, You lose" possibilities, was developed by von Neumann and economist Oskar Morgenstern in their classic 1944 book *Theory of Games and Economic Behavior.*[dclix] The 1945 review of this book by Arthur Copeland began "Posterity may regard this book as one of the major scientific achievements of the first half of the 20th century." It has! (We have seen the work of John von Neumann popping up in many places: here in game theory, the architecture of computers (Section 7.1), pseudorandom numbers (Section 8.3), and Monte Carlo simulations (Section 15.7). This is truly amazing!)

The von Neumann-Morgenstern theory largely deals with "cooperation" between the players, in which they can commit to potential coalitions, irrevocable threats, side payments, and enforceable agreements, and this is called cooperative game theory. In real life, the parties often cannot cooperate and make binding agreements. When this is true, and all players have correctly assessed the strategies and outcomes, non-cooperative game theory predicts that there is at least one stable outcome, called the Nash equilibrium. With this outcome, no player can improve his or her own outcome by choosing a different strategy, and will not regret such a decision. (This is similar to each player pursuing a minimax strategy, but is not limited to constant-sum games.[dclx]) This equilibrium is named after American mathematician John Nash who distinguished between cooperative and non-cooperative game theory, and mathematically proved the existence of this equilibrium in non-cooperative games. His life was portrayed in the book and the popular movie *A Beautiful Mind.*[dclxi] He shared the Nobel Memorial Prize in Economic Science in 1994 with John Harsanyi, who showed how this equilibrium can be determined even when there is only incomplete information available to assess the other player, and Reinhard Selten, who showed how to select the most likely outcome when a game has more than one Nash equilibrium. Nash equilibrium analysis explains collective irrationality even when each player acts rationally for him or herself, such as in some international trade wars and threats to the global environment by large emission of pollutants, and the evolution concept of natural selection, both within and between species.[230,dclxii,dclxiii]

Chase scenes in many movies and books are replete with decisions made by the chaser and the one being chased that are "real-life" enactments of game theory. The classic train chase of Sherlock Holmes by Professor Moriarty in *The Final Problem* (by Arthur Conan Doyle) is one without a saddle point (below) and Nash equilibrium. Each can get off at the first stop of Canterbury or the second stop of Dover, which is the desired final destination. Holmes has a head start, but Moriarty can take a faster train, and he would certainly kill Holmes if they both arrived

---

[230] The winner in approval voting is often the one at a Nash equilibrium, which is stable in the sense that voters have no reason to depart from this choice.

at the same station. So, if they both get off at Canterbury or Dover, Holmes has a 0% chance of survival. If Moriarty gets off at Canterbury, and Holmes at the later stop of Dover, he is so far ahead that his chance of survival may approach 100%. If Holmes gets off at Canterbury, and Moriarty at Dover, the chase continues, so Holmes' chance of survival might be roughly 50%. Mathematically, this last possibility turns out to be choice most likely chosen by game theory (see below), and is the choice that the author chose in his story, in part so the chase could continue.[dclxiv]

Some basic games can be phrased in the simple terms of two players, A and B, each with two potential strategies, for a total of four potential situations (2 × 2) and so four possible outcomes. The outcome for each situation is expressed by the outcome for each player, which can be ranked 1, 2, 3, 4 for the worst to the best outcome for each. So an outcome labeled (4,4) would be the optimal outcome for each individually, (1,1) the worst for each, (3, 2) the next-to best outcome for player A (outcome 3) and the next-to worst outcome for player B (outcome 2), and so on. Using only relative rankings and not absolute impacts is not perfect. For example, in a particular game outcome 2 may be just a little worse than outcomes 3 and 4, so on a relative basis it is not preferred but on an absolute basis it could be fine, while in another it could be much worse, and almost as bad as that for outcome 1. We consider the simpler version with only relative rankings.

Because these scenarios can be expressed by a set of 2 × 2 boxes, this format is called a *matrix* in the area of math involving linear equations and relationships called linear algebra, and so this matrix is called a *payoff matrix*, as shown in the diagram for *Prisoner's Dilemma* game we consider next (Figure 19.1). Each strategy of player A corresponds to two boxes next to each other (in what are called *rows*) and each of B is on top of each other (in *columns*). There are several approaches or "solutions" to the game. The mathematical framework can become more complicated, but we will largely stick to the simpler classic examples. An element in a general matrix with relative or absolute values that is a minimum in its row and a maximum in its column (or vice versa) is called a saddle point (Section 18.1), and represents a Nash equilibrium.

In 1950, Merrill Flood and Melvin Dresher devised the classic *Prisoner's Dilemma* (or Prisoners' Dilemma) game, so named and formalized by Albert W. Tucker. Two suspects in a crime, A and B, are arrested and then kept apart, and are questioned by prosecutors who need evidence to win the case against them. The sentence of a given prisoner each could be 0, 1, 5 or 10 years (corresponding to outcome ranks 4 (the best outcome), 3, 2, or 1 (the worst) respectively) depending on the choices they each make. This depends on whether either or both betray the other, by confessing, or cooperate with each other, without explicit discussion, by remaining silent. The strategies of A are betraying and cooperating, which, in this example are the same as those for B. If both betray each other, each serves 5 years in prison (outcome (2,2)). If both remain silent, each serves 1 year in prison on a lesser charge, since the prosecutors lack evidence for the more serious crime or (3,3), so this is a better outcome but may not be the best one. If A betrays B, who remains silent, the betrayer is set free and the one who is betrayed serves 10 years or (4,1), and *vice versa* leading to (1,4). Because each prisoner makes his or her own decision, it is really a prisoners' rather than a prisoner's dilemma because it depends on the decisions by the two prisoners (Figure 19.1).[dclxv]

	B betrays	B cooperates (is silent)
A betrays	A 5 yrs/B 5 yrs (2,2)*	A 0 yrs/B 10 yrs (4,1)
A cooperates (is silent)	A 10 yrs/B 0 yrs (1,4)	A 1 yr/B 1 yr (3,3)**

* Dominant
** Compromise-with cooperation

Figure 19.1: Prisoner's dilemma payoff matrix.

Analysis is often phrased in terms of whether it is better for the players to cooperate, as in remaining silent here, or not. Classic game theory would say that each will betray the other in the *Prisoner's Dilemma.* A would think: "If B stays silent, my sentence will be shorter for betrayal (0 years) (4,1), than for remaining silent (1 year) (3,3), so I should betray B. If B betrays me, then my sentence will still shorter for betrayal (5 years) (2,2), than for remaining silent (10 years) (1,4), so again I should betray B." Either way, A would think it better to betray B, and vice versa. So, both would choose their own *dominant strategy* and betray each other, and serve 5 years for this Nash equilibrium outcome (2,2). However, this is worse for both than the compromise approach (3,3), for which each would serve 1 year. This could happen if they were able to talk with each other to consider cooperation.[dclxvi] Such scenarios have been incorporated into the plots of several TV shows and movies, including in *Law and Order: Special Victim's Unit.*[dclxvii]

When two race car drivers are on a collision course, the *Chicken* scenario, either can decide to swerve away to avoid a collision (and be "chicken") and lose some time, or decide not to swerve and risk a crash. Both do not want to swerve and hope that the other swerves to avoid a crash. The worse outcome for both is when both decide not to swerve and they crash (1,1). The best outcome for A is for A not to swerve (and save time and "face") while B swerves, leading to (4,2), and for B it is the reverse, leading to (2,4); both are (pure strategy) Nash equilibria. When both swerve the outcome is a bit sub-optimal for both (3,3) because they both lose some time (Figure 19.2).[dclxviii]

	B swerves	B does not swerve
A swerves	No crash (3,3)	No crash A loses time (2,4)
A does not swerve	No crash B loses time (4,2)	Crash (1,1)

Figure 19.2: Chicken payoff matrix.

Chicken is one model of the Cuban Missile Crisis in 1962, in which the U.S. wanted to respond to the installation of nuclear-arms ballistic missiles in Cuba targeting a large part of the U.S. The U.S. strategies were a Naval Blockade and a "Surgical" Air strike of the missiles, with both possibly followed by stronger action. The Soviet strategies were Withdrawal and Maintenance of the missiles. The worst outcome would have been for Air strike/Maintenance U.S./Soviet action that would have led to nuclear war (1,1). Air strike/Withdrawal could have led to near-term U.S. victory/Soviet defeat (4,2), while Blockade/Maintenance could have led to near-term Soviet victory/U.S. defeat (2,4). Blockade/Withdrawal would have been the Compromise outcome (3,3)—which, fortunately, is what essentially happened.

There are 78 distinct such 2×2 games with outcomes that are simply relatively ranked. In symmetric games, the payoff is the same for each player in similar circumstances. If both players cooperate, they have the same outcome CC (C for cooperating). If both players defect (or do not cooperate), they have the same outcome DD (D for defecting). If one cooperates and the other defects, the outcome for the cooperator is CD and that for the defector is DC. CC, DD, CD, and DC can be ranked relative to each other. If their outcomes are unequal (and there are no ties), there are 4 possibilities for the best one, 3 for the next best, 2 for the next to worst and then 1 remaining one for the worst. So there are $4 \times 3 \times 2 \times 1 = 4!$ or 24 distinct, symmetric 2×2 games, including Prisoner's Dilemma and Chicken.

In many social games, you want your opponent to cooperate because you will be better off no matter what you decide to do. When you cooperate the CC outcome is better than the CD outcome, which we will express as CC > CD, and when you defect DC > DD. This occurs for 6 of the 24 possibilities:

(1)	CC > CD > DC > DD	
(2)	CC > DC > CD > DD	
(3)	CC > DC > DD > CD	*Stag Hunt*
(4)	DC > CC > CD > DD	*Chicken*
(5)	DC > CC > DD > CD	*Prisoner's Dilemma*
(6)	DC > DD > CC > CD	*Deadlock*

In the first two, there is no incentive to defect, either when the other player cooperates, because CC > DC, or when the other player defects, because CD > DD. There is an incentive to defect when either DC > CC or DD > CD, or both, as in the last four games.

Prisoner's Dilemma is #5, with both DC > CC and DD > CD, so each individual sees more rational good in defecting, even though cooperation may lead to a better outcome. In *Deadlock* (#6) both conditions are also met and the Nash equilibrium is when both defect (DD). In *Chicken* (#4) only the first condition is met. In *Stag Hunt* (#3) only the second condition is met and the Nash equilibrium is when both cooperate (CC).[dclxix]

Each of these six cases involves each player choosing his or her best strategy among the listed *pure strategies*. However, in the Holmes-Moriarty chase, neither Holmes nor Moriarty can definitively choose their best strategies among the four pure strategies of either exiting at either stop, because there is no saddle point in the payoff matrix. But, by infusing the math of probability into game theory, there is a *mixed strategy* that is optimal for both, in which each chooses a stop, Canterbury or Dover, with a given probability; this is optimal because no other choice improves their position.[dclxx]

Figure 19.3 shows within the matrix the percent probability that Holmes survives for the 4 pure strategies that each exits at a given stop, along with the probabilities that gives them their best mixed strategies. Say Holmes decides that he will leave at Canterbury with a 2/3 probability and at Dover with a 1/3 probability. He could do this by deciding in advance to adopt the first strategy if he rolls a die and gets a 1, 2, 3, or 4 and the second strategy if he rolls a 5 or 6. Given this, if Moriarty then chooses Canterbury, Holmes should expect a payoff of 0% survival probability for a choice he would make with a 2/3 likelihood plus a 100% survival probability for a choice he would make with a 1/3 likelihood, leading to a (2/3)(0%) + (1/3)(100%) = 1/3 chance of survival for this weighted average probability. If Moriarty chooses Dover, Holmes should expect a payoff (2/3)(50%) + (1/3)(0%) = 1/3 chance of survival. So, on average his chance of survival does not depend on what Moriarty decides. You can show that these probability choices give

Holmes the best chance of survival for this payoff matrix.[231] This 1/3 chance of survival is better from Holmes' perspective than the best he could assure for himself by adopting only pure strategies, which is the maximin result of 0%.

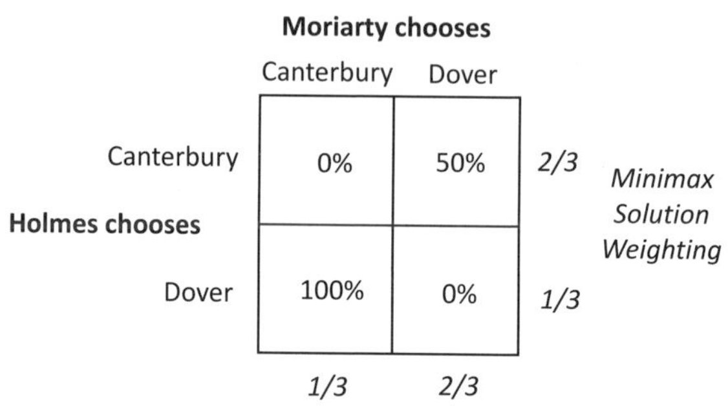

Figure 19.3: Holmes-Moriarty chase game of mixed strategy.[dclxxi]

If Moriarty decides to leave at Canterbury with 1/3 probability and at Dover with 2/3 probability, the chance that Holmes survives if he leaves at Canterbury is (1/3)(0%) + (2/3)(50%) = 1/3 and if he leaves at Dover it is (1/3)(100%) + (2/3)(0%) = 1/3. These are equal and these probabilities give Holmes the smallest chance of survival,[232] so from Moriarty's perspective it is the best he could assure himself of Holmes not surviving, and better than that possible by using only pure strategies, which is the minimax result of 50%.

---

[231] If Holmes decides that he will leave at Canterbury with a probability p and at Dover with a probability $1 - p$, he would expect a survival payoff of $(p)(0\%) + (1 - p)(100\%) = 1 - p$ if Moriarty chooses Canterbury and $(p)(50\%) + (1 - p)(0\%) = p/2$ if Moriarty chooses Dover. These are equal when $p = 2/3$, so $1 - p = 1/3$. For any other choice of p, these two payoffs are unequal and one of them will give Holmes a survival rate < 33.33%, which he would not want.

[232] If Moriarty decides that he will leave at Canterbury with a probability q and at Dover with a probability $1 - q$, he would expect a Holmes survival payoff of $(q)(0\%) + (1 - q)(50\%) = (1 - q)/2$ if Holmes chooses Canterbury and $(q)(100\%) + (1 - q)(0\%) = q$ if Moriarty chooses Dover. These are equal when $q = 1/3$, so $1 - q = 2/3$. For any other choice of q, these two payoffs are unequal and one of them will give Holmes a survival rate > 33.33%, which Moriarty would not want.

Game theory can proceed by analyzing such payoff matrices in this static manner, for which both sides choose options simultaneously and there is no interaction between them, but more quantitatively.[dclxxii] It can also proceed via a "tree" (as in the tree "branching" analysis in the discussion of risks in Chapter 20) in which A decides on 1 of 2 courses of action, and then B chooses from two possible actions, followed again by A with two choices, and so on (Figure 20.4 below).[233,dclxxiii]

Real-life situations can involve commitment, promises, and threats, and these can be reflected in game theory models that use more detailed math. In one of the strategies in *Chicken*, each player commits to staying on course without swerving no matter what. A political candidate vying for a party's nomination may vow to remain in a race until the end of the primary election season, and other candidates will plot their strategies assuming the candidate is or is not telling the truth. A player (or political candidate) can make promises of side deals or concessions, as part of a "cooperation" bargaining procedure. A player can make threats of doing this (such as supporting a different candidate) unless you do that (appointing someone to a certain political position if you are elected). A game can be about one of these strategies or it can be transformed because of them.[dclxxiv]

To optimize our control over our lives we need to evaluate the uncertainties of risk in light of expected rewards, such as good health and financial security.

---

233 A dynamic game theory, called the *Theory of Moves* combines these two modes within a payoff matrix. (The reference is in the text.)

Chapter 20

The Math of Risk

You assess the potential relative risk and reward of hundreds of actions daily, most with very little thought and very little direct math. But at times you incur a greater risk by not using the math of risk to help guide some aspects of your life. This subworld of math assesses and optimizes the potential upsides and downsides of an action on the basis of the uncertainties of risk in the face of known rewards.

A typical 65-year old man in the U.S. wonders whether he should buy term life insurance for one year. An insurance company is willing to sell him a one-year term policy for $1,650 that would pay his heirs $100,000 if he were to die in that period, and which would not be subject to taxes. Do the risk, cost, and payout (or reward) make it reasonable for him to accept the risk of paying the premium? The insurance company offers him this policy because the probability he would die in the coming year is 1.6%, from the life expectancy tables. The premium needed to cover only payouts would be on 1.6% × $100,000 = $1,600. Because the company sells so many policies, fluctuations in which policies pay out that year cancel and this is what they would pay out on average. The company needs to make some money in the deal to pay for expenses and provide some profit, so they will add, say, $50 to the price; it will make a little more money by putting the $1,650 in safe investments, such as bonds. By "thinking math" and checking the online actuarial tables, the man should be able to see that this policy is financially rational and may be worth the risk.

Some consider risk negative and avert it; others embrace it. In the 1939 film classic *The Wizard of Oz*, Dorothy, while "dreaming" in her unconscious state, readily accepted the numerous and ever-changing risks on her trek to the Emerald City because the potential benefit of returning home trumped all. The famous quote "Tis better to have loved and lost, Than never to have loved at all," from the Alfred, Lord Tennyson poem *In Memoriam A.H.H.*, expounds on one example where the benefit is judged to be greater than the risk.

We have already examined risk and the risk-related factors associated with the cost of mortgages (Section 13.2.1), the outcomes of gambling (Sections 15.1 and 15.7), weather predictions (Section 15.3), reliability of medical testing (Section 15.6), and the use of game theory to make decisions (Chapter 19). Risk is intimately related to probability. In gambling, the winning probability and payout are set so that on the average the house wins. If the odds are even, the winning payout is a bit less than the bet itself. If the odds are a bit in favor of the house, the payout can equal the bet. In this chapter, we will focus on risks to well-being due to health involving long-term environmental exposure and accidents and those due to risks in the world of finance. Risk analysis is proving to be essential, or at least prevalent, in other areas as well, as in ensuring the reliability of the electric power grid, assessing the risk of "upcoming" earthquakes, and implementing the varying levels of benefits and risks in new construction and in retrofitting old buildings.[dclxxv]

Each time you fly in a plane you are trusting the basic physics of flight, that of lift, and the safety risk analysis performed by a myriad of engineers and operations personnel. You have a gut qualitative feeling about absolute and relative risk and about the probability of a negative outcome, be it in flying, health, economics, business, engineering safety, military risk, dying, and so on. Risk analysis was once random guesswork, but the development of the mathematics of risk has introduced a degree of rationality that has underpinned much human advancement.

Risk analysis began with the development of probability theory and statistical analysis, and led to the development of other mathematics, such as game theory. The wonderful book *Against the Gods: The Remarkable Story of Risk* by Peter Bernstein "tells the story of a group of thinkers whose remarkable vision revealed how to put the future at the service of the present. By showing the world how to understand risk, measure it, and weigh its consequences, they converted risk-taking into one of the prime catalysts that drives modern Western society."[dclxxvi] *Risk–Benefit Analysis* by physicists Richard Wilson and Edmund Crouch broadly examines the math of evaluating one type of risk, that to health and human welfare.[dclxxvii] Risk is usually evaluated in light of the potential benefits, at least implicitly and qualitatively. The result is called the risk-benefit ratio or the benefit-risk ratio, depending on the community, though it is not always explicitly presented as an arithmetic ratio.[dclxxviii,dclxxix]

There is a systematic part and a stochastic component of the uncertainty of risk. For example, there is a baseline uncertainty of contracting lung cancer for the general population, but also a systematic additional component if you have added risk due to being a smoker. These are systematic "average" factors. Everyone does not "succumb" to the risk factor. It is not deterministic. This is the inherent stochastic uncertainty of risk.[dclxxx,dclxxxi]

Risk is assessed from statistics derived from data, projections from models, and perceptions. Sometimes these distinct, but related assessments are presented as a merged result. The mathematics of risk vs. benefit entails the assessment of potential scenarios, including game theory and conditional probability theory, as well as the math of how people make decisions on the basis of these risks.

This is what the automobile company Ford did in a classic blunder when they assessed cost vs. benefit in their risk vs. reward analysis to decide whether or not to correct a design flaw in the 1971 Ford Pinto, a flaw that could have led to gas tank explosions in rear-end collisions. They estimated improvements to avoid tank explosions, the better safety option, would be $11 for each of 11 million cars and 1.5 million light trucks, for an overall cost of $137 million. This was much greater than their estimated costs of potential liability due to explosions after rear-end crashes if improvements were not made. These were $200,000 for each of the estimated 180 burn deaths + $67,000 for each of the 180 estimated serious burn injuries + $700 for each of the 2,100 burned vehicles, for a total of $49.15 million. By doing nothing, Ford risked $49.15 million, and benefitted by $137 million – $49.15 million = $88 million, which they considered their reward. This analysis became a major embarrassment for Ford, because it was widely viewed as being immoral since it cost lives.[dclxxxii] I am sure such analyses are still prevalent, but are kept more secret. In any case, the liability estimates in this type of estimate would now be much higher than the repair costs.

The medical assessment of disease risk and treatment continues to advance with the aid of statistical evidence. For example, those thought to have a high risk of a coronary disease due to genetic predisposition have a ≈10.7% probability of having a coronary event within 10 years if they have adopted an unfavorable lifestyle. This means they have only 0 or 1 of the following positive factors: not smoking cigarettes, not being obese, performing weekly physical activity, and having a healthful diet. This probability decreases to ≈5.1% if they instead adopt a favorable lifestyle by having 3 or 4 these factors. Even those at low genetic risk can decrease their risk from ≈5.8% to ≈3.1% by adopting a favorable lifestyle.[dclxxxiii]

By "thinking math," you can interpret these data as: the rate of a coronary event is halved if you have good rather than bad genes (which you cannot control) independent of lifestyle, and it is also halved if you change from an unhealthful to a healthful lifestyle (which you can control) whatever your genes. The other math lesson is how useful it is to bin data into proper and helpful groups for analysis, interpretation, and presentation, and how essential it is to do so fairly. Otherwise, interpretations can be of little value and perhaps outright misleading.

Vaccinations are usually risks worth taking, both for individuals and society. For example, the cost of manufacturing and distributing the annual influenza vaccine outweighs the cost of hospitalization and other medical costs for those who are not vaccinated and contract the flu (when the vaccination side effects are rare).[dclxxxiv] It is reassuring that influenza vaccination efforts are win-win propositions both from a human and dollars perspective. In some years, the prevalent flu strains are predicted better than in others, so the success of immunization varies annually and is never perfect.[dclxxxv] One study shows that in years when the vaccine is "thought to be effective," 1.2% of the vaccinated contract the flu, compared to 3.9% of those who were not vaccinated. So 3.9% – 1.2% = 2.7% more of the unvaccinated get the flu, meaning an additional ≈ 1 of 37 people (≈ 1/2.7% of them) are protected by vaccination. This means that 8 million fewer people will contract the flu in a country of 300 million (the U.S. population in 2006) if all are vaccinated than if all are not. When the vaccine is "thought to be ineffective," vaccination protects fewer, yet still millions more people. The success of immunization is even greater for children.[dclxxxvi]

Society continually faces the tradeoffs of the added cost of new and supposedly better drugs vs. their benefits. Siddhartha Mukherjee has wondered whether it is better to prescribe Brilinta to prevent clot formation in the arteries of the heart (for ~$6.50 a pill) rather than the ~25× cheaper generic version of Plavix (~$0.25 per pill) when its performance is only a bit better. In a trial, 10% who took Brilinta died in a year from vascular causes, heart attacks, or stoke, compared to 12% who took Plavix, so "only" ~1 in 50 more people die with this generic.[dclxxxvii]

Risk analysis advances come with better data acquisition and analysis, along with improved diagnostics and treatments. For example, using results from new tests of the protective IgG antibodies may improve the assessment of risk factors for heart attacks and strokes.[dclxxxviii] Roughly 1 in 2 people are at greater risk for high blood pressure or a heart attack because they have variants of the gene CYP1A2 that make them metabolize caffeine very slowly. ~2 out of 3 people are at greater risk of heart disease and stroke, because their variants of the MTHFR gene convert the folate in green vegetables to the useful form very slowly. ~7 out of 10 people are at greater risk of high blood pressure because of how their versions of the gene ACE produce the enzyme that regulates the response to sodium intake. ~1 in 5 people are at greater risk of being overweight and developing cardiovascular disease because of how their version of the gene GLUT2 regulates glucose from sweets.[dclxxxix] Moreover, these risk values depend on the population group. With DNA testing, these gene variants can be determined individually and those at risk should be able to modify their diet and lifestyle, and lower their own risk.[dcxc] Such beneficial personalized medicine via DNA testing will advance,[dcxci] but our medical and life insurance rates may eventually be determined by our gene-based likelihood of contracting diseases (despite current protestations to the contrary).

Risk quantification and analysis can be controversial in other ways as well, as in the now common quantification and subsequent use of risk-evaluation factors for prisoner recidivism, such as COMPAS developed by Northpointe, Inc. They are used in pretrial detention, criminal sentencing, and parole decisions. Northpointe claims that their risk score is ~70% accurate. Aside from it being wrong a whopping ~30% of the time, this procedure seems unfair and to violate due process.[dcxcii] States use widely varying and often unclear algorithms to predict recidivism.[dcxciii]

Statistical risk levels can evolve with time because of experience. Fewer people are smoking than decades ago because of its now-known risks, and so the overall risk of lung cancer in society has plummeted. Improvements in auto technologies have made driving much safer. Unfortunately, proper action does not always occur even when a large risk is clear, as is illustrated by the Space Shuttle Challenger disaster in 1986. The Space Shuttle engineers told their employers and NASA not to launch because the launch temperature would be much lower than that at which key parts had been tested and shown to be usable, so this was deemed to be extremely risky. But, due to financial and political pressures, the shuttle was launched, followed by disaster. The human tragedy was likely amplified because there was no emergency escape system on the Challenger spacecraft even though the 1967 cabin fire in Apollo 1 during launch rehearsal had caused the deaths of its three astronauts.[dcxciv] This led to improvements in the design of subsequent Apollo spacecrafts, including an improved door hatch for improved emergency escape, so the risk of not having an escape system, should have been well appreciated. The Challenger had no escape system; one was added before subsequent Space Shuttle flights.

As important as the math of risk in betting and other activities is the math of the perception of risk and decision-making, which we examine later. The perception of risk also changes with time. Riding in elevators and in planes was once perceived to have significant risk, though now they do not.

20.1 Evaluating Risks from Hazards: Environmental Exposure and Accidents

Health hazards can come from long-term events, such as chronic exposure to chemicals and radioactivity, and short-term events, such as accidents, including acutely high levels of exposure. Both can lead to a range of outcomes from minor health issues, such as minor disease or injury, to major issues, and to death.

Risk can be quantified to understand hazards and make decisions on taking the appropriate course of action.[dcxcv,dcxcvi] It increases when the source of danger or hazard becomes worse and decreases when safeguards to avoid danger improve. Risk increases when either the uncertainty of the outcome (its probability on the basis of statistical analysis) or the potential damage of an event (its consequences) increases.[dcxcvii]

In health and safety, risk can be evaluated in terms of death or injuries by age or cause, and per person year or lifetime, event occurrence, or exposure of say a pollutant. Alternatively, it can be assessed in somewhat different terms, such as life expectancy or working days lost.[dcxcviii] One general way of characterizing risk is as the product of a hazard probability and its severity, but there are variations in this quantification as we will see. With multiple sources, the total risk is the product of the probability, risk, and a weighting factor, summed over the different sources, when all are uncorrelated. The probability of causation (POC) is the ratio of the risk of a disease from one given factor, say radioactivity from medical treatment, to the overall risk from all causes, such as radioactivity from this treatment and that from air travel and living at high elevation, and the risks from water pollution, air pollution, and so on.[dcxcix] The math used to evaluate the risk from environmental exposure and from accidents differs because the risk in former generally comes from continuous influx of stimuli over a period of time, while that in the latter it comes from an event or a discrete series of events. Consequently, we will look at them separately.

20.1.1 Health risk from long-term environmental exposure

The health risk from exposure is determined by the continuous nature of the exposure and the *baseline level of risk*, which is the risk to health with no exposure or apparent cause.[dcc] Smokers have a risk of contracting lung cancer of ~8% (~80 out of 1,000 contract it in a lifetime), which is ~10× the baseline risk for non-smokers, ~0.8% (~8 out of 1,000). Living with a smoker increases this baseline risk for non-smokers contracting lung cancer by ~1.3×, to an overall risk ~0.8% × 1.3 = ~1.04% (~10.4 out of 1,000), which is an additional risk of 2.4 cases for 1,000 people.[dcci]

An exposure model was chosen to help understand the consequences of smoking once it was realized that cumulative smoking exposure increases the risk of lung cancer, and then a specific, quantitative type of a model was created to fit the available data. In this model, the risk of humans contracting a disease, such as cancer, from smoking or from environmental exposure to chemicals and radioactive materials is evaluated using probability principles. This risk can be modeled as a product of three factors: (1) the risk to humans per unit dose, called the *potency, potency slope, cancer slope factor* or *unit risk*; (2) the dose or exposure; and (3) a threshold factor which is a function of dose. This factor equals 1 if the risk is always proportional to the dose; such *linearity* or proportionality is expected in the probability of random, uncorrelated events. If there is a threshold dose level below which cancer does not occur or occurs at a much reduced rate (than above any background rate), this factor is 0 below the threshold dose and then above it, it increases and becomes 1 at high doses (Figure 20.1).[234,dccii] Whether such thresholds exist is still controversial, and this is particularly so for exposure to ionizing radiation from radioactivity.[dcciii,dcciv] In any case, models should generally include the possibility of thresholds,[dccv] though for simplicity we will usually assume linearity, with no threshold.

There are two ways to increase the scope and usefulness of this model. One is to use tests on animals to determine risks factors in humans. Tests of the risk potency measured in tested animals are used to determine the human risk by multiplying it by the ratio of human potency to that of the animal, which is assumed to be near 1 if the daily dose

---

[234] This risk is $R = \beta dE(d)$, where β is the potency, potency slope, cancer slope factor or unit risk; d is the dose or exposure; and $E(d)$ is a threshold factor, which is a function of dose.

given for a lifetime is expressed as a fraction of body weight.[235,dccvi] The other way is to make it more realistic by including the baseline or background risk even with no exposure.

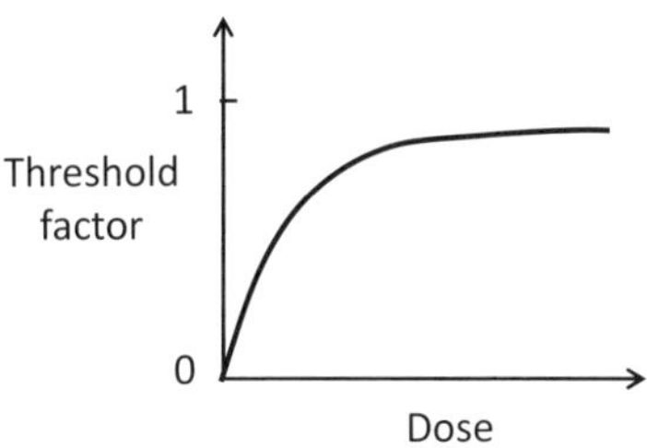

Figure 20.1: Example of the threshold factor in exposure model.

The data from statistical epidemiological studies can be incorporated, such as the results of the American Cancer Society Cancer Prevention Studies I (CPS-I, started in 1959) and II (CPS II, 1982) for lung cancer. We will use this as an example.[dccvii] This approach led to a model that shows how risk depends on age (scaled relative to reference age of 50 years), exposure potency (in days/cigarette), the dose rate (self-reported cigarettes/day), how risk depends on exposure (the exposure-response shape), and the effect of long-term exposure (the duration-response shape) in terms of the fraction of one's period of exposure (in years).[236,237,dccviii]

---

235 The human risk R_h is obtained from $R_h = K_{ah}\beta_a dE(d)$, where β_a is the risk potency measured for tested animals, K_{ah} is the ratio of human potency to that of the animal (or β_h/β_a), and the other parameters are described in the previous footnote.

236 The mortality rate from lung cancer, in deaths per 100,000 per year, is $r = a(t/t_0)^{\alpha}(1 + bf(d)\,g(\tau/t))$, where the age, t, in years is scaled relative to the reference age t_0, 50 years; the background rate, a, is in deaths/100,00 per year; the power law dependence is α; the exposure potency, b, is in days/cigarette; the dose rate, d, is the self-reported cigarettes/day; the dose- or exposure-response shape, $f(d)$, is how the risk depends on exposure; the exposure time is τ; and the duration-response shape, $g(\tau/t)$, is the effect of long-term exposure. f is normalized to 1 for 20 cigarettes/day. g is normalized so it is 0 and 1 for τ/t equal to 0 and 1; for a 60-year old who started smoking at 15, τ/t would be $45/60 = 0.75$. f increases linearly with dose up to the "break-point" d_b of 20 cigarettes/day,

Models should also include changes in the level of dose or exposure to include, for example, how risk might decrease after someone stops smoking. The excess level of risk of lung cancer of a former smoker above baseline or background levels decays exponentially after smoking has stopped. This decay constant from the CPS-II study is 0.11018/year for previous cigarette consumption of 20 or fewer cigarettes/day and 0.08839/year for over 20 cigarettes per day for men (and 0.22573/year and 0.09873/year for females), so the risk of lung cancer for former smoker decreases roughly 10% per year, and even faster for females who were "lighter" smokers.

This risk model is entirely a statistical analysis of data and not based on the science of how cigarettes can cause lung cancer and death. It is only as good as the quality of the data collection, and so it relies on the reliability of self-reporting by people of how many cigarettes they smoke or smoked a day.

Another well-known potential risk to health is the risk of cancer due to ionizing radiation (radioactivity). Both the real risk due to exposure, including the uncertain threshold effects, and the public perception of the risk (the perceived risk) are significant. Radiation exposure is now measured in terms of millisieverts; historically, it was measured in terms of millirems, where 1 millisievert equals 100 millirems. Assuming a linear, no-threshold model, a dose of 1 millisievert over any period will eventually lead to five fatal cancers in a population of 100,000, and this increases proportionately with dose.[dccix,dccx] There are several sources of radiation that lead to a total "natural" background level of ~1.5-3 millisieverts/year[dccxi] and an effective total background level of typically ~3-6 millisieverts/year.[238,dccxii]

---

where $f = 20$, and then increases as the square root of the does above it. $g(x) = (e^{\beta x} - 1)/(e^{\beta} - 1)$.

[237] For the CPS-II study, a is 5.2967 per 100,000 per year for males (and 3.6193 for females), α is 5.1877, β as given is 2.8154, b is 2.6123 per cigarette/day for males (and 2.6226 for females), and d_b is 8.6409 cigarettes/day for males (and 26.537 for females).

[238] One source of baseline exposure from radioactivity is due to cosmic radiation from charged particles from outer space. This dose per person at sea level in the U.S. is 0.40 millisieverts/year and it increases to 0.65 millisieverts/year in the mile-high city of Denver due to the higher elevation (and consequently the less screening by air). Added to

Being near Chernobyl after the nuclear accident in 1986 gave the average nearby resident a dose of ~50 millisieverts/year the first year. Statistical studies have shown that nuclear workers generally exposed occupationally to protracted ionizing radiation have a ~1.8× increased risk of leukemia and a ~0.3-1.0× increased risk to other cancers (above baseline levels); this is significant but not extraordinarily large. For those doing cleanup at Chernobyl, the risk of leukemia increased by much more, by 3.4-4.8×.[dccxiii]

When the data on risk vs. exposure are not certain, as is common for air pollution, standards for safe exposure can be changed with little justification. For example, more or less arbitrarily lowering the standards for the level of safe exposure to, say, air particulates would lead to fewer predicted premature deaths for the same conditions, but this numerical change would be unjustifiable.[dccxiv]

20.1.2 Health risk from accidents and other short-term, high-level events

We accept the risk of accidents whenever we travel. We noted earlier that the risk of being on a plane flight that resulted in at least one fatality was one in several million, on the basis of the frequency of past events. We automatically use this knowledge and ignore this very small risk of flying on commercial airlines, as well as the concomitant risk due to increased radiation exposure.

The risk of flying in the U.S. Space Shuttle is clearly much higher. It is unbelievable to me that NASA once predicted that the shuttle would fail only once in every 100,000 flights.[dccxv] Was this realistic or wishful thinking for this cutting-edge, low-volume form of travel? In 1983, before the Space Shuttle Challenger accident in 1986, Colglazier and

---

this baseline dose is that from exposure from the potassium 40 that naturally exists in our body, which is an additional 0.30 millisieverts/year for all of us, and that from other sources. So, the total "natural" background level of ~1.5-3.0 millisieverts/year. There is an additional dose of 0.02 millisieverts/year for each transcontinental trip by air per year and so 4 millisieverts/year for an aircrew exposed 60 hours per month. In 2000, medical exposures from x-rays typically added 0.4 millisieverts/year. The dose from of a single chest x-ray has decreased from 9 to 0.07 millisieverts from 1945 to 2000, so the risk from such x-ray examinations has decreased much.

Weatherwax predicted a much higher failure rate, with ~1 of 35 Space Shuttle flights ending in failure, and this has been closer to that seen by experience.[dccxvi] In 1988, physicist (and Nobel laureate) Richard Feynman estimated that 1% of the shuttle flights would fail.[dccxvii] (He was a member of the Rogers Commission that investigated the Challenger accident.) A NASA-sponsored study in 1995 found a median estimate of shuttle failure as ~1 in 145 = ~0.69%.[dccxviii] More recent NASA retrospective risk studies suggest that the first 9 shuttle flights had a 1 in 9 chance for disaster, which is 10× the current risk.[dccxix] Despite the thorough preparation at NASA, space flight has significant risks.[dccxx] Astronaut Jim Lovell, has said, "Spaceflight is risky. People who want to become astronauts, accept the risk."[dccxxi] Lovell was the commander of the Apollo 13 flight in 1970 that was aborted after a spark, fire, and subsequent loss of electrical power in the main module; it returned to Earth safely "defying" the very small estimated probability of safe return. Two of the 135 Space Shuttle flights (as of May 2019)[dccxxii] led to the death of the astronauts, shuttle flights 25 (Challenger in 1986) and 113 (Columbia in 2003), which is a frequency of 2 in 135, or 1 in 67.5 ≈ 1.48%.[dccxxiii]

You can estimate the probability of future Space Shuttle accidents by using this frequency of past fatal shuttle accidents as the probability of failure. The three remaining space shuttles (Discovery, Atlantis, Endeavor) have flown 97 times. If each one flies no more than 50 times, there could be 150 – 97 = 53 remaining flights. If each flies 100 times in its operational lifetime, then 300 – 97 = 203 flights could remain. What is the likelihood of at least one more fatal event in the remaining flights if for each flight life you assume a flight risk estimate of either 1 per 67.5 flights, from the past frequency, or 1 per 145 flights, from the above estimate? For a probability of a fatal accident of 1 in 67.5, the probability of no failure is 66.5/67.5 in one flight, 66.5/67.5 × 66.5/67.5 in two flights, … and $(66.5/67.5)^{53}$ in 53 flights. So the probability of one or more accidents is $1 - (66.5/67.5)^{53} \approx 0.547$. This is the same as the binomial distribution math (Section 15.1.2). The estimated probabilities of one or more accidents with at least one fatal event in the 53 remaining flights are ≈54.7% and ≈30.7% for the two higher and lower chosen flight risks. For 203 remaining flights, these probabilities are ≈95.2% and

≈75.5%, respectively. Because the risk per event is much less than 1, the Poisson distribution limit of the binomial distribution can also be used to obtain good estimates (Section 15.1.2).[239] Unless equipment and procedures are improved, future deaths in Space Shuttle flights would not be surprising.

More generally, to assess the risk of potential accidents the hazards need to be identified, the potential consequences must be determined along with their probabilities, and the frequency of each hazardous event needs to be estimated.[dccxxiv] The risk of a hazardous event, R, is given by the product of consequences of the event, C, and the frequency or probability of the event, p (as in the number per year), so $R = Cp$. The *risk index* is defined as $\log R$ (in base 10), which equals $\log C + \log p$, because the log of a product equals the sum of the logs. C is 1 for hazards causing minor damage, 2 for moderate damage, 3 for major damage, 4 for severe loss, and 5 for a catastrophe. To keep all numbers positive, 5 is added to $\log p$ and it used instead of $\log p$. The frequency is often presented for a range of frequency possibilities. It is 5 for the probability $p = 1\text{-}10$ per year (because log 1 is 0, so $5 + 0 = 5$), 4 for $p = 0.1\text{-}1$ per year, 3 for $p = 0.001\text{-}0.1$ per year, 2 for $p = 10^{-5}\text{-}10^{-3}$ per year, and 1 for $p = 0\text{-}10^{-5}$ per year. (Note that for this definition and assignment, the ranges of p are not the same factors and the changes of p and its logarithm across the ranges are not always consistent.)

This risk index increases with accident severity and accident frequency, and could be the same for two different accidents, one being more severe and the other being more frequent. For fire with smoke in a subway train in a tunnel, $\log C$ is 5, indicating a catastrophe. Such events are expected only once every 10-100 years, so it falls with the 3 frequency category and the risk index is $5 + 3 = 8$. This is high in the risk index range from 2 to 10. (The lowest range is $1 + 1 = 2$ and the highest one is $5 + 5 = 10$.)[dccxxv]

---

239 The probability of no catastrophic event occurring in the first example is $e^{-(\text{probability of an event in one trial})\times(\text{number of trials})} = e^{-(1/67.5)53} \approx e^{-0.785} \approx 0.456$, so the probability of at least one event occurring is $\approx 1 - 0.456 = 54.4\%$, which is very close to the exact result using the binomial distribution. (The four probabilities in this limit are ≈54.4%, 30.6%, 95.1%, and 75.3%, respectively, which are close to the "exact" binomial distribution results.)

This evaluation of risk lumps together all of the consequences into one parameter C. The range of scenarios that can occur for a given event can be addressed by evaluating the triplet of the scenario type, the probability of the scenario occurring, and the possibly multiple consequences of that scenario, and then analyzed as the risk frequency vs. severity. This is a probability density function (PDF, f; as in Section 15.2.1). The overall estimated probability of the event consequences is often presented as a risk curve or table that is *cumulative* over the severity of the outcomes, and so the frequencies for a particular outcome or one that is worse. Consequently, this is a cumulative distribution function (CDF, F; Section 15.2.1). Such plots are called *Farmer curves* (Figure 20.2).

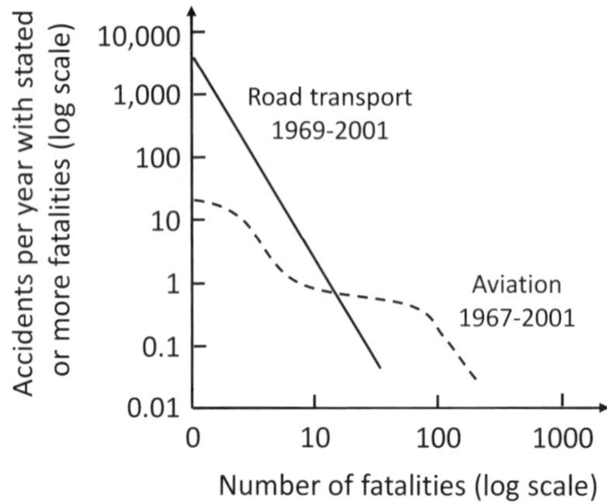

Figure 20.2: Log-log dependence of Farmer curves, of the frequency of events or worse vs. severity.[dccxxvi]

For example, there is a range of the severity of potential injuries that could occur during chemical plant operation each year. If the annual risk of an injury per worker that is minor is 0.0050 or 0.50%, moderate 0.0025, major 0.0005, and fatal 0.0001, then the risk of an injury that is considered minor or worse is $0.0050 + 0.0025 + 0.0005 + 0.0001 = 0.0081$, moderate or worse is $0.0025 + 0.0005 + 0.0001 = 0.0031$, major or worse is $0.0005 + 0.0001 = 0.0006$, and fatal is 0.0001. This risk relation is characterized by the limits of the probability that any damage or injury can occur (0.0081 here) and the maximum possible damage or injury (or death here). Figure 20.3 presents this frequency as a bar graph histogram. Such a trend of smaller rate of accidents for increasing severity is common. A study of almost 1 million accidents in Britain up to 1975 showed that for every fatal or serious injury, there were 3 minor injuries (with the victim absent from work for up to 3 days), 50 injuries requiring only first-aid, 80 incidents with only property damage, and 400 accidents with no injuries or damage, or that were near misses.[dccxxvii,dccxxviii] Figure 20.2 shows that such trends also occur for road and aviation accidents. Moreover, it shows there are fewer aviation accidents than those on the road, but when they occur they tend to have larger numbers of fatalities (which makes sense). Such statistics are usually displayed as log-log plots because of the many decades in the variation of the values of the frequency and severity data. Straight lines on such log-log plots indicate (as noted in Section 6.2.2) that the cumulative frequency varies as the number of accidents raised to a power, here a negative power.

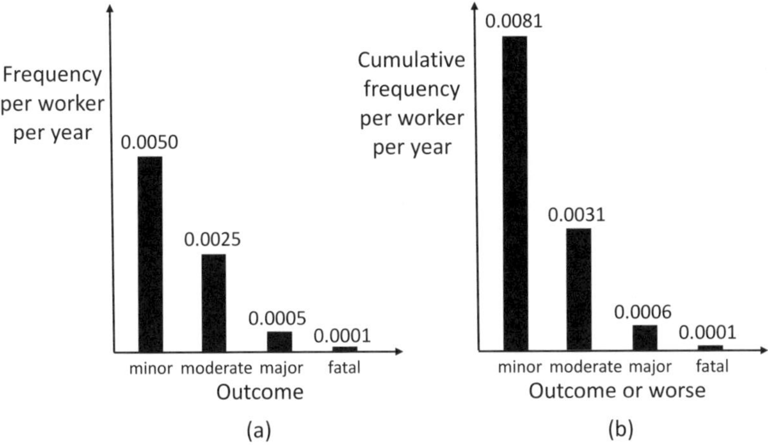

Figure 20.3: Example of frequency (or risk) vs. outcome and cumulative frequency vs. outcome or more severe outcome.

Such data are also presented in other ways. The *individual risk per annum* or IRPA is the frequency that an individual is killed per year, sometimes for a given type of hazard. For example, 1,989 people were killed in accidents in Norway in 2009. Given the Norwegian population of 4,900,000, the IRPA was $1{,}989/4{,}900{,}00 \approx 4 \times 10^{-4}$, so about 4 out of 10,000 people. More specific is the localized individual risk (LIRA), which is the probability that an average unprotected person always present in a specific location is killed per year due to an accident associated with that potentially hazardous location.[240,dccxxix]

Whatever the outcome, the frequencies of accidents to people can be characterized as occurring at a rate p. More generally, component failures (or "accidents") in products, machines, systems (such as underground pipelines[dccxxx] and Space Shuttles), factories, and terrorism[dccxxxi] can also sometimes be characterized as occurring at a rate p (failures per time of use). So, the product of this rate and the time they are in use t, pt, is the average number of accidents (or failures) in this unit time. If the accident rate is very small, as one would hope and so pt would be much smaller than 1, and if the accidents and failures are not correlated with each other, they are random events. Then, the probability of having 0, 1, 2, 3, 4, or more accidents or failures in this time t is given by the Poisson distribution, as we noted in our discussion of Space Shuttle accidents and in Section 15.1.2.[241,dccxxxii]

Several technology hardware failures are not totally random. As such, they can be described by the probability density function, the *Weibull distribution*, named after the Swedish mathematician Waloddi

---

[240] The LIRA may represent observed frequencies or design targets. In Australia, the upper limits of LIRA in the design of hospitals and schools is 5×10^{-7} (1 person per year in two million), for commercial developments, such as offices and restaurants, it is 5×10^{-6}, and for industrial sites it is 5×10^{-5}. Another statistic for risk exposure is the *fatal accident rate* (FAR), which is the number of fatalities per 100 million working hours. In Nordic countries from 1980 to 1989 it was 0.7 for banking and insurance, 5.0 for building and construction, and 10.5 for raw material extraction.

[241] For n failures, this probability is $((pt)^n/n!)e^{-pt}$. This notation differs a bit from that used earlier.

Weibull who detailed it in 1951.[242,dccxxxiii] (As seen in the footnote, it is a combination of exponentials and powers.) Depending on the values of the parameters that characterize it, it can describe a failure rate that is constant over time and then it is the usual exponential decay; one with a failure rate that decreases with time, so it characterizes situations where defective items fail quickly and the remaining ones are more robust; and one in which the failure rate increases with time, so it characterizes aging.

20.1.3 Risk analysis of rare events and assessing new technologies

Risk is often assessed on the basis of historical frequencies. This is often not possible in assessing new technologies, such as for flying in its early days, so estimates need to be made. When guesses by experts are not valued to do this, you need to turn to several semi-quantitative ways of estimating the probability.[dccxxxiv]

Fault tree analysis, so named because the branching of the decision paths looks like branches of a tree, is used to evaluate risks that are not well understood from past frequencies, including those of rare events. If you know that event Z can occur if first A occurs, and then B, and then C, the probability of Z occurring will be the product of the probabilities of A occurring, B occurring given A occurred, C occurring given B, and then Z occurring given C occurred (if they were uncorrelated). If these probabilities can be estimated fairly well and they are small, the overall probability for a potentially rare and possibly catastrophic event can be very small. If Z can also occur after distinct events D, E, F, and G occur, the overall probability for Z is the sum of the probabilities of these two sequences or paths (Figure 20.4).[dccxxxv]

You can also estimate risk probabilities by scaling the known probability of a similar, reference process by using *similarity judgments.* The first step is to enumerate the number of features the two processes have in common, say 2 of them, the number possessed only by the

---

[242] The Weibull probability density function PDF is $f(x) = (k/\lambda)(x/\lambda)^{k-1}e^{-(x/\lambda)^k}$ for variable x (such as time, ≥ 0) and two positive parameters, k and λ (for scaling). When $k = 1$ the failure rate is constant over time and this is the usual exponential decay; when $k < 1$, the failure rate decreases with time; and when $k > 1$, the failure rate increases with time.

comparison case, say 4, and the number possessed only by the process of interest, say 3. The latter two are multiplied by weighting factors that sum to 1, say 0.2 and 0.8 respectively, to arrive at $0.2 \times 4 = 0.8$ and $0.8 \times 3 = 2.4$, and then summed, $0.8 + 2.4 = 3.2$. The scaled probability is the reference probability, say 0.001, times the ratio of the number of elements in common divided by the sum of this and the weighted sum of non-overlapping factors, $2/(2+3.2) = 2/5.2 \approx 0.385$, to arrive a probability of 0.385×0.001, or ~ 0.0004. Such binning of factors and the weighting may seem a bit arbitrary.

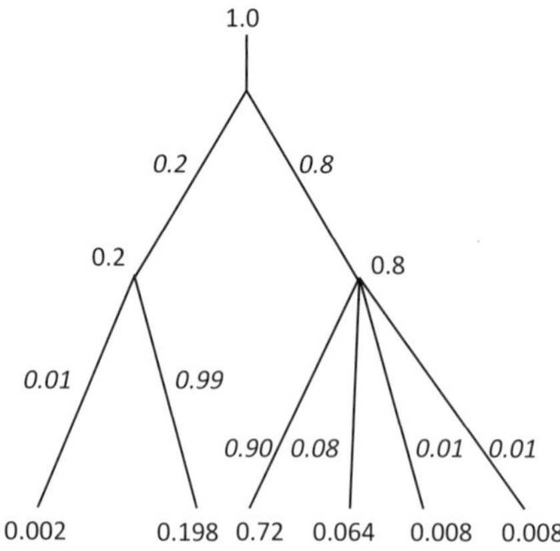

Figure 20.4: Branching or fault tree estimate of probability.

Monte Carlo simulations, another tool of random event analysis, are also used to study critical events in accident risk, along with other math approaches, such as Bayesian networks (which employ Bayes conditional probability analysis) and so-called Petri net diagrams of the propagation of risks and faults.

The understanding and perception of risk can evolve with time, especially for new technologies. For example, we now have a relatively good idea of the risks of radioactivity for long-term and accidental high-level exposures, though it is still a bit controversial. The risks of radioactivity exposure were unknown to Nobel-Prize winning physicist Wilhelm Röntgen when he developed x-rays in 1895 and then used them to image the hand of his wife Anna Bertha, with what we now know to be extremely high levels of x-rays.[dccxxxvi] Women painting watch dials with paint that was luminous because it contained radioactive radium were told to make points on the brushes with their lips to save time and materials. They were told that the paint was harmless and posed no risk, and this was known to be a lie at the time; later many died from this.[dccxxxvii] The potential of risk was clear when the first plutonium production facilities were being built in Hanford, Washington by the U.S. government during World War II, working with DuPont, but no one knew what the exact risks were, the duration or effects of potential hazards, and or the possibility and extent of any major catastrophes. Because private insurers would need to know these risks to provide insurance and they could not be quantified, the federal government set up a fund for DuPont to cover the risk.[dccxxxviii]

20.1.4 Acceptance of risk from exposure and accidents

Determining the risk frequencies and probabilities and establishing risk metrics is one thing. Accepting a given level of risk is another. The upper limits of risk levels for new nuclear reactors have been set by some to be 10^{-5} events per reactor year for significant core damage and for small releases of radioactive materials that may require temporary evacuation of the local population. It has been set to be 10^{-6} events per reactor year for large releases of radioactive materials that may require long-term relocation of the local population. As with the Ford Pinto case, the acceptance of such risk is tied to the assessed dollar and the moral and political values of human well-being.[dccxxxix] This is closely tied to the perception of risk, which we will address later this chapter.

20.2 Insurance: The Interplay of the Risks to Your Physical and Financial Health

Insurance companies set rates using data-driven statistics to arrive at risk levels and to determine how these risks might be pooled or shared. You need to understand this to optimize what insurance is best for you. Earlier this chapter we saw a back-of-the envelope way of using the actuarial (or life or mortality) tables noted in Sections 13.4 and 16.3 to set life insurance rates.

Automobile insurance rates for young men have often been set higher than for other groups because they entail higher risk in insurance claims. Medical (or health) insurance typically covers treatment from usual health diseases and disorders, including those that might result from risks of environmental exposures and accidents. Such medical insurance for older people would be expected to cost much more than for younger people because their use for medical care is much higher. When these risks are pooled in a controlled way, as in a company plan, older people might pay a bit less than they would otherwise and younger people a bit more (with the understanding that younger people would be paying relatively low rates in later years). If this relative risk is mispriced, younger people may not pay the higher rates in a voluntary or semi-voluntary plan, even one with modest incentives to join. Then, without the participation of all age groups, the insurance companies will lose money and likely opt out of providing such insurance plans. In 2016 this was a challenge in the insurance program of the U.S. Affordable Care Act. Apparently, sixty-four year olds each then consumed ~6× more medical care than 21-year olds, on average, but at that time the act allowed insurance companies to charge them only 3× more.[dccxl] However, even when the available actuarial risks are properly included in the math, medical insurance rates have been known to increase faster than had been expected, not only due to increasing medical costs, but also to poor projections of longevity and lapsing policies.[dccxli]

Medical insurance is often defined by the premium that you need to pay monthly, copayments per visit, the deductible (the total amount you need to pay in a year before insurance starts paying at all), and the maximum out-of-pocket expenses you are asked to pay in a year. When the deductible is relatively high, you might not go for treatment at all (which is good for the insurance company, but bad for you) or you might try to opt out of having insurance and paying any premiums (which is bad for all). Each strategy entails risks and tradeoffs. Which plan available to you is best for you? Methodically "do the math" and make a risk-benefit decision that is best for you based on cost and your health projections.

First, consider the relatively simple example in which you are told to choose one of the two plans offered: *Plan A*, with an annual premium of

$1,400 and a deductible of $500, and *Plan B* with a premium of $800 and a deductible of $1,000. In this plan, the insurance pays all expenses after you have paid the deductible. With no expenses, only the premium is paid so *Plan A* costs $1,400 and *Plan B* $800. The difference decreases as deductible expenses increase until the deductible has been paid off. The maximum cost of *Plan A* is ($1,400 + the maximum payment of $500 =) $1,900 and of *Plan B* is ($800 + $1,000 =) $1,800 and so *Plan B* is the cheaper plan, surprisingly independent of your medical expenses, and your decision is obvious.[dccxlii]

In reality, medical insurance plans are usually more complicated. Say the choice is between a *Plan 80*, in which you pay monthly premiums of $60 (or 12 × $60 = $720 annually), all expenses up to the annual deductible of $600, and then 20% up to the maximum out-of-pocket expense of $3,750 exclusive of the premiums (while the company pays the other 80% in this "coinsurance"), and then pay no more, and *Plan 100*, in which you pay monthly premiums of $320 (or 12 × $320 = $3,840 annually), all expenses up to the annual deductible of $200, and then no more (while the company pays 100%). (However, in both plans you will likely need to pay a copayment ("copay") per visit that is not included in these maxima or reimbursements.) If you expect to have no medical expenses during the year, you pay only the premiums, so *Plan 80* would be better ($720 < $3,840).

What happens if you expect to have expenses? Analyze the possible scenarios in stages, recognizing the "cross-over" points where a plan changes from one regime to another or when the payments in the two plans are equal:

– For up to $200 in total expenses (a crossover point in *Plan 100*) you pay all expenses in both plans, and so for $200 expenses you pay $720 + $200 = $920 in *Plan 80* and $3,840 + $200 = $4,040 in *Plan 100*, which is the maximum expense for this plan.

– From $200 to $600 in total expenses (a crossover point in *Plan 80*) you pay all the expenses in the first plan and nothing in excess of $200 in the second, and so for $600 expenses you pay $720 + $600 = $1,320 in *Plan 80* and still $4,040 in *Plan 100*.

– For expense amounts in excess of \$600 you pay 20% in *Plan 80* (up to a limit, as we will see) and nothing extra in *Plan 100*. So, up to \$14,200 in annual medical expenses, including premiums, you pay less in *Plan 80* than in *Plan 100*, and at \$14,200 you pay an equal amount: \$720 + \$600 + (\$14,200 – \$600) × 0.2 = \$720 + \$600 + \$13,600 × 0.2 = \$720 + \$600 + \$2,720 = \$720 + \$3,320 = \$4,040 in *Plan 80* (with \$3,320 out of pocket), and still \$4,040 in *Plan 100*. (This crossover or tipping point of equal cost can be obtained by algebra[243] or by plugging in a trial medical expense number, then by increasing the expenses if *Plan 80* costs are lower and decreasing them if *Plan 80* costs are higher, until the cost of both plans are equal.)

– For annual total expenses beyond \$14,200, you now pay more in *Plan 80* than in *Plan 100* and you pay more until you reach the maximum out-of-pocket expenses of \$3,750, which occurs for \$16,350,[244] for which the total, including premiums, is \$720 + \$600 + (\$16,350 – \$600) × 0.2 = \$720 + \$600 + \$15,750 × 0.2 = \$720 + \$600 + \$3,150 = \$720 + \$3,750 = \$4,470 in *Plan 80* and still \$4,040 in *Plan 100*.

So, unless you expect very large expenses (>\$14,200), *Plan 80* is the financially wiser and less risky deal. The graph in Figure 20.5 makes this very clear. In any case, the message is: (Methodically) *Do the Math!*

Life and medical insurance are supposed to be actuarially fair and at the fair market value, but this is not true for all types of insurance. People once bought flight insurance much more than they do now and paid premiums that were much higher than the actuarially fair market value. In contrast, historically they have tended not to buy flood insurance even

---

[243] Algebraically, the equation of interest is \$720 + \$600 + (x – \$600) × 0.2 = \$4,040. It is solved by subtracting \$720 + \$600 = \$1,320 from both sides of the equation [to get (x – \$600) × 0.2 = \$2,720], dividing both sides by 0.2 (to get x – \$600 = \$13,600), and then adding \$600 to both sides (to get x = \$14,200).

[244] Algebraically, the equation of interest is \$600 + (x – \$600) × 0.2 = \$3,750. It is solved by subtracting \$600 from both sides of the equation [to get (x – \$600) × 0.2 = \$3,150], dividing both sides by 0.2 (to get to get x – \$600 = \$15,750), and then adding \$600 to both sides (to get x = \$16,350).

when it was subsidized and so it was available at much less than the fair market value.[dccxliii]

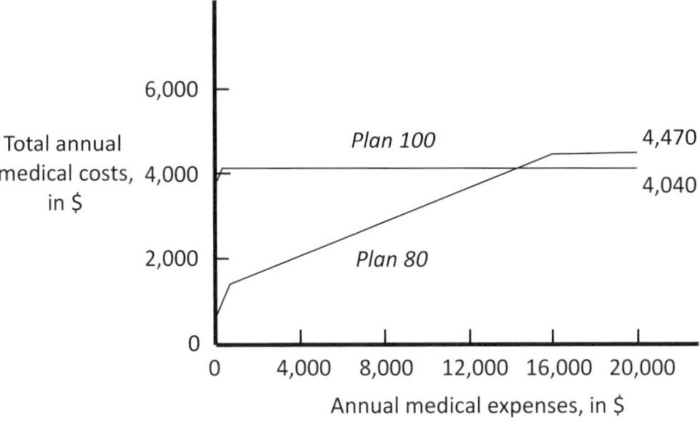

Figure 20.5: Comparing the total costs of the example Plans 80 and 100 (note the different horizontal and vertical scales).

So, perception affects how we view insurance and risk in ways that may be mathematically inconsistent. A study published in 1993 presented the results of a poll that asked how much those polled would pay for $100,000 of flight life insurance that would cover either any act of terrorism, any non-terrorism related mechanical failure, or any reason.[dccxliv] On average they were willing to pay $14.12, $10.31, and $12.03 respectively. The first two categories do not overlap and the third last category includes the first two and other possibilities as well, so actuarially the third premium would be expected to be set a bit higher than the sum of the first two (because it includes additional categories or risk). However, fear of terrorism then and how the choices were framed led people to make poor choices and, in effect, be willing to pay more for less coverage. This is not logical. We will revisit such math of risk perception and decision making later this chapter.[dccxlv]

20.3 Minimizing Risk in Options and other Investments in the Financial Market

20.3.1 Risk in managing market investments

Financial investments at all levels entail evaluating the tradeoff of risk vs. the desired reward and managing that risk,[dccxlvi,dccxlvii] including your tolerance for this risk.[dccxlviii] We have examined this for insurance. Common advice in minimizing risk in your total personal financial assets is to diversify your assets. This lessens variations in the value of your total portfolio due to random changes, so a decrease in the price of one stock you own in one company perhaps due to an unexpected strike would *be expected* to be cancelled by random positive changes in others. It is relatively unlikely that all random changes will be negative at the same time if the overall market is fairly steady. Diversification can also help manage risk due to broader market changes. The goal of mutual or investment funds is to professionally manage a set of investments for specified targets of risk and reward. Mathematical models and projections have been used increasingly to do this and to address other aspects of finance as well.

Those developing and implementing mathematical models for quantitative analysis of finance are sometimes referred to as quants.[dccxlix] The models and equations they develop may seem complex, but their underlying math concepts are easy to understand. They can be solved exactly (analytically) in some cases, but are usually analyzed using numerical, computational analysis. They are examples of stochastic math using random walk analysis. Quants devise and rely on algorithms based on past and essentially real-time data to constantly make investments decisions, rather than on their instincts, and are becoming evermore successful and important in the investment world.

Quantitative hedge fund accounts increased from ~14% of U.S. stock trading in 2013 to ~27% in 2017. They have been managed by algorithms that are strongly based on mathematical models, but are increasingly being controlled by neural nets within the realm of machine learning. Such a strategy is not better *per se*, but if done right, it can be. Financial meltdowns, as in 2007, can occur when many of the algorithms used are too similar—and so they react to market conditions the same way and accentuate trends.[dccl] (See *Black Swans* below.)

Risk is evaluated in light of how the value of a financial asset with a given initial price would increase through a geometric progression at the return rate per period, which may be assumed to be fixed, to arrive at the *future value* after a given number of time periods. Sometimes a proposed investment is compared for a known fixed interest rate. If the interest rate is 4% per year, or 1% per quarter, after the 3 years, or 12 quarters, the future value of \$100 would be $\$100 \times (1 + 0.01)^{12} \approx \$100 \times 1.1268 =$ \$112.68.[245] The *discounting or present value* of a future payoff in a given number periods is just the reverse, so for a future value of \$100 the present value would be $\$100/(1+0.01)^{12} \approx \$100/1.1268 \approx \$88.74$ for the same length of time.[246]

Risk is also evaluated in terms of the market *volatility* of an asset, which is its variability about the expected mean. This can be assessed in terms of the past performance of the asset, characterized in terms of percentage changes over a series of periods and by the average and standard deviation of these changes. This standard variation, the square root of the variance, is often equated to the volatility. For example, in their book, *Preparing for the Worst: Incorporating Downside Risk in Stock Market Investments*, Rick Vinod and Derrick Reagle tracked the changes in the Standard and Poor's (S&P) 500 Index annual return from 1990 to 2000, over which it ranged from -10.14% to 34.11%.[dccli] Its average annual return was 13.74% and its standard deviation was 14.96% (over a year). Over this same period, the annual return from U.S. three-month treasury bills ranged from 3.00% to 7.50%, its average was 4.94%, and the standard deviation was 1.15% (for the year). The index had more average growth, larger volatility, and more (though occasional) periods of losses.

The *value at risk* (VaR) is an assessment of the highest expected loss of an investment assuming past return averages and volatility. It is sometimes calculated using a normal distribution of the return, given your choice of the probability of a given loss that may be acceptable. If

---

[245] After N periods, the future value is $P_N = P_0(1 + r_N)^N$ for a given initial price, P_0, increasing through a geometric progression at the fixed return rate of r_N per period.

[246] The present value is $P_v = P_N/(1 + r_N)^N$, N periods before the future value of P_N for a fixed return rate of r_N per period.

you start with \$100, a mean return of 4% over the time period, and a volatility or standard deviation of 2% (for this time period), half of the time you would expect to leave with an average rate of return or more, or \$104. The cumulative probability in the normal distribution is ≈68.27% within one standard deviation of the mean, and half of it that is between the mean and the mean minus one standard deviation, so the cumulative probability above a 4% – 2% or 2% gain is ≈68.27%/2 + 50% or ≈84.1%; consequently, the probability of you leaving with \$102 or more would be expected to be ≈84.1%. Since the cumulative probability in the normal distribution is ≈95.45% within two standard deviations of the mean, the cumulative probability above a 4% – (2 × 2%) = 0% gain is ≈95.45%/2 + 50% or ≈97.7%, so the probability that you will leave with \$100 or more would be ≈97.7%; consequently, the probability that you would expect to lose any money would be ~2.3%. Because the cumulative probability in the normal distribution is ≈99.73% within three standard deviations of the mean, the cumulative probability above or 4% – (3 × 2%) = -2% is ≈99.73%/2 + 50% or ≈99.865%. So, the probability of you leaving with \$98 or more (and therefore lose no more than \$2 of your initial investment) would be expected to be ≈99.86%. The probability would be ≈0.14% or ≈1 in 700 that you will lose as much as \$2. The amount you could lose, or the value at risk, would be \$100 – \$98 = \$2, which can occur a fraction ≈0.14% of the time (or ≈1 out of 700 times), <u>if</u> the input parameters are correct and <u>if</u> the distribution is the normal distribution exactly. Values at risk are often cited for 1%, 5%, and 10% fractions of initial investments, and in practice are usually calculated assuming a mean gain of zero.

How does math show that diversification partially counters risk? The math of variances, covariances, and correlations from Section 16.2.1 shows why diversification lowers risk: random changes in different investments tend to cancel to some degree.[247] (Warning: The wordiness of this discussion shows why algebraic notation can simplify the analysis

---

[247] For different securities with prices x_1 and x_2, average returns μ_1 and μ_2, and variances σ_1^2 and σ_2^2, a portfolio with a fraction w in security in 1 and $1 - w$ in security 2, has a value $wx_1 + (1 - w)x_2$. The mean return of the portfolio is $w\mu_1 + (1 - w)\mu_2$ and its variance is $w^2\sigma_1^2 + (1 - w)^2\sigma_2^2 + 2w(1 - w)$ cov(1,2). The correlation coefficient $\rho = \text{cov}(1,2)/\sigma_1\sigma_2$.

and discussion! See the footnote about this if you like.[248]) Consider two different portfolios of stocks, 1 and 2, each with the same initial value. Each evolves with time and its own variance can be determined; we assume the variances are equal. The covariance of the two portfolios can also be tracked. The correlation coefficient is this covariance divided by the square root of the product of their individual variances, which is the individual variance here, and so this coefficient could range from -1 (for perfect anticorrelation) and 1 (for perfect correlation). So, the covariance can range between minus and plus the variance of the portfolio. Now consider a portfolio of the same initial value that is composed of half of the stocks that were initially in portfolio 1 and half of those in portfolio 2. The total variance is now one quarter of the sum of the variances of the original portfolios 1 and 2, which is half of either one of them, plus half the covariance of the two original portfolios. For perfect correlation, this sum is the original portfolio variance and for perfect anticorrelation, it is zero. In general, it can range between these two limits. For no correlation, the total variance is half the original individual variance. For any non-perfect correlation, the total variance is still less than that of either original portfolio and so diversification has lowered the variances and the risk. These conclusions also apply to combining unequal portfolios and to combining many of them. Such diversification helps with asset-specific, random risk, but cannot eliminate systematic market risks, as from disasters, wars, currency crises, worldwide recessions, and extreme weather.

The standard deviation of the returns in a typical stock portfolio decreases with an increasing number of stocks from different firms, first rapidly down from the ~16% standard deviation for 1 stock, then slowly to ~6% with 30 stocks (assuming the same average correlation), and then much slower with additional stocks because the remaining market or systematic risk tends to affect all stocks to some degree.[dcclii]

---

[248] Using the previous footnote, with initial equal amounts $w = 1 - w = 0.5$ and with the same variances σ^2, the variance of the total portfolio is $0.25\sigma^2 + 0.25\sigma^2 + 0.5\ \text{cov}(1,2)$. With $\text{cov}(1,2) = \rho\sigma^2$, the variance of the total portfolio is $0.5\sigma^2 + 0.5\rho\sigma^2 = 0.5(1 + \rho)\sigma^2$, which ranges between σ^2 for $\rho = 1$ and 0 for $\rho = -1$.

Let us look at combinations of different financial investments, which could include individual stocks and bonds and bills, and stock market funds, and parallel your own financial portfolio. Vinod and Reagle examined the monthly returns from January 1997 to September 1999 for the Fidelity Magellan and Vanguard 500 funds, and for three-month T-bills.[dccliii] They had averages, variances and standard deviations of: 1.94803%, 26.42334%, and 5.220064% for Magellan, 1.926667%, 23.86518%, and 4.960944% for Vanguard, and 0.410879%, 0.000718%, and 0.027211% for T-bills. The covariances and correlations were 24.40821% and 0.971986 between the Fidelity and Vanguard funds, -0.02557% and -0.18563 between Fidelity and T-bills, and -0.01697% and -0.12964 between Vanguard and T-bills. The two stock funds tracked either other very well (the correlation coefficient was very nearly 1), so combining the two funds lowered risk only a bit. Combining either stock fund with T-bills lowered risk due to the imperfect (and in fact negative) correlation, but it also lowered average return.

Diversification even among reasonably correlated investments helps limit losses (!), albeit with the concomitant limitations in gains, as has been noted by Meir Statman.[dccliv] The annual correlation between U.S. and foreign stocks is "generally" high, ~0.90. In 2008, U.S. stocks lost ~37% of their value while foreign stocks lost ~43%, so the loss would have been ~40% for someone investing equal value in each at the beginning of the year. In 2009, U.S. stocks gained ~26% of their value while foreign stocks gained ~32%, so the gain would have been ~29% for someone starting with equal value in both categories at the beginning of 2009. Those investing equally in both at the beginning of the 2-year period, lost less in the first year, but gained less in the second year than those investing only in foreign stocks, while they lost more in the first year and gained more in the second year than those investing only in U.S. stocks.

However, "the devil is in the (investment and math) details." Those starting 2008 by investing $10,000 would have ended 2009 with $7,938 in this recovering market if they had invested in U.S. stocks only (=$10,000 × (1.0 – 0.37) × 1.26), with $7,524 in foreign stocks only (=$10,000 × (1.0 – 0.43) × 1.32), and with $7,731 (= ($7,938 + $7,524)/2) for initial \$5,000 investments in each. In this last case, the total investment remaining at the end of 2008 (=\$5,000 × (1.0 – 0.37) + \$5,000 × (1.0 – 0.43) = \$6,000) could have been purposely reinvested equally in U.S. and foreign stocks at the beginning of 2009 (so, "rebalancing" the investment). Then, at the end of 2009, the diversified portfolio would be different and slightly higher in this specific case, \$7,740 (= \$3,000 × 1.26 + \$3,000 × 1.32), compared to \$7,731 without rebalancing. This is why some recommend you at times *rebalance* a diversified portfolio rather than just sticking with it.

What is the math behind managing and minimizing the risk of financial assets, other than by diversification? The price of a stock or other financial asset can be modeled, in part, as a random walk that is a function of time, to determine its rate of return. Part of the change in its value over time is on average expected to be deterministic, and equal to (the stock value) × (a given *drift* or mean rate, in dollars per unit time, such as years) × (a small, differential time).[249] Because this is thought to be a random walk process, the possible values of the stock price at later times will be a log-normal distribution, with this average value, but with a variance that increases quadratically as the *volatility* and linearly with time.[250] This part of the change in the stock value is (the stock value) × (this volatility) × (the differential change in a variable that varies randomly, which randomly weights the volatility factor).[251] The differential describing the evolution of the change in stock price is the sum of these two changes, the average and the stochastic parts in this model of stock returns.[252,dcclv]

How does this variation compare with changes you would expect from more secure investments? In the financial market, you can buy or sell commodities or assets right now at the current or *spot price*, or in the future by buying an option, and paying the option seller a fee or *premium*

---

249 So, this differential change or drift of the stock price S, dS, in a small time, dt, would be $dS = S\mu dt$ for *drift* of mean rate μ.

250 Then, the volatility or standard deviation will have a characteristic width that increases as the square root of time.

251 So, this random part of the change in stock price dS is $S\sigma dW$, where the volatility is σ and the differential stochastic variable that simulates a random walk is dW.

252 So, this model of stock returns is $dS = S\mu dt + S\sigma dW$.

for this right. The buyer of a *call option* can, but need not, buy a commodity from the option seller for the agreed-upon *strike price* at (for a *European option*) or before (for an *American option*) the option expiration date. The option buyer hopes that the commodity price will increase beyond the *strike price*, to make money by buying it at the strike price and then selling it. If it does not, the option buyer can do nothing and lose only the fee. The buyer of a *put option* can, but need not, sell a commodity to the option seller for the agreed-upon *strike price* at (*European option*) or before (*American option*) the option expiration date. The put option buyer hopes that the commodity price will decrease below the *strike price*, to make money by selling it at the strike price and then buying it. If it does not, the buyer can do nothing and lose only the fee.

Though these may seem to be risky propositions, they are really designed to control the level of risk. For a call option, if the stock price does not increase by at least the premium you do not exercise your option and all you lose is the premium, but if it increases by more you reap the great increase in stock price (less the premium). So-called financial *instruments*, such as these options, whose value is *derived* from some other asset are called *derivatives*. (They are distinct from the mathematical operation of finding the derivative of a function in calculus.) A *hedger* uses market instruments to reduce risks, while a *speculator* uses them to increase risks and achieve greater returns. An *arbitrageur* seeks to buy and sell risk at different prices to make profits without incurring any risk.[dcclvi]

The value of a stock in the future held for a time T can be compared to what its value would be at that later time if its current value were instead invested in risk-free bonds. (This increase would be at a continuously compounding rate r, so this later value would be $e^{rT} \times$ its current value.) No variation in the stock price is assumed for now, but the stock value would be enhanced by dividends. (We assume this occurs at a continuously compounding rate d, so this value would end up being $e^{dT} \times$ its current value.) However, stock prices do change, and there is a risk and the price of an option needs to reflect these variations in light of asset changes due to bond interest and stock dividends.

You would expect the price change of the option would be proportional to the expected change in stock price, and that the stock price change would vary as the price of the stock and the degree of variability or volatility during this time.[253] Volatilities should be smaller for shorter periods and, for a random walk, it should be smaller by the square root of this fractional period, so the volatility for 3 months = 0.25 years would be that for a year × the square root of this period, or $(0.25)^{1/2} = 0.5$ here. If the spot and strike prices are the same, say $100, the annual volatility is 10%, and the time factor for this volatility for an option expiring after 3 months is 0.5, one could expect that the option price change would be the product of these three factors times a factor on the order of 1. Better calculations, including one using the below *Black-Scholes equation*, show that this is true and that this factor is very nearly 0.4,[254] so the call option price change would be $100 \times 0.1 \times (0.25)^{1/2} \times 0.4$ = $2. The cost of money (i.e. interest) and differences in the spot and strike prices would modify this. (The time scales used in such calculations are usually shorter than 3 months.)

Such analysis is used to follow the price of a call option on a stock, which is a function of both the stock price and time. This is done by seeing how small changes in the call option price (the differential of the option price) can be caused by small changes in the stock price and time (the differentials of both).[255] Our model of stock returns is then used to replace this differential of the stock price, and so the differential of the call option price now has a random, stochastic component. If a fund

---

[253] The value of the option, the premium, depends on: (1) the time before the option expires (the longer the duration, the more valuable the option), (2) the difference in the current and strike prices (strike prices lower than current prices makes the call option more valuable), (3) the income the buyer can make on that person's current money (due to interest) and the seller can make on a current asset (such as, dividends) (the more these are the more the option price), and, most importantly, (4) the expected volatility (uncertainty in asset value) during the option period (the more the volatility, the more the option).

[254] So, this product would be $0.4S\sigma T^{1/2}$ for time period T.

[255] Using what is known as the Taylor series expansion, this can be expressed as $dC = (\partial C/\partial t)dt + (\partial C/\partial S)dS + (1/2)(\partial^2 C/\partial S^2)(dS)^2$, where S is the stock price and C is the price of a call option on the stock (which depends on the stock price and time, so is $C(S,t)$).

owns both stocks and options, risk can be removed from it, at least mathematically, by cancelling out the stochastic part of the changes in the total fund holdings by using this equation and the model of stock returns, and by judiciously choosing and managing the relative amount of stock and options the fund has.[256] If the entire portfolio is indeed free of risk, then its total value must be free of risk, and the drift in the total holdings must grow at the risk-free interest rate r. This results in the well-known *Black-Scholes Equation* for setting option prices. (The seemingly endless wordiness of this discussion is yet another example of the value of math notation and equations; the relevant equations are presented in the footnotes. Admittedly, the level of math needed to obtain them would challenge any college student who had not taken a few terms of calculus.)[257,dcclvii,dcclviii]

The Black-Scholes equation is solved by using analytic and, more usually, numerical computing methods, and it is evaluated at the option expiration. If you decide to ignore the risk-free interest rate, which the fund would not do, the Black-Scholes equation becomes identical to the usual equations describing particle and heat diffusion. They are also governed by random walk mathematics.[dcclix]

This Black-Scholes mathematical model and equation for option pricing are used by investment banks and hedge funds to set prices for options to *hedge the option* (and more generally for determining economic valuations). The goal is to minimize risk for European options (and, with a modified version, for American options), by buying and selling the asset in question in the way the model suggests. This method was developed by Fischer Black and Myron Scholes in 1973 and the math underpinning option pricing was furthered by Robert C. Merton in the same year. Merton and Scholes shared the 1997 Nobel Memorial Prize in Economic Sciences for this work; Black had died in 1995 and so

---

[256] If the fund owns α stocks for each option, you can show that the change in the holdings is proportional to $d(C + \alpha S)$ and the stochastic differential variable dW part cancels out when α is set equal to $-\partial C/\partial S$.

[257] The Black-Scholes equation is $(\partial C/\partial t) + (r-d)S(\partial C/\partial S) + (1/2)\sigma^2 S^2(\partial^2 C/\partial S^2) = rC$, where r and d are the interest and dividend rates.

was ineligible for the prize, but was noted as a contributor in the award information.

The financial crash of 2008 has often been attributed to several factors, including the widespread use of the Black-Scholes and similar equations. This is questionable. However, if it did play a role, it would not have been due to the use of the equation, but its misuse and the misuse of other math of risk, through the use of unjustified model parameters and glaringly simplistic, improper and unrealistic assumptions that market risk was much smaller than it really was.[dcclx,dcclxi] Part of the problem was that some did not fully recognize that statistics entails distributions and that points within of 2.33 standard deviations of the average in a normal distribution are expected to occur only 99% of the time. The 1% of points outside the distribution (really the half of them in the lower tail) may lead to *failure* and cannot be ignored (Figure 15.2). Such unlikely events, or *Black Swans*, are more probable than some in the financial market had thought[dcclxii,dcclxiii] and the market systems should have been designed to 99.99...% certainty.[dcclxiv] Moreover, the assumption that the relevant distributions were precisely normal distributions was not well-founded; the outer limits on the wings may be more probable than these distributions would suggest.

The math used to set insurance rates is similar to that used in option pricing, to minimize the risks taken by the insurance company.

20.3.2 Risk in retirement financial planning

Given the increasing scarcity of pensions and the limited funds from Social Security, you are risking your future in the U.S. if you do not perform long-term financial planning for retirement through savings and investments, 401K plans, and the like. The math of financial planning entails deterministic math based on past averages of performance projected into the future—which is itself a risk—and stochastic math based on random events and unexpected long-term trends.

Retirement planning is risk planning. It entails analysis of Social Security payments in the U.S. and pension payments to you and other potential sources of income, which may depend on the money you and your company put in your retirement accounts and on your other savings. This analysis also needs to incorporate when you expect to retire, how long you are expected to live past retirement (using guidance from actuarial tables, with their concomitant uncertainty), and how much you want to live on. Online calculators can give you some guidance,[dcclxv] but only if you understand the key math issues. Some guidance comes from rules of thumb based on experience, such as you can live on 70-80% of your pre-retirement income,[dcclxvi] but with your extra free time you might really want to live on 130% of this income;[dcclxvii] other guidance might be based on the below estimates. (I am offering no financial advice, just noting you should be comfortable with some of the relevant "math thinking.")

What can you expect for the future based on the past? Jeremy Siegel and others have argued that stocks are the best investment in the long term.[dcclxviii] He claims the annualized return from 1802 to 2012 has been 8.1% for stocks (including dividends), 5.1% for long- and short-term U.S. Treasury bonds, 4.2% for bills (or interest rates on long- and short-term bonds, treasury bills/T-bills), 2.1% for gold, and 1.4% for the U.S. dollar. The real annualized return, adjusted for inflation over that period was 6.6% for stocks, 3.6% for bonds, 2.7% for bills, 0.7% for gold, and -1.4% for the U.S. dollar. He also showed that there are shorter- and longer-term fluctuations in these rates that can strongly impact your specific investing and retirement decisions. Though the 15-year periods from 1966 to 1981 and 1982 to 1999 the real stock returns averaged -0.4% and 13.6% per year, respectively, very different returns, their average, 6.6%, is not far from the average return of stocks over even longer periods. You would be unhappy if your critical years were during the former period.

From 1871-2012, stocks have been much riskier than bonds for 1- and 2-year holding periods. This means that they have had possibilities of larger losses for such durations, but these differences largely vanish for 5-year holdings. Then they are progressively less risky for longer holding periods. Stocks outperformed bonds 61.3% of the time and T-bills 66.9% of the time over 1-year holding periods. This increased to 69.0% and 74.6% respectively over 5-year holding periods and to 99.3% and 100.0% over 30-year holding periods. *Will the future see the same trends?* For you, the crux may be to manage short-term fluctuations, because if you do not manage such risks you may be unhappily surprised while trying to maximize your longer-term interests. As we just saw, diversification lowers risk. Retirees and near-retirees are often advised to have a portfolio rich in bonds because of the shorter-term fluctuations in stocks. Part of your planning is to assess the tradeoff of having enhanced lower risk to ensure a steady income or enhanced higher risk for possible higher future income, and also to decide whether or not your planning includes leaving some funds for your heirs. Professional analysis of your retirement portfolio entails simulations of many scenarios that are built on your goals, the level of risk you are comfortable with, and the historical performance of the market.[dcclxix]

What is the math behind how you should be using your savings? In Section 13.2 we saw that money in a fund that pays fixed interest increases by compounding interest and that mortgage payments include partial payment of the principal and interest on the remaining principal, which compounds every month. In a funds withdrawal plan, you receive income for an extended time from a fund that continues to accrue interest, which may or may not be guaranteed. The standard goal is to receive a good income protected from inflation, with very little risk of zeroing out the principal from your retirement account during your generously-calculated expected life span after retirement. According to actuarial tables, the life expectancy of those turning 65 in 2016 was 17.92 years for men and 20.49 for women,[dcclxx] but planning should be for longer periods, perhaps for 30 years to be on the safe side. Zeroing out your reserves could happen if you extract too much money each year, if market interest rates are low and become lower, and if inflation becomes higher. If you extract little, you lessen the probability of it zeroing out and you may have funds to leave for your heirs, but you may prefer to live with higher expenditures.

If the "interest" rate of your financial portfolio were fixed over the duration and your lifespan were exactly known, you could use your financial resources to pay yourself a fixed amount for the rest of your life. This math is exactly the same as that of fixed-rate mortgages, except in reverse. Instead of an amount you owe, there is an amount you own in your financial portfolio. Instead of your remaining loan balance increasing at a fixed interest rate, your remaining bank account increases at a fixed interest rate. Instead of you paying back part of your loan each month, you withdraw a fixed amount each month. You could easily account for a fixed inflation rate and adjust your withdrawals accordingly. Though the rates and payments in a fixed-rate mortgage are deterministic, the actual interest rate, how inflation affects your needed expenditures, and your actual lifespan are stochastic.

In a funds withdrawal plan, you can manage your resources in some sort of static or dynamic manner by investing your resources and managing your withdrawals, possibly helped by simulations. Alternatively, you can purchase annuities that take these risk factors into account and offer you a known *streaming income*, but your heirs may not get any benefit from potential leftover funds. You will derive more financial benefit from your annuity than you put in if you have a longer-than-expected lifespan and less if you have a shorter-than-expected lifespan. All approaches involve uncertainty. We will look at the core numbers in common strategies that try to account for different tradeoffs between fund growth, fund withdrawals, and risk when you manage your own nest egg. Please (!) check with your own financial advisor before making any decisions.

One standard protocol is to receive 4.0% of the initial fund amount per year, and to spend this same amount each year, with increases for inflation over the 30-year timeframe of the withdrawals. With no fund investment, this exact initial amount would last for exactly (100/4% =) 25 years. Simulations using a range of past market and inflation scenarios (interest rates, stock market changes, and so on) predict that this 4.0% strategy plus inflation strategy is "safe" and can provide stable income for the duration with very low risk, with the right blend of stocks and bonds (often with at least half invested in bonds). Supposedly safe withdrawals rates are higher for shorter periods, 5.0-5.5% over 20 years, and lower for longer periods, 3.5% over 40 years. Some projections claim a 95% level of confidence, which means that your money would be expected to run out before you do in as many as 5% of the scenarios they have chosen. This is a risk. Some, as the ones just presented, have higher confidence levels and are called "safe," but they still have risk.[dcclxxi] This safety factor means that much money remains in the fund after 30 years in most scenarios, which is bad if you want to extract as much income as possible because *You Can't Take it With You* (the title and theme of

George S. Kaufman's play and movie). However, this is fine if you want to lessen risk and pass a large fund to your heirs.

Some think the 4% strategy is far too cautious and suggest alternatives. Starting at the larger fixed rate of 5% per year would work 90% of the time for 30 years, based on historical trends with 60% stocks and 40% bonds. Can you accept this 10% risk of failure and adjust to less money down the line to adjust for inflation and market fluctuations to continue this for 30 years? A more aggressive strategy starting with 6.5%, with adjustments only for inflation, would have failed in 50% of the sample 30-year periods since the late 19th century. Though the 4% strategy would have worked even during the Great Depression, some think it is still too risky on the basis of simulations using future projections made early in 2018 (rather than historical performance) and they recommend a 3% strategy.[dcclxxii]

A more dynamic approach allows larger withdrawals when possible by managing the risk annually. You need to be comfortable with the math to do this. In one of several strategies, your actual withdrawals can increase in years your fund increases, by up to a fixed % ceiling, say 5%, but they will decrease in years your fund decreases, by at most a fixed %, say 2.5%, so there can be more income with set ceilings and floors that can change annually.[dcclxxiii] In one safe strategy that adjusts to large increases and decreases in the fund, in the first year you withdraw 5% and then annually you increase your most recent withdrawal for inflation. But, if after this adjustment your withdrawal would be less than 4% of your remaining funds you can increase your withdrawal to 10% above this amount and if it is more than 6% you need to decrease your withdrawal to 10% below this amount.[dcclxxiv] (Again, I am not endorsing any particular strategy, but only that you should explore the math and the risk.)

One such retirement income vehicle that controls many of our futures in the U.S. is the "tax-deferred" individual retirement account (IRA). The specific rules governing these accounts change in time, so we will consider a plausible scenario. (Check the current rules!) Say you are required to withdraw money from them starting at age 72; you may use the withdrawn funds or invest or save them in a different type of account. Each year the minimum you must withdraw is the remaining amount at a specified time divided by your *life expectancy factor* in years (also called the distribution period) as is specified by the U.S. government in widely available tables. Say this factor starts at 25.6 at the age of 72. (Usually, it is very roughly twice your actual life expectancy in years[258] and initially decreases by roughly 0.9 years for every year you advance.[dcclxxv]) If your account is initially $100,000, you must withdraw at least $100,000/25.6 = $3,906 the first year and you will have $96,094 left for the next year if you withdrew the minimum, assuming no growth or loss in the investment or loss due to IRA management fees—as is assumed here. At age 73, say the factor is 24.7, so you would need to withdraw $3,890 minimum and if you do so you would have $92,204 left, and so on. The minimum amount you need to withdraw is currently roughly constant from year to year if you do not take out more than this minimum.

What happens if your remaining lifetime greatly exceeds that expected from actuarial tables? What happens if your needs for expenditures skyrocket? What happens if there is a great inflation? What happens if your expectations about a stable income from Social Security are wrong? This is all part of the risk in life and the math of the risk. Luckily, there is no risk of the math itself being wrong!

During your path to retirement, at times you may need to pay off debts as well as to save. Whether it is better to pay off debts, such as student loans, before or after you begin to start longer-term savings, as in IRAs, is a longstanding question that depends on the math of compound interest and the math of risk. For example, some thought that in the financial climate of 2017 it was better to pay off debt quickly for debt interest rates above 3.75% and to make only minimum payments for debt for interest rates lower than roughly 3.75%.[dcclxxvi] (Once again, I am not endorsing any particular strategy, but only that you should "Do the math!".)

---

[258] In 2016, the life expectancies in years were 14.40 for men and 16.57 for women at age 70; 13.73 and 15.82 at age 71; 13.07 and 15.09 at age 72; 12.43 and 14.37 at age 73, and so on. At the same time, when the age for required withdrawing began at age of 70½, the life expectancy factors were 27.4 at 70, 26.5 at 71, 25.6 at 72, 24.7 at 73, and so on.

20.4 The Math of Risk Perception and Associated Decision Making

After seeing a mathematical risk analysis, what decision will you make? Your perception of risk and decision-making can also be analyzed with math tools, as has been known for some time. In 1738 Daniel Bernoulli noted that people are usually averse to risk and that this aversion decreases with increasing wealth.

Different groups perceive risk levels differently, as uncovered by Paul Slovic. In 1987, college students and the League of Women Voters rated nuclear power as the number 1 risk of 30 selected activities and technologies, while experts ranked it only number 20. Experts ranked x-rays the number 7 risk, but these two groups ranked them as only numbers 17 and 22. All groups ranked vaccinations near the lowest in risk level.[dcclxxvii] Risk levels and perception change in time. Several in the general public now think vaccinations are quite risky, given the more prevalent misinformation about their real risks vs. real benefits, but the experts would still rank them as having low risk.[dcclxxviii]

Slovic also noted how hazard risks can be characterized on two distinct scales (described by two different variables), one from known to the unknown risk and the other from not dreaded risk to dreaded risk, and that there is little correlation between these two. For example, different unknown risks can be not dreaded and dreaded to different degrees. Non-experts perceive greater risk when the dread is great, while experts see it higher with higher expected mortality (a known risk). He also found that the more voluntary an activity, the higher the level of risk that is accepted.[dcclxxix,dcclxxx]

Sometimes, our decisions appear to be based on how the mathematics of the risk is posed and our perceptions of the outcomes, even for the same average numerical amount of risk. In 1979 Daniel Kahneman and Amos Tversky began studying how we make decisions for given risks, which they called *Prospect Theory*,[dcclxxxi] and learned that people generally avoid risks in decisions involving sure gains (suggesting that a bird in the hand is worth two in the bush). They found that 80% of those they polled preferred the certainty of winning 3,000 to the alternative 80% chance of the bigger payoff of winning 4,000 and a 20% chance of winning nothing, even though the latter had the higher average expected gain of 3,200. (The money units were Israeli pounds, at a time when the median Israeli monthly income was ~3,000 Israeli pounds.) People made different decisions when there was less certainty in the choices. 65% preferred a 20% chance of winning 4,000 and an 80% chance of winning nothing to a 25% chance of winning the smaller payoff of 3,000 and a 75% chance of winning nothing. The prospect of greater winnings when all options were less than likely was apparently the deciding factor, and not the slightly larger average expected winning in the first option (800 vs. 750 in the second option). Mathematically, these choices can be framed in terms of value functions weighted by decision weights. People generally accepted risks in decisions involving sure losses (going for broke). In a mirror image of our first example, 92% preferred a 20% chance of no loss and an 80% chance of losing $4,000 to a definite loss of $3,000. Clearly, the value function is not symmetric for losses and gains.[dcclxxxii]

How a choice is framed can also change a decision. To combat the expected death of 600 people from a disease, Kahneman and Tversky proposed two paths in which 200 people on average would be saved.[dcclxxxiii] In one path, 200 people would be saved and in the alternative there was a 1/3 probability that 600 people would be saved and a 2/3 probability that no one would be saved. They found that 72% preferred the first option. When faced with exactly the same options phrased differently, one choice where 400 people would die and in the alternative where there was a 1/3 probability that no one would die and a 2/3 probability that 600 would die, 78% approved the latter option instead of the 28% that had with the earlier phrasing. Such life and death decisions are often made by military officers selecting alternative ways of sending soldiers into battle armed with intelligence reports.

Such framing of options also affects our medical decisions. A broad spectrum of people without cancer were given options of hypothetically being treated for operable lung cancer by either surgery or radiation. They were given the statistics of the life and death outcomes, with surgery having the poorer short-term outcomes but the better long-term outcomes, and no information about side effects and so on. More than 40% chose radiation when the statistics were phrased in terms of mortality, but this fraction was cut in half when the same statistics were phrased in terms of survival.[dcclxxxiv]

Sometimes it is not clear how people rate risks and benefits, so they are not able to compare them when needed. Are the risks of pain endured, say for medical treatment, and the benefits of joyous events, such as vacations, determined by their average level during the event? Fredrickson and Kahneman found evidence that both are determined by the peak value and the ending value during the event (*Peak-end Rule*), rather than their average values.[dcclxxxv] Moreover, in 1994 Tversky and Derek Koehler showed that generally risk may be more acceptable when presented as the entire risk, rather than the risk associated with each component of the process.[dcclxxxvi]

Kahneman shared the 2002 Nobel Prize in Economic Sciences, with Vernon L. Smith, for the work he did in collaboration with Tversky on these new ways of evaluating risk. (Tversky had died six years earlier, and so was not eligible for the award; in his Nobel Prize lecture, Kahneman noted his work with joint with Tversky.)[dcclxxxvii] Analysis of how humans assess risk has continued, in part by Richard Thaler, who won the 2017 Nobel Prize in Economic Sciences for advancing this area of behavioral economics to see how human traits such as "limited rationality, social preferences, and the lack of self-control" affect decisions.[dcclxxxviii]

Perception also affects the choice of retirement plans. Annuity payments that start at 65 years of age currently pay monthly very roughly 0.5% of the lump sum, so the options of direct income from $1 million from a lump payment and monthly payments of $5,000 from a $1 million annuity are roughly the same on average. However, some perceive the large lump sum to be the more adequate route in providing their retirement needs, while others perceive the monthly payment mode to be more adequate. The former group has been said to "suffer" from the "illusion of wealth" and members of this group may undersave because they may overestimate the impact of the resources they have, while members of the latter group may suffer from the "illusion of poverty" and may worry excessively about running out of money. The lump sum route is generally more likely preferred with smaller initial funds and the annuity monthly payment mode is generally preferred with larger initial funds. But of course, the higher the beginning fund the more adequate the resources are deemed.[dcclxxxix] Perception enters the world of annuities in another way as well. They are less popular than expected, because many think it is unfair that the company issuing the annuity keeps any unused funds after they pass away (even though using the unused funds of those with shorter than expected lifespans to support those with longer than expected lifespans is what makes the annuity system work).[dccxc]

Sometimes people make decisions that seem to defy logic to some, even when the facts are clearly presented. *Would you participate in a betting scheme that returned only ~60% of the money bet to the bettors?* Of course not. The effective odds are terrible, and much worse than the relatively poor ≈94.7% returned on average per amount bet in American Roulette. Still, the answer is apparently yes, because this is the typical return of the very successful state-run lotteries. Of course, some may rationalize that this very small return is acceptable because the rest is apparently used to pay for education and the like and, in any case, this betting is a form of entertainment. *Would you participate in a betting scheme that returned ~55% of the money bet back to the bettors, when this money is distributed to very, very few people and the probability of the really big payout is 1 in more than 250 million?* One might think, of course not, especially when the outcomes and risks are phrased this way because the probability of you winning a bet would be so very small. Yet the answer is apparently yes, given the great participation in the Powerball and Mega Millions lotteries. For example, fewer than 4% of the participants won anything in the July 30, 2016, Powerball drawing, and 78% of these winners won $4. One ticket won $2 million and the big winner won a $487 million payout (before taxes).[259,dccxci,dccxcii]

Because this returned money goes to very few winners, it very unlikely that you will ever see your bet money again even if 100% of all bet money were returned. Jeff Sommer has suggested that given the risk of getting no return, you could be better off by investing $1,000 a year,

---

[259] To win this $1,000,000 Powerball Prize, the second biggest prize, you needed to choose 5 numbers correctly out of 69 (white balls), which as we have seen, has odds of 1 out of $69!/(64!5!)$ = 1 out of $(68 \times 67 \times 66 \times 65)/(5 \times 4 \times 3 \times 2 \times 1)$, or 1 out of ≈11.2 million. To win the biggest Powerball Prize, you needed to choose 5 numbers correctly out of 69 (white balls) <u>and</u> the correct number from the 26 hot-red "Powerballs", which is 1 out of ≈11.2 million × 26, or 1 out of ≈292 million. Lesser prize money was awarded for matching fewer balls.

the typical annual lottery loss for regular lottery players, from the age of 20 to 65. With 5% annual return it would produce $150,000.[dccxciii]

Closing Thoughts

You are now ready to count to ten:

1. Feel comfortable whenever you encounter math.
2. Accept that you probably know and understand more math than you think.
3. Realize that you are able to analyze math situations.
4. Ask questions about any new math you encounter.
5. Access and then learn any math method you need.
6. Make math an integral part of your mode of thinking.
7. Derive joy in thinking quantitatively.
8. See the math in the world around you.
9. Take control of the math that rules your life and "Do the math!" when needed.
10. Welcome home to your world of math!

Bibliography

[i] Queens College Professor Believes Algebra Is Unnecessary Stumbling Block Forcing Millions Of Students To Drop Out, *CBS New York/AP*, March 27, 2016, 1:27 PM, https://newyork.cbslocal.com/2016/0 3/27/andrew-hacker-algebra/; Karr, Jane. Who Needs Advanced Math? Not Everybody, *New York Times*, Feb. 5, 2016, https://www.nytimes.com/2016/02/07/education/edlife/who-needs-advanced-math-not-everybody.html.

[ii] Markarian, James. Who Needs Calculus? Not High-Schoolers, *Wall Street Journal*, May 14, 2018, 6:49 PM ET, https://www.wsj.com/articles/who-needs-calculus-not-high-schoolers-1526338152.

[iii] Wing, Jeannette M. (2008), "Computational thinking and thinking about computing," *Phil. Trans. R. Soc. A*, 366, 3717–3725; Wing, Jeannette M. Computational thinking, 10 years later, *Microsoft Research Blog*, March 23, 2016, https://www.microsoft.com/en-us/research/blog/computational-thinking-10-years-later/.

[iv] Hawking, Stephen, editor (2007, 2nd edition 2007) *God Created the Integers: The Mathematical Breakthroughs That Changed History* (Running Press Adult, Philadelphia).

[v] Conway, John H. and Guy, Richard K. (1998) *The Book of Numbers* (Copernicus Books, Germany).

[vi] Scheinerman, Edward R. (2017) *The Mathematics Lover's Companion: Masterpieces for Everyone* (Yale University Press, USA).

[vii] Gamow, George (1947, revised 1961) *One, Two, Three…Infinity* (Viking Press, New York).

[viii] *God Created the Integers.*

[ix] Stewart, Ian (2012) *In Pursuit of the Unknown: 17 Equations That Changed the World* (Basic Books, New York).

[x] Suri, Manil. The Importance of Recreational Math, *New York Times*, October 12, 2015, https://www.nytimes.com/2015/10/12/opinion/the-importance-of-recreational-math.html.

[xi] Martin Gardner (1914-2010), http://www.martin-gardner.org/, downloaded Dec. 13, 2018.

[xii] Rota, Gian-Carlo; Palombi, Fabrizio, editor (1997) *Indiscrete Thoughts* (Birkhauser, Boston-Basel-Berlin, 1997), pg. 89.

[xiii] Hoffman, Jan "Generations of math fears," *New York Times*, Aug. 24, 2015, 1:53 PM, https://well.blogs.nytimes.com/2015/08/24/square-root-of-kids-math-anxiety-their-parents-help/.

[xiv] Klass, Perri "Fending off math anxiety," *New York Times*, April 24, 2017, https://www.nytimes.com/2017/04/24/well/family/fending-off-math-anxiety.html.

[xv] Antonick, Gary "John Conway's wizard puzzle," *New York Times*, August 10, 2015, 12:00 PM, https://wordplay.blogs.nytimes.com/2015/08/10/feiveson-1.

[xvi] "Flight delayed when math mistaken for terrorism by passenger," May 8, 2016, *Fox News*, http://www.foxnews.com/us/2016/05/08/flight-delayed-when-math-mistaken-for-terrorism-by-passenger.html; Rampell, Catherine "Ivy League economist ethnically profiled, interrogated for doing math on American Airlines flight," *The Washington Post*, May 7, 2016, updated 11:20 AM, https://www.washingtonpost.com/news/rampage/wp/2016/05/07/ivy-league-economist-interrogated-for-doing-math-on-american-airlines-flight/.

[xvii] Groom, Winston (2015) *The Generals: Patton, MacArthur, Marshall, and the Winning of World War II* (National Geographic, Washington D. C.), pp. 72-73.

[xviii] Chernow, Ron (2017) *Grant* (Penguin Press, New York), pg. 23, Chapter 2, citing Eaton, *Grant, Lincoln and the Freedom*, pg. 256.

[xix] Ouellette, Jennifer (2010) *The Calculus Diaries: How Math Can Help You Lose Weight, Win Vegas, and Survive a Zombie Apocalypse* (Penguin, London).

[xx] Each quote is from "Doron Zeilberger's Collection of Quotes," updated Oct. 19, 2018, http://www.math.rutgers.edu/~zeilberg/quotes.html.

[xxi] Samuel, Alexandra "How I beat math phobia—and became a better entrepreneur," *Wall Street Journal*, Nov. 26, 2017, 10:06 p.m. ET, https://www.wsj.com/articles/how-i-beat-math-phobiaand-became-a-better-entrepreneur-1511751960.

[xxii] "The Lives They Lived: Maryam Mirzakhani," *New York Times*, https://www.nytimes.com/interactive/2017/12/28/magazine/the-lives-they-lived-maryam-mirzakhani.html. (The New York Times Magazine, Sunday, Dec. 31, 2017, pp. 22-23.)

[xxiii] Nasar, Sylvia (1998) *A Beautiful Mind* (Simon & Schuster, New York).

[xxiv] Lewis, Michael (2004) *Moneyball: The Art of Winning an Unfair Game* (W. W. Norton & Company, New York).

[xxv] McGinty, Jo Craven "Calculators in class: Use them or lose them?," *Wall Street Journal*, May 23, 2016, 2:10 PM ET, https://www.wsj.com/articles/calculators-in-class-use-them-or-lose-them-1463756482.

[xxvi] Willingham, Daniel T. "You still need your brain," *New York Times*, May 19, 2017, https://www.nytimes.com/2017/05/19/opinion/sunday/you-still-need-your-brain.html.

[xxvii] Schumacher, Carol S.; Siegel, Martha J. Co-Chairs, Zorn, Paul Editor (2015) *2015 CUPM Curriculum Guide to Majors in the Mathematical Sciences* (The Mathematical Association of America, USA), pp. 31 and 74.

[xxviii] Byers, William (2010) *How Mathematicians Think: Using Ambiguity, Contradiction, and Paradox to Create Mathematics*, 6th edition (Princeton University, Princeton).

[xxix] Levitin, Daniel J. (2016) *A Field Guide to Lies* (Dutton, Florida).

[xxx] Arbesman, Samuel "The misinformation age," *Wall Street Journal*, Sept. 16, 2016, 4:56 PM ET, https://www.wsj.com/articles/the-misinformation-age-1474059411.

[xxxi] *2015 CUPM Curriculum Guide to Majors in the Mathematical Sciences*, pg. 12.

[xxxii] Feynman, Richard (1967) *The Character of Physical Law* (The MIT Press, Cambridge, MA), 1967, Chapter 2.

[xxxiii] *The Character of Physical Law*, pg. 40.

[xxxiv] *The Character of Physical Law*, pg. 58.

[xxxv] Rosenberg, Jonathan, as quoted in "Doron Zeilberger's Collection of Quotes," updated Oct. 19, 2018, http://www.math.rutgers.edu/~zeilberg/quotes.html.

[xxxvi] Wigner, Eugene P. (1960) "The unreasonable effectiveness of mathematics in the natural sciences," *Communications on Pure and Applied Mathematics*, 13, 1-14.

[xxxvii] Dabbaghian, Vahid "What is mathematical modeling?," https://www.sfu.ca/~vdabbagh/Chap1-modeling.pdf, downloaded Dec. 13, 2018.

[xxxviii] Westlake, Donald E. (2013) *Dancing Aztecs* (Harry N. Abrams, New York).

[xxxix] International classification of diseases, tenth revision, Clinical modification (ICD-10-CM), *CDC/Centers for Disease Control and Prevention/National Center for Health Statistics*, https://www.cdc.gov/nchs/icd/icd10cm.htm.

[xl] Rosenthal, Elisabeth "Those indecipherable medical bills? They're one reason health care costs so much," *New York Times*, March 29, 2017, https://www.nytimes.com/2017/03/29/magazine/those-indecipherable-medical-bills-theyre-one-reason-health-care-costs-so-much.html.

[xli] Cooke, Patrick "Raising the bar," *Wall Street Journal*, July 29, 2016, 3:49 PM ET, https://www.wsj.com/articles/raising-the-bar-1469821771.

[xlii] Gellman, Lindsay and Baer, Justin "Goldman Sachs to stop rating employees with numbers," *Wall Street Journal*, May 26, 2016, 6:53 PM ET, https://www.wsj.com/articles/goldman-sachs-dumps-employee-ranking-system-1464272443.

[xliii] "Pareto principle," *Wikipedia*, https://en.wikipedia.org/wiki/Pareto_principle, downloaded Dec. 13, 2018.

[xliv] Federal Register, Vol. 44, No. 43, Friday, March 2, 1979; Question and answer #11, *The U.S. Equal Employment Opportunity Commission*, https://www.eeoc.gov/policy/docs/qanda_clarify_procedures.html.

[xlv] Bialik, Carl "Milestone figures grab attention, but their impact is hazy," *Wall Street Journal*, Feb. 12, 2010, 12:01 AM ET, https://www.wsj.com/articles/SB10001424052748703382904575059862880464510.

[xlvi] "deep 6," *Urban Dictionary*, https://www.urbandictionary.com/define.php?term=deep+6, downloaded Dec. 14, 2018.

[xlvii] "dressed to the nines," *Urban Dictionary*, https://www.urbandictionary.com/define.php?term=dressed%20to%20the%20nines, downloaded Dec. 14, 2018; Dressed to the Eights, *Urban Dictionary*, https://www.urbandictionary.com/define.php?term=Dressed%

20to%20the%20Eights, downloaded Dec. 14, 2018.

[xlviii] Quote 1,171 attributed to Sal Davino, as quoted in Byrne, Robert (2012) *The 2,548 Wittiest Things Anybody Ever Said* (Touchstone, New York).

[xlix] Verghese, Abraham (2009) *Cutting for Stone* (Vintage, New York), pg. 361.

[l] "Puzzle 324. Self-descriptive numbers," *Problems & Puzzles: Puzzles*, http://www.primepuzzles.net/puzzles/puzz_324.htm, downloaded Dec. 14, 2018.

[li] Weisstein, Eric W. "Self-descriptive number, From *MathWorld--A Wolfram Web Resource*," http://mathworld.wolfram.com/Self-DescriptiveNumber.html, downloaded Dec. 14, 2018.

[lii] "Six degrees of separation," *Wikipedia*, https://en.wikipedia.org/wiki/Six_degrees_of_separation, downloaded Dec. 13, 2018; Bialik, Carl, We're Far Removed From Proof of 'Six Degrees' Theory, *Wall Street Journal*, Aug 6, 2008, 3:53 PM ET, https://blogs.wsj.com/numbers/were-far-removed-from-proof-of-six-degrees-theory-391.

[liii] TV Show *JAG,* Season 10, Episode 2, "Fair Winds and Following Seas," Aired April 29, 2005 on CBS.

[liv] TV Show *Barney Miller*, Season 4, Episode 21, "Evaluation," Aired 8:00 PM, May 4, 1978 on ABC.

[lv] Bialik, Carl "Number-crushing: When figures get personal," *Wall Street Journal*, Oct. 28, 2009 11:59 PM ET, https://www.wsj.com/articles/SB125668948820711987

[lvi] Bialik, Carl "When numbers add up to more than math," *Wall Street Journal*, Oct 27, 2009, 9:04 PM ET, https://blogs.wsj.com/numbers/when-numbers-add-up-to-more-than-math-829; "Number-crushing: When figures get personal"; Bialik, Carl "Fearing Friday the 13th," *Wall Street Journal*, Apr 13, 2007, 3:23 pm ET, https://blogs.wsj.com/numbers/fearing-friday-the-13th-83; Dudley, Underwood (1997) *Numerology* (Math. Assoc. Am., USA).

[lvii] Gartenberg, Chaim "Happy Tau Day, the true circle constant," *The Verge*, updated June 28, 2018, 9:00 AM EDT, https://www.theverge.com/tldr/2018/3/14/17119388/pi-day-pie-math-tau-circle-constant-mathematics-circumference-diameter-radius-holiday-truth, downloaded Dec. 14, 2018; Bartholomew, Randyn Charles "Let's use tau—It's easier than pi," *Scientific American*, June 25, 2014, https://www.scientificamerican.com/article/let-s-use-tau-it-s-easier-than-pi/, downloaded Dec. 14, 2018; Hartl, Michael "The Tau Manifesto," *Tau Day*, Tau Day, 2010; updated Tau Day, 2018, https://tauday.com/tau-manifesto, downloaded Dec. 14, 2018.

[lviii] Bialik, Carl "Happy Square Root Day," *Wall Street Journal*, March 3, 2009 1:00 PM ET, https://blogs.wsj.com/numbers/happy-square-root-day-612; Inan, Aziz "Palindrome dates in the 21st century," http://faculty.up.edu/ainan/palindromedates21stcentury2011.pdf, downloaded Dec. 14, 2018; Bialik, Carl. Happy Pi Day, *Wall Street Journal*, March 14, 2007, 12:01 AM ET, https://blogs.wsj.com/numbers/happy-pi-day-60/.

[lix] *The Book of Numbers*, pg. 151.

[lx] Victor, Daniel; Stevens, Matt "United Airlines passenger is dragged from an overbooked flight," *New York Times*, April 10, 2017, https://www.nytimes.com/2017/04/10/business/united-flight-passenger-dragged.html.

[lxi] The Editorial Board "America needs a bigger house," *New York Times*, corrected November 15, 2018, https://www.nytimes.com/interactive/2018/11/09/opinion/expanded-house-representatives-size.html.

[lxii] "International language environments guide: Decimal and thousands separators," *Oracle*, 2010, https://docs.oracle.com/cd/E19455-01/806-0169/overview-9/index.html; "Decimal separator," *Wikipedia*, https://en.m.wikipedia.org/wiki/Decimal_mark, downloaded Dec. 13, 2018.

[lxiii] "Examples for equation solving," *WolframAlpha*, https://www.wolframalpha.com/examples/EquationSolving.html, downloaded Dec. 14, 2018.

[lxiv] *In Pursuit of the Unknown: 17 Equations That Changed the World*, Equation 9: Fourier Transform.

[lxv] Jebb, Andrew T.; Tay, Louis; Diener, Ed; Oishi, Shigehiro (2018) "Happiness, income satiation and turning points around the world," *Nature Human Behaviour*, 2, 33-38.

[lxvi] This is inspired by: Rattner, Steven "2016 in Charts. (And can Trump deliver in 2017?)," *New York Times*, Jan. 3, 2017, https://www.nytimes.com/2017/01/03/opinion/2016-in-charts-and-can-trump-deliver-in-2017.html.

[lxvii] Wang, Samuel, as quoted in Bialik, Carl "When polls turn up the wrong number," *Wall Street Journal*, Jan. 6, 2012, 9:35 pm, https://blogs.wsj.com/numbers/when-polls-turn-up-the-wrong-number-1109.

[lxviii] Jahren, Hope (2016) *Lab Girl* (Vintage, New York), pg. 3.

[lxviii] *The Road to Reality*, pg. 103.

[lxix] *In Pursuit of the Unknown: 17 Equations That Changed the World*, Equation 3: Calculus.

[lxx] Penrose, Roger (2006) *The Road to Reality: A Complete Guide to the Laws of the Universe* (Knopf, New York), pg. 103.

[lxxi] *The Road to Reality*, pg. 103.

[lxxii] *The Mathematics Lover's Companion*, Chapter 12; Cheng, Eugenia "Algorithms aren't just for computers," *Wall Street Journal*, May 9, 2019, 9:04 AM ET, https://www.wsj.com/articles/algorithms-arent-just-for-computers-11557407055.

[lxxiii] Reichl, L. E. (1980) *A Modern Course in Statistical Physics* (University of Texas Press, Austin, TX), pg. 138.

[lxxiv] "Counting, permutations, and combinations," *Khan Academy*, https://www.khanacademy.org/math/statistics-probability/counting-permutations-and-combinations, downloaded Dec. 18, 2018.

[lxxv] Permutation, *Wikipedia*, https://en.wikipedia.org/wiki/Permutation, downloaded Dec. 13, 2018.

[lxxvi] *The Book of Numbers*, pp. 65ff.

[lxxvii] *The Book of Numbers*, pp. 70, 91ff.

[lxxviii] *Discrete Mathematics and Its Applications*, pp. 314ff.; Zhou, Xinfeng (2008) *A Practical Guide to Quantitative Finance Interviews*, 1st edition (CreateSpace) pp. 20-21.

[lxxix] Dexter, Colin (1999) *The Remorseful Day* (Macmillan, London) pg. 186, in this the last Inspector Morse mystery novel.

[lxxx] Roberts, Siobhan (2015) *Genius At Play: The Curious Mind of John Horton Conway* (Bloomsbury, New York), pg. 247.

[lxxxi] Author Unknown, as quoted in Quotations about Math & Numbers, *The Quote Garden*, http://www.quotegarden.com/math.html, downloaded Dec. 14, 2018.

[lxxxii] TV Show *Scorpion,* Season 3, Episode 15, "Sharknerdo," Aired 10:00 PM, Feb. 6, 2017 on CBS.

[lxxxiii] TV Show *House,* Season 1, Episode 1, "Pilot, Everybody Lies," Aired Nov. 16, 2004 on FOX; House MD - 1.01 Pilot, *(Unofficial) House Transcripts*, http://clinic-duty.livejournal.com/385.html, downloaded Dec. 14, 2018.

[lxxxiv] Groves, Leslie M. (1962) *Now It Can Be Told* (Da Capo Press, Boston), pg. 312.

[lxxxv] Quote 32 attributed to Steven Wright, as quoted in Byrne, Robert (2012) *The 2,548 Wittiest Things Anybody Ever Said* (Touchstone, New York).

[lxxxvi] Barbanel, Josh "Manhattan apartment sales hit speed bump," *Wall Street Journal*, March 17, 2016, 9:17 PM ET, https://www.wsj.com/articles/manhattan-apartment-sales-hit-speed-bump-1458262300.

[lxxxvii] "Demographics of New York City," *Wikipedia*, https://en.wikipedia.org/wiki/Demographics_of_New_York_City, downloaded Dec. 13, 2018.

[lxxxviii] "Demographics of New York City," *Wikipedia*; Koeske, Zak. "Staten Island by the numbers: Income, jobs and education data by ZIP code," *silive.com*, updated Dec. 12, 2018, http://www.silive.com/specialreports/index.ssf/2015/01/staten_island_by_the_numbers.html.

[lxxxix] Cheng, Eugenia "The mysteries of means and medians," *Wall Street Journal*, July 13, 2017, 12:26 PM ET, https://www.wsj.com/articles/a-mathematicians-guide-to-means-and-medians-1499963169.

[xc] "Labor force statistics from the current population survey," *United States Department of Labor-Bureau of Labor Statistics*, http://www.bls.gov/cps/cps_htgm.htm, downloaded Dec. 14, 2018.

[xci] McGinty, Jo Craven "Gas guzzlers' gains in fuel efficiency go farther than fuel sippers' do," *Wall Street Journal*, April 21, 2017, 9:00 AM ET, https://www.wsj.com/articles/gas-guzzlers-gains-in-fuel-efficiency-go-farther-than-fuel-sippers-do-1492779603; Larrick, Richard P.; Soll, Jack B. (2008), "The MPG illusion," *Science*, 320, 1593-1594.

[xcii] Powell, Robert "What investors need to understand about 'Investment Return'," *Wall Street Journal*, April 12, 2017, 9:18 PM ET, https://www.wsj.com/articles/what-investors-need-to-understand-about-investment-return-1491790501.

[xciii] Flynn, James R. "Are we really getting smarter?," *Wall Street Journal*, Sept. 21, 2012, 9:10 PM ET, https://www.wsj.com/articles/SB10000872396390444032404578006612858486012; Intelligence quotient, Wikipedia, https://en.wikipedia.org/wiki/Intelligence_quotient, downloaded Dec. 13, 2018.

[xciv] Castelvecchi, Davide " 'We hate math,' Say 4 in 10 -- a Majority of Americans," Scientific American, Oct 15, 2011, https://blogs.scientificamerican.com/degrees-of-freedom/we-hate-math-say-4-in-10-a-majority-of-americans/.

[xcv] Lester, Will; Associated Press Writer "Poll shows America's love-hate relationship with math," LJWorld.com Lawrence Journal-World, Aug. 17, 2005, http://www2.ljworld.com/news/2005/aug/17/poll_shows_americas_lovehate_relationship_math/; Savage, Sam. Americans Love to Hate Math, Poll Shows, *RedOrbit*, Aug. 17, 2005, http://www.redorbi
t.com/news/education/211362/americans_love_to_hate_math_poll_shows/.

[xcvi] "Hall of Famers," *National Baseball Hall of Fame*, http://baseballhall.org/hall-of-famers/bbwaa-rules-for-election, downloaded Dec. 14, 2018.

[xcvii] Nurnberg, Maxwell; Rosenblum, Morris (1949) *How to Build a Better Vocabulary* (Warner Books, New York), Chapter 8, Count Off!.; Chrisomalis, Stephen. Numerical Prefixes, *The Phrontistery*, http://phrontistery.info/numbers.html, downloaded Dec. 14, 2018; *The Book of Numbers*, Chapter 1.

[xcviii] *The Mathematics Lover's Companion*, Chapter 16.

[xcix] "Decade," *Wikipedia*, https://en.m.wikipedia.org/wiki/Decade, downloaded Dec. 13, 2018.

[c] "Decade," *Wikipedia.*

[ci] "Biannual, biennial, semiannual," *GrammarBook.com*, https://www.grammarbook.com/homonyms/biannual-biennial-semiannual.asp, downloaded Jan. 1, 2019.

[cii] "Score," *Wikipedia*, https://simple.wikipedia.org/wiki/Score, downloaded Dec. 13, 2018.

[ciii] "Lustrum," *Wikipedia*, https://en.wikipedia.org/wiki/Lustrum, downloaded Dec. 13, 2018.

[civ] "Dozen," *Wikipedia*, https://en.wikipedia.org/wiki/Dozen, downloaded Dec. 13, 2018.

[cv] "Dime (United States coin)," *Wikipedia*, https://en.wikipedia.org/wiki/Dime_(United_States_coin), downloaded Dec. 13, 2018.

[cvi] Author Unknown, as quoted in Quotations about Math & Numbers, *The Quote Garden*, http://www.quotegarden.com/math.html, downloaded Dec. 14, 2018.

[cvii] TV Show *Bull,* Season 2, Episode 7, "No Good Deed," Aired Nov. 7, 2017 on CBS.

[cviii] decimate, *Merriam-Webster*, https://www.merriam-webster.com/dictionary/decimate, downloaded Feb. 18, 2019.

[cix] Noll, Landon Curt "The English name of a number - How high can you count?," *Landon Curt Noll homepage*, Revision: 7.9, Sept. 30, 2015, 2:00:29,

http://www.isthe.com/chongo/tech/math/number/howhigh.html, downloaded Dec. 14, 2018.

[cx] Krulwich, Robert "Which Is greater, the number of sand grains on Earth or stars in the sky?," *Krulwich Wonders - Robert Krulwich On Science*, *National Public Radio*, Sept. 17, 2012, 10:19 AM ET, https://www.npr.org/sections/krulwich/2012/09/17/161096 233/which-is-greater-the-number-of-sand-grains-on-earth-or-stars-in-the-sky, downloaded Dec. 14, 2018.

[cxi] "Shannon number," *Wikipedia*, https://en.wikipedia.org/wiki/Shannon_number, downloaded Dec. 13, 2018; Observable universe: Matter content, *Wikipedia,* https://en. wikipedia.org/wiki/Observable_universe#Matter_content, downloaded Dec. 13, 2018.

[cxii] Bialik, Carl "There could be no Google without Edward Kasner," *The Wall Street Journal*, updated June 14, 2004, 12:01 AM ET, https://www.wsj.com/articles/SB1085759 24921724042; Flannery, S.; Flannery, D. (2000) *In Code: A Mathematical Journey* (Profile Books, London), pp. 112-113; Kasner, Edward; Newman, James R. (1989) *Mathematics and the Imagination* (Tempus Books, Redmond, WA), pp. 20-27; "Googol," *WolframMathWorld*, http://mathworld.wolfram.com/Googol.html, downloaded Dec. 14, 2018.

[cxiii] Koller, David "Origin of the name "Google"," Stanford University, Jan. 2004, Archived from the original on July 4, 2012, http://www.webcitation.org/68ubHzYs7. Retrieved Jan. 26, 2016.

[cxiv] *The Book of Numbers*, pp. 248ff.

[cxv] *In Pursuit of the Unknown: 17 Equations That Changed the World*, Equation 2: Logarithms.

[cxvi] "Earthquake magnitude, energy release, and shaking intensity," *USGS Earthquake Hazards Program*, https://earthquake.usgs.gov/learn/topics/mag-intensity/, downloaded Dec. 14, 2018.

[cxvii] "Magnitude (astronomy)," *Wikipedia*, https://en.wikipedia.org/wiki/Magnitude_(astro nomy), downloaded Dec. 13, 2018.

[cxviii] "Specific intervals," *musictheory.net*, https://www.musictheory.net/lessons/31, downloaded Dec. 14, 2018.

[cxix] "Musical tuning," *Wikipedia*, https://en.wikipedia.org/wiki/Musical_tuning, downloaded Dec. 13, 2018.

[cxx] *The Book of Numbers*, pp. 256-257.

[cxxi] Author Unknown, as quoted in Quotations about Math & Numbers, *The Quote Garden*, http://www.quotegarden.com/math.html, downloaded Dec. 14, 2018.

[cxxii] Gertner, Jon (2013) *The Idea Factory: Bell Labs and the Great Age of American Innovation* (Penguin Books, London), pg. 129.

[cxxiii] "Explain Half Adder and Full Adder with Truth Table," *elprocus*, https://www.elprocus.com/half-adder-and-full-adder/, downloaded Jan. 22, 2019.

[cxxiv] "Binary multiplier," *Wikipedia*, https://en.wikipedia.org/wiki/Binary_multiplier, downloaded Dec. 13, 2018.
[cxxv] "Alan Turing and Hawking," *Wikipedia*, https://en.wikipedia.org/wiki/Alan_Turing and Hawking, downloaded Dec. 13, 2018; Computer, *Wikipedia*, https://en.wikipedia.org/wiki/Computer, downloaded Dec. 13, 2018.
[cxxvi] *The Book of Numbers*, pp. 248ff.
[cxxvii] "*e* (mathematical constant)," *Wikipedia*, https://en.wikipedia.org/wiki/E_(mathemati cal_constant), downloaded Dec. 13, 2018; *The Book of Numbers*, pp. 248-256.
[cxxviii] *The Idea Factory*, Chapter 7.
[cxxix] *The Idea Factory*, pg. 129.
[cxxx] *The Idea Factory*, pg. 131.
[cxxxi] "Nyquist–Shannon sampling theorem," Wikipedia, https://en.wikipedia.org/wiki/Nyquist%E2%80%93Shannon_sampling_theorem, downloaded Jan. 24, 2019.
[cxxxii] *In Pursuit of the Unknown: 17 Equations That Changed the World*, Equation 15: Information Theory.
[cxxxiii] Kiersz, Andy "These equations changed the course of history," *World Economic Forum*, April 4, 2016, https://www.weforum.org/agenda/2016/04/the-17-equations-that-changed-the-world.
[cxxxiv] Brillouin, Léon (1962) *Science and Information Theory,* 2nd edition (Academic Press, New York).
[cxxxv] *The Idea Factory*.
[cxxxvi] "Data compression," *Wikipedia*, https://en.wikipedia.org/wiki/Data_compression, downloaded Dec. 13, 2018; Mahoney, Matt. Data Compression Explained, updated April 15, 2013, http://mattmahoney.net/dc/dce.html.
[cxxxvii] "Hamming code," *Wikipedia*, https://en.wikipedia.org/wiki/Hamming_code, downloaded Dec. 13, 2018.
[cxxxviii] Rosen, Kenneth H. (2005) *Elementary Number Theory and Its Applications*, 5th Ed. (Pearson/Addison Wesley, Boston), pg. 209. *Elementary Number Theory and Its Applications*, 5th Ed. (Pearson/Addison Wesley, Boston), pg. 209.
[cxxxix] "Error detection and correction," *Wikipedia*, https://en.wikipedia.org/wiki/Error_det ection_and_correction, downloaded Dec. 13, 2018.
[cxl] *In Pursuit of the Unknown: 17 Equations That Changed the World*, Equation 15: Information Theory.
[cxli] "Red-Solomon error correction," *Wikipedia*, https://en.wikipedia.org/wiki/Red%E2%80%93Solomon_error_correction, downloaded Dec. 13, 2018.
[cxlii] "Come again?: A surprisingly simple test to check research papers for errors," *The Economist*, June 16, 2016, http://www.economist.com/news/science-and-technology/21700620-surprisingly-simple-test-check-research-papers-errors-come-again; Brown,

Nicholas J. L. and Heathers, James A. J. (2016) "The GRIM Test: A simple technique detects numerous Anomalies in the reporting of results in psychology," *Social Psychological and Personality Science*, https://doi.org/10.1177/1948550616673876.

[cxliii] Taubman, William (2017) *Gorbachev: His Life and Times* (Simon & Schuster Ltd, New York), pg. 500.

[cxliv] Rich, Tracey R. "The Jewish Calendar: A Closer Look," *Judaism 101*, updated Sept. 25, 2017, http://www.jewfaq.org/calendr2.htm.

[cxlv] *The Book of Numbers*, pg. 176-177.

[cxlvi] "Floating-point arithmetic," Wikipedia, https://en.wikipedia.org/wiki/Floating-point_arithmetic, downloaded Jan. 22, 2019.

[cxlvii] "The Patriot Missile failure," University of Minnesota, http://www-users.math.umn.edu/~arnold/disasters/patriot.html, downloaded Jan. 25, 2019.

[cxlviii] "Double-precision floating-point format," Wikipedia, https://en.wikipedia.org/wiki/Double-precision_floating-point_format, downloaded Jan. 22, 2019.

[cxlix] McGinty, Jo Craven "Crowd estimates easy to fight over but harder to verify," *Wall Street Journal*, Jan. 27, 2017, 9:00 AM ET, https://www.wsj.com/articles/crowd-estimates-easy-to-fight-over-but-harder-to-verify-1485525601.

[cl] "Experts say Times Square NYE crowd is smaller than people think," *New York Post*, from the *Associated Press*, updated Dec. 31, 2018, 2:54 PM, https://nypost.com/2018/12/31/experts-say-times-square-nye-crowd-is-smaller-than-people-think/.

[cli] Bialik, Carl "Americans stumble on math of big issues," *Wall Street Journal*, Jan. 7, 2012, httpswww.wsj.com/articles/SB10001424052970203471004577144632919979666.

[clii] From Joshua Clinton, as reported in Bialik, Carl "When polls turn up the wrong number."

[cliii] Brancazio, P. J. (1984) *Sport Science: Physical Laws and Optimum Performance* (Touchstone, New York); Katz, J. Sylvan; Katz, Leon (1999), "Power laws and athletic performance," *Journal of Sports Sciences*, 17, 467-476; Heisler, Kevin "Modeling lower bounds of world record running times," *Stetson University*, 2009, https://www2.stetson.edu/~efriedma/research/kheisler.pdf.

[cliv] *Problematical Recreations: Tenth in a Series, Litton Industries,* 1968, problem 37.

[clv] West, Geoffrey (2017) *Scale: The Universal Laws of Growth, Innovation, Sustainability, and the Pace of Life in Organisms, Cities, Economies, and Companies* (Penguin Press, USA).

[clvi] Herman, Irving P. (2016) *Physics of the Human Body*, 2nd edition (Springer, Heidelberg).

[clvii] Fienberg, S. E. (1971) "Randomization and Social Affairs: The 1970 Draft Lottery," *Science*, 171, 255-261.

[clviii] "Random number generation," *Wikipedia*, https://en.wikipedia.org/wiki/Random_n

umber_generation, downloaded Dec. 13, 2018; Pseudorandom number generator, *Wikipedia*, https://en.wikipedia.org/wiki/Pseudorandom_number_generator, downloaded Dec. 13, 2018.

[clix] Rosenbaum, David E. "Statisticians Charge Draft Lottery Was Not Random," *New York Times*, Jan. 4, 1970, https://www.nytimes.com/1970/01/04/archives/statisticians-charge-draft-lottery-was-not-random.html; Draft lottery (1969), *Wikipedia*, https://en.m.wikipedia.org/wiki/Draft_lottery_(1969), downloaded Dec. 13, 2018. Fienberg, S. E. (1971), "Randomization and Social Affairs: The 1970 Draft Lottery," *Science*, 171, 255-261.

[clx] "Randomization and Social Affairs: The 1970 Draft Lottery."

[clxi] *Indiscrete Thoughts*, pg. 89.

[clxii] Sondheimer, Ernst; Rogerson, Alan (1981) *Numbers and Infinity: A Historical Account of Mathematical Concepts* (Cambridge University Press, Cambridge).

[clxiii] This is similar to the quote "Culture Eats Strategy for Breakfast," as attributed to Peter Drucker, in "Culture eats strategy for breakfast," *The Management Centre*, https://www.managementcentre.co.uk/culture-eats-strategy-for-breakfast/, downloaded Dec. 16, 2018.

[clxiv] For more on this topic see *The Book of Numbers*.

[clxv] *The Book of Numbers*, pp. 157ff.

[clxvi] *The Road to Reality: A Complete Guide to the Laws of the Universe*; *The Book of Numbers*, pp. 176ff.

[clxvii] *The Book of Numbers*, pp. 228-229.

[clxviii] *In Pursuit of the Unknown: 17 Equations That Changed the World*, Equation 1: Pythagoras's Theorem.

[clxix] *The Book of Numbers*, pp. 183ff.

[clxx] de Campos, Deivis; Malysz, Tais; Bonatto-Costa, João Antonio; Jotz, Geraldo Pereira; Lino Pinto de Oliveira Junior, Lino Pinto; da Rocha, Andrea Oxley (2015) "More than a neuroanatomical representation in *The Creation of Adam* by Michelangelo Buonarroti, a representation of the *Golden Ratio*," *Clinical Anatomy*, 28, 702-705. The Golden Ratio: Phi, 1.618, *The Golden Number*, https://www.goldennumber.net/, downloaded Jan. 1, 2019.

[clxxi] *The Book of Numbers*, Chapter 9.

[clxxii] Adhikari, Ani; DeNero, John *Computational and Inferential Thinking: The Foundations of Data Science*, https://www.inferentialthinking.com/chapters/intro, downloaded Oct. 5, 2017, pp. 70 and 75.

[clxxiii] "Square root of 2," *Wikipedia*, https://en.wikipedia.org/wiki/Square_root_of_2, downloaded Dec. 13, 2018.

[clxxiv] Weisstein, Eric W. "e Continued Fraction," *MathWorld--A Wolfram Web Resource*, http://mathworld.wolfram.com/eContinuedFraction.html, downloaded Dec. 14, 2018.

[clxxv] *The Book of Numbers*, pg. 186.

[clxxvi] *The Road to Reality: A Complete Guide to the Laws of the Universe.*

[clxxvii] "Dedekind cut," *Wikipedia*, https://en.wikipedia.org/wiki/Dedekind_cut, downloaded Dec. 13, 2018.

[clxxviii] Weisstein, Eric W. Pi Approximations, *MathWorld--A Wolfram Web Resource*, http://mathworld.wolfram.com/PiApproximations.html, downloaded Dec. 14, 2018.

[clxxix] *In Pursuit of the Unknown: 17 Equations That Changed the World*, Equation 5: The Square Root of Minus One.

[clxxx] *The Book of Numbers*, Chapter 8.

[clxxxi] Martinez, Albert A. (2006) *Negative Math: How Mathematical Rules Can Be Positively Bent* (Princeton University Press, Princeton).

[clxxxii] *The Book of Numbers*, pp. 254-256.

[clxxxiii] "North American numbering plan," *Wikipedia*, https://en.wikipedia.org/wiki/North_American_Numbering_Plan, downloaded June 12, 2019.

[clxxxiv] Rosen, Kenneth H. (2003) *Discrete Mathematics and Its Applications*, 5th Ed. (McGraw Hill, New York).

[clxxxv] McGinty, Jo Craven "Losing your old area code? You're not alone," *Wall Street Journal*, Aug. 4, 2017, 9:00 AM ET, https://www.wsj.com/articles/losing-your-old-area-code-youre-not-alone-1501851600.

[clxxxvi] This quote was a take-off of "Beyond the infinite" that appears as a title card in the 1968 movie *2001: A Space Odyssey*. "Buzz lightyear," *Wikipedia*, https://ipfs.io/ipfs/QmXoypizjW3WknFiJnKLwHCnL72vedxjQkDDP1mXWo6uco/wiki/Buzz_Lightyear.html, downloaded March 5, 2019.

[clxxxvii] *God Created the Integers*; *One, Two, Three...Infinity*; *The Book of Numbers*, Ch. 11; *Numbers and Infinity: A Historical Account of Mathematical Concepts*, Ch. 11.

[clxxxviii] *Discrete Mathematics and Its Applications*, pp. 233-236.

[clxxxix] *Discrete Mathematics and Its Applications*, pp. 234-236.

[cxc] Duarte, Gustavo "Counting Infinity," *Many But Finite Tech and science for curious people.*, Jan. 6, 2009, https://duartes.org/gustavo/blog/post/counting-infinity/.

[cxci] Strogatz, Steven H. (2015) *Nonlinear dynamics and chaos: With applications to physics, biology, chemistry, and engineering* (Westview Press, Boulder, CO), 2nd edition.

[cxcii] Knuth, Donald E. (1974) *Surreal Numbers: How two ex-students turned on to pure mathematics and found total happiness* (Addison-Wesley, Reading, Massachusetts).

[cxciii] "Surreal number," *Wikipedia*, https://en.wikipedia.org/wiki/Surreal_number, downloaded Dec. 13, 2018; *The Book of Numbers*, pg. 283.

[cxciv] *Indiscrete Thoughts*, pp. 217-218.

[cxcv] As quoted in the "Apophthegms, sentiments, opinions and occasional reflections" of Sir John Hawkins (1787-1789) in Johnsonian Miscellanies (1897), vol. II, p. 2, George Birkbeck Hill, editor.

[cxcvi] Clark, Carol "Mathematicians find 'Magic Key' to drive Ramanujan's Taxi-cab Number," *Emory University eScience Commons*, Oct. 14, 2015, http://esciencecommons.blogspot.com/2015/10/mathematicians-find-magic-key-to-drive.html.

[cxcvii] *Elementary Number Theory and Its Applications*, pp. 535-536.

[cxcviii] "Examples for Numbers," *WolframAlpha*, https://www.wolframalpha.com/examples/mathematics/numbers/, downloaded Dec. 14, 2018.

[cxcix] TV Show *Futurama*, respectively Season 2, Episode 2 "Brannigan, Begin Again", Aired November 28, 1999 and Season 2, Episode 4 "Xmas Story", Aired December 19, 1999, both on Fox.

[cc] Weisstein, Eric W. "Taxicab Number," *MathWorld--A Wolfram Web Resource*, http://mathworld.wolfram.com/TaxicabNumber.html, downloaded Dec. 14, 2018; "1729 (number)," *Futurama Wiki*, http://futurama.wikia.com/wiki/1729_%28number%29, downloaded Dec. 16, 2018; Singh, Simon; Oakes, Kelly "The 12 Geekiest Maths Jokes Hidden In Futurama," *BuzzFeed*, Oct. 18, 2013, 5:20 AM, https://www.buzzfeed.com/simonsingh/the-12-geekiest-maths-jokes-hidden-infuturama?utm_term=.bqrjRYB8z#.ixlxAQdPk, downloaded Dec. 16, 2018.

[cci] *The Book of Numbers*, pg. 106.

[ccii] Caldwell, Chris "The prime pages: Prime number research, records and resources," *The University of Tennessee at Martin*, https://primes.utm.edu/index.html, downloaded Dec. 16, 2018; *The Mathematics Lover's Companion*.

[cciii] *Discrete Mathematics and Its Applications*, pp. 154-158; *The Book of Numbers*, Ch. 5.

[cciv] *Elementary Number Theory and Its Applications*, pp. 77-86.

[ccv] *The Book of Numbers*, pp. 132ff.

[ccvi] *The Book of Numbers*, pp. 136-137.

[ccvii] Hardy, G. H. (1940) *A Mathematician's Apology* (Cambridge University Press, Cambridge).

[ccviii] *Discrete Mathematics and Its Applications*, pg. 155.

[ccix] Lenstra, Jr., Hendrik W. *Applied Number Theory*, https://www.math.leidenuniv.nl/~hwl/PUBLICATIONS/1990h/art.pdf, downloaded Jan. 7, 2019.

[ccx] Caldwell, Chris "Why do people find these primes?," *The University of Tennessee at Martin*, https://primes.utm.edu/notes/faq/why.html, downloaded Dec. 16, 2018.

[ccxi] Curtis, Matthew; Tularam, Gurudeo Anand (2011), "The importance of numbers and the need to study primes: The prime questions," *J. Math. Stat.*, 7, 262-269.

[ccxii] Webb, G. F. (2001) "The prime number periodical cicada problem," *Discr. Cont. Dyn. Sys., Series B*, 1, 3987-399; Tanaka, Yumi; Yoshimura, Jin; Simon, Chris; Cooley, John R.; Tainaka, Kei-ichi. (2009) "Allee effect in the selection for prime-numbered cycles in periodical cicadas," Proc. Nat. Acad. Sci., 106, 8975–8979; Gould, Stephen Jay (1977) *Ever Since Darwin: Reflections in Natural History* (Norton, New York); "Periodical cicadas," *Wikipedia*, https://en.wikipedia.org/wiki/Periodical_cicadas, downloaded Dec. 13, 2018.

[ccxiii] "Pythagorean prime," *Wikipedia*, https://en.wikipedia.org/wiki/Pythagorean_prime, downloaded Dec. 13, 2018.

[ccxiv] Ribet, Kenneth A. "The five fundamental operations of mathematics: addition, subtraction, multiplication, division, and modular forms," *UC Berkeley Math*, March 31, 2008 at Trinity University, https://math.berkeley.edu/~ribet/trinity.pdf; *The Book of Numbers*, pp. 146-147.

[ccxv] "8675309," *WolframAlpha*, https://m.wolframalpha.com/input/?i=8675309, downloaded May 9, 2017.

[ccxvi] "8675309," *WolframAlpha.*

[ccxvii] "8675309," *WolframAlpha.*

[ccxviii] *Area Code Listing, by Number*, http://www.bennetyee.org/ucsd-pages/area.html, as of March 21, 2017, downloaded June 6, 2017.

[ccxix] "Examples for numbers," *WolframAlpha*, https://www.wolframalpha.com/examples/mathematics/numbers/, downloaded May 9, 2017.

[ccxx] "Examples for Numbers," *WolframAlpha.*

[ccxxi] Ribet, Kenneth, as quoted in Bialik, Carl "Number-crushing: When figures get personal," *Wall Street Journal*, Oct. 28, 2009, 11:59 PM ET, https://www.wsj.com/articles/SB125668948820711987.

[ccxxii] *Discrete Mathematics and Its Applications*, pg. 189.

[ccxxiii] Caldwell, Chris "Lists of small primes," *The University of Tennessee at Martin*, https://primes.utm.edu/lists/small/. Downloaded Dec. 16, 2018.

[ccxxiv] Caldwell, Chris "The largest known primes--A summary," *The University of Tennessee at Martin*, https://primes.utm.edu/largest.html, downloaded Dec. 16, 2018.

[ccxxv] Erdős, Paul; Surányi, János K. (2003) *Topics in the Theory of Numbers*, 2nd edition (Springer, Heidelberg).

[ccxxvi] Rassias, Michael Th. (2011*) Problem-Solving and Selected Topics in Number Theory: In the Spirit of the Mathematical Olympiads* (Springer, Heidelberg).

[ccxxvii] "Prime number," *Wikipedia*, https://en.wikipedia.org/wiki/Prime_number, downloaded Dec. 13, 2018.

[ccxxviii] *The Book of Numbers*, pg. 143.

[ccxxix] *The Book of Numbers*, pg. 143.

[ccxxx] Rankin, R. A. (1938), "The difference between consecutive prime numbers," *J. London Math. Soc.*, 13, 242-247.

[ccxxxi] Klarreich, Erica "Mathematicians make a major discovery about prime numbers," *Wired*, Dec. 22, 2014, 6:30 AM, https://www.wired.com/2014/12/mathematicians-make-major-discovery-prime-numbers/.

[ccxxxii] Klarreich, Erica. Mathematicians Make A Major Discovery About Prime Numbers.

[ccxxxiii] *The Book of Numbers*, pp. 127-130.

[ccxxxiv] *The Book of Numbers*, pg. 134.

[ccxxxv] *The Book of Numbers*, pp. 137-139.

[ccxxxvi] *Great Internet Mersenne Prime Search (GIMPS)*, https://www.mersenne.org/primes/, downloaded June 7, 2019.
[ccxxxvii] *Great Internet Mersenne Prime Search (GIMPS).*
[ccxxxviii] *The Book of Numbers*, pp. 135-136.
[ccxxxix] *Great Internet Mersenne Prime Search (GIMPS).*
[ccxl] *The Book of Numbers*, pp. 135ff.
[ccxli] "Examples for Numbers," *WolframAlpha*, https://www.wolframalpha.com/examples/mathematics/numbers/, downloaded December 14, 2018.
[ccxlii] *Elementary Number Theory and Its Applications; Discrete Mathematics and Its Applications; The Book of Numbers*, pp.130-132.
[ccxliii] *The Book of Numbers*, pp. 28-29.
[ccxliv] McGinty, Jo Craven "Time to change global clock management? It's under debate," *Wall Street Journal*, Jan. 16, 2015, 12:54 p.m. ET, https://www.wsj.com/articles/planned-june-leap-second-stirs-a-timely-debate-1421430888; Hershber, Brian "Time check: Markets fret a leap second," *Wall Street Journal*, May 19, 2015, 10:04 AM ET, https://blogs.wsj.com/numbers/time-check-markets-fret-a-leap-second-2072; McGinty, Jo Craven "Leaping lizards! A timely look at seconds," *Wall Street Journal*, Jan 16, 2015, 2:58 PM ET, https://blogs.wsj.com/numbers/leaping-lizards-a-timely-look-at-seconds-1908; Donahue, Michelle Z. "2016 will be one second longer than expected," *National Geographic*, July 8, 2016, https://news.nationalgeographic.com/2016/07/leap-second-added-year-december-time-clocks-earth-science/.
[ccxlv] *Elementary Number Theory and Its Applications*, pg. 196.
[ccxlvi] *Elementary Number Theory and Its Applications*, pg. 200.
[ccxlvii] *Elementary Number Theory and Its Applications*, pg. 202; *Discrete Mathematics and Its Applications*, pp. 163-164.
[ccxlviii] *Elementary Number Theory and Its Applications*, pg. 209; Cheng, Eugenia "Secret codes built on very large numbers," *Wall Street Journal*, Oct. 3, 2018, 11:12 AM ET, https://www.wsj.com/articles/secret-codes-built-on-very-large-numbers-1538579523.
[ccxlix] *Elementary Number Theory and Its Applications*, Ch. 8; *Discrete Mathematics and Its Applications*, pp. 165-166, 191-194; Swenson, Christopher (2008) *Modern Cryptanalysis: Techniques for Advanced Code Breaking* (John Wiley & Sons, New York); Kaufman, Charlie; Perlman, Radia; Speciner, Mike (2002) *Network Security: Private Communication in a Public World*, 2nd edition (Prentice Hall, Upper Saddle River, New Jersey).
[ccl] *Elementary Number Theory and Its Applications*, pg. 297; "Cryptology and data secrecy: The Vernam Cipher," *Protechnix*, http://www.pro-technix.com/information/crypto/pages/vernam_base.html., downloaded Dec. 16, 2018; The Vernam Cipher, *Crypto Museum*, https://cryptomuseum.com/crypto/vernam.htm, downloaded Dec. 16, 2018.
[ccli] "RSA (cryptosystem)," *Wikipedia*, https://en.wikipedia.org/wiki/RSA_(cryptosyst

em), downloaded Dec. 13, 2018; *Discrete Mathematics and Its Applications*, pp. 191-194; *Elementary Number Theory and Its Applications*, pp. 310-314.

[cclii] Mims, Christopher "The day when computers can break all encryption is coming" *Wall Street Journal*, June 4, 2019, 7:12 AM ET, https://www.wsj.com/articles/the-race-to-save-encryption-11559646737.

[ccliii] *Discrete Mathematics and Its Applications*, pp. 164-165.

[ccliv] Zetter, Kim "How a crypto 'backdoor' pitted the tech world against the NSA" *Wired*, Sept. 13, 2014, https://www.wired.com/2013/09/nsa-backdoor/; "Cryptographically secure pseudorandom number generator," Wikipedia, https://en.wikipedia.org/wiki/Cryptographically_secure_pseudorandom_number_generator, downloaded Jan. 23, 2019; "Random number generator attack," Wikipedia, https://en.wikipedia.org/wiki/Random_number_generator_attack, downloaded Jan. 23, 2019; "Dual_EC_DRBG," Wikipedia, https://en.wikipedia.org/wiki/Dual_EC_DRBG, downloaded Jan. 23, 2019.

[cclv] "The On-Line Encyclopedia of Integer Sequences® (OEIS®)," The OEIS Foundation, https://oeis.org/, downloaded Dec. 16, 2018; *Discrete Mathematics and Its Applications*, pp. 225-233.

[cclvi] Antonick, Gary "John Horton Conway, the Genius at Play," *New York Times*, August 3, 2015, 12:00 PM, https://wordplay.blogs.nytimes.com/2015/08/03/conway; Roberts, Siobhan (2015) *Genius at Play: The Curious Mind of John Horton Conway* (Bloomsbury, New York), pp. 76-77; *The Book of Numbers*, pp. 208-209.

[cclvii] Queen, Ellery (1949) *Cat of Many Tails* (Little, Brown & Company, New York).

[cclviii] Moore, Gordon E. "Lithography and the future of Moore's Law," reprinted in https://ieeexplore.ieee.org/stamp/stamp.jsp?arnumber=4785861 and chttp://www.lithoguru.com/scientist/CHE323/Moore1995.pdf. Initially: *Proc. SPIE*, 2437, pp. 2-17.

[cclix] "American wire gauge *Wikipedia*, https://en.wikipedia.org/wiki/American_wire_gauge, downloaded Dec. 13, 2018.

[cclx] "Pyramid scheme," *Wikipedia*, https://en.wikipedia.org/wiki/Pyramid_scheme, downloaded Dec. 13, 2018.

[cclxi] Strogatz, Steven H. (2015) *Nonlinear dynamics and chaos: With applications to physics, biology, chemistry, and engineering* (Westview Press, Boulder, CO), 2nd edition.

[cclxii] *The Mathematics Lover's Companion*, Chapter 17.

[cclxiii] Peitgen, Heinz-Otto; Jürgens, Hartmut; Saupe, Dietmar (1992) *Fractals for the Classroom: Part One Introduction to Fractals and Chaos* (Springer, Heidelberg).

[cclxiv] "Loan," *Wikipedia*, https://en.wikipedia.org/wiki/Loan, downloaded Dec. 13, 2018; "Monthly payment formula," *Wikipedia*, https://en.wikipedia.org/wiki/Mortgage_calculator#Monthly_payment_formula, downloaded Dec. 13, 2018; "Compound interest," *Wikipedia*, https://en.wikipedia.org/wiki/Compound_interest, downloaded Dec. 13, 2018.

[cclxv] Elkins, Kathleen "Here's what happens when you only pay the minimum on your credit card balance," *CNBC make it*, July 24, 2018, 12:37 PM ET, https://www.cnbc.com/

amp/2018/07/23/what-happens-when-you-only-pay-the-minimum-on-your-credit-card-balance.html.

[cclxvi] Powell, Robert "What investors need to understand about 'Investment Return'" *Wall Street Journal*, updated April 12, 2017, 9:18 PM ET, https://www.wsj.com/articles/what-investors-need-to-understand-about-investment-return-1491790501.

[cclxvii] "Compound interest," *Wikipedia*.

[cclxviii] "*e* (mathematical constant)," *Wikipedia*.

[cclxix] Hoffmann, Peter "Physics makes aging inevitable, not biology," *Nautilus*, May 12, 2016, http://nautil.us/issue/36/aging/physics-makes-aging-inevitable-not-biology.

[cclxx] McGinty, Jo Craven "The genius behind accounting shortcut? It wasn't Einstein," *Wall Street Journal*, June 16, 2017, 9:00 a.m. ET, https://www.wsj.com/articles/the-genius-behind-accounting-shortcut-it-wasnt-einstein-1497618000.

[cclxxi] "U.S. Department of Health and Human Services study *Health, United States, 2017," National Center for Health Statistics. Health, United States, 2017: With special feature on mortality. Hyattsville, MD. 2018*, https://www.cdc.gov/nchs/data/hus/hus17.pdf, Tables 17, 21-24, downloaded June 7, 2019.

[cclxxii] Kolata, Gina "A medical mystery of the best kind: Major diseases are in decline," *New York Times*, July 8, 2016, https://www.nytimes.com/2016/07/10/upshot/a-medical-mystery-of-the-best-kind-major-diseases-are-in-decline.htm; Welch, H. Gilbert; Robertson, Douglas J. (2016) "Colorectal cancer on the decline - Why screening can't explain it all," *N. Engl. J. Med.*, 374, 1605-1607; Jones, David S.; Greene, Jeremy A. (2016) "Is dementia in decline? Historical trends and future trajectories," *N. Engl. J. Med.*, 374, 507-509; Satizabal, Claudia L.; Beiser, Alexa S.; Chouraki, Vincent; Chêne, Geneviève; Dufouil, Carole; Seshadri, Sudha (2016), "Incidence of dementia over three decades in the Framingham Heart Study," *N. Engl. J. Med.*, 374, 523-532; "Longer life, disability free: Increases in life expectancy accompanied by increase in disability-free life expectancy, study shows, *Science Daily*, June 6, 2016, https://www.sciencedaily.com/releases/2016/06/160606120039.htm; Chernew, Michael; Cutler, David M.; Ghosh, Kaushik; Landrum; Mary Beth "Understanding the improvement in disability free life expectancy in the U.S. elderly population, from insights in the economics of aging" (2017), David A. Wise, editor, by *National Bureau of Economic Research*, pp. 161-201, http://www.nber.org/chapters/c13631.pdf.

[cclxxiii] "Backgrounder on tritium, radiation protection limits, and drinking water standards," *U.S. Nuclear Regulatory Commission*, http://www.nrc.gov/reading-rm/doc-collections/fact-sheets/tritium-radiation-fs.html, downloaded Dec. 17, 2018; Singh, V. P.; Pai, R. K.; Veerender, D. D.; Vishnu, M. S.; Vijayan, P.; Managanvi, S. S.; Badiger, N. M.; Bhat, H. R. (2010) "Estimation of biological half-life of tritium in coastal region of India." *Radiation Protection Dosimetry*. 142, 153–159.

[cclxxiv] Strogatz, Steven H. (2015) *Nonlinear dynamics and chaos: With applications to physics, biology, chemistry, and engineering* (Westview Press, Boulder, CO), 2nd edition.

[cclxxv] Sharov, Alexei "Logistic model," *University of Texas*, https://www.ma.utexas.edu/users/davis/375/popecol/lec5/logist.html, downloaded Dec. 17, 2018.
[cclxxvi] Gleick, James (1987) *Chaos: Making a New Science* (Viking Books, New York); Alligood, Kathleen T.; Sauer, Tim D.; Yorke, James A. (1996) *Chaos: an introduction to dynamical systems* (Springer, New York); Chaos theory, *Wikipedia*, https://en.wikipedia.org/wiki/Chaos_theory, downloaded Dec. 13, 2018; Weisstein, Eric W. "Chaos," *MathWorld--A Wolfram Web Resource*, http://mathworld.wolfram.com/Chaos.html, downloaded Dec. 14, 2018; "Butterfly effect," *Wikipedia*, https://en.wikipedia.org/wiki/Butterfly_effect, downloaded March 6, 2019.
[cclxxvii] *Chaos: An introduction to dynamical systems*
[cclxxviii] *In Pursuit of the Unknown: 17 Equations That Changed the World*, Equation 16: Chaos Theory.
[cclxxix] *The Mathematics Lover's Companion*, Chapter 21.
[cclxxx] Dabbaghian, Vahid "What is Mathematical Modeling?," https://www.sfu.ca/~vdabbagh/Chap1-modeling.pdf, downloaded Dec. 13, 2018.
[cclxxxi] McGinty, Jo Craven "The math behind flu season," *Wall Street Journal*, Feb. 23, 2018, 8:00 AM, https://www.wsj.com/articles/the-math-behind-flu-season-1519390800.
[cclxxxii] Oster, Emily; Kocks, Geoffrey "After a debacle, how California became a role model on measles," *New York Times*, Jan. 16, 2018, https://www.nytimes.com/2018/01/16/upshot/measles-vaccination-california-students.html.
[cclxxxiii] "Compartmental models in epidemiology," *Wikipedia*, https://en.m.wikipedia.org/wiki/Compartmental_models_in_epidemiology, downloaded Dec. 13, 2018.
[cclxxxiv] "Preparing for, detecting and responding to infectious disease threats," *Midas Models Of Infectious Disease Agent Study*, http://www.epimodels.org/drupal-new/, downloaded Dec. 17, 2018; Smith, Robert "Simple epidemic models," University of Ottawa, https://mysite.science.uottawa.ca/rsmith43/MAT4996/Epidemic.pdf, downloaded Dec. 17, 2018; Smith, David; Moore, Lang "The SIR Model for Spread of Disease - The Differential Equation Model," *Mathematical Association of America*, http://www.maa.org/press/periodicals/loci/joma/the-sir-model-for-spread-of-disease-the-differential-equation-model, downloaded Dec. 17, 2018; Stroyan, Keith "Using calculus to model epidemics," The University of Iowa, http://homepage.math.uiowa.edu/~stroyan/CTLC3rdEd/3rdCTLCText/Chapters/Ch2.pdf, downloaded Dec. 17, 2018; Mahaffy, Joseph M. "Epidemic models," San Diego State University, http://www-rohan.sdsu.edu/~jmahaffy/courses/f09/math636/lectures/epidemics/epidemics.pdf, downloaded Dec. 17, 2018.
[cclxxxv] Bergsman, L. D., Hyman, J. M., Manor, C. A. (2016), "A mathematical model for the spread of West Nile virus in migratory and resident birds," Math. Biosci. Eng., 13,401-424.
[cclxxxvi] McGinty, Jo Craven "How math helps fight epidemics like Zika," *Wall Street Journal*, Jun 3, 2016, 1:19 pm ET, https://blogs.wsj.com/numbers/how-math-helps-fight-

epidemics-like-zika-2207; McGinty, Jo Craven "Zika draws U.S. researchers into a race for understanding," *Wall Street Journal*, June 3, 2016, 10:13 AM ET, https://www.wsj.com/articles/zika-draws-u-s-researchers-into-a-race-for-understanding-1464956736.

[cclxxxvii] Billings, Molly "The Influenza Pandemic of 1918," *Stanford University*, updated Feb. 2005, https://virus.stanford.edu/uda/.

[cclxxxviii] Bozic, Martha "Have you heard? The maths of rumour spreading," *+plus Magazine*, Sept. 26, 2016, https://plus.maths.org/content/have-you-heard-maths-rumour-spreading.

[cclxxxix] *Discrete Mathematics and Its Applications*, pp. 259, 403; *Elementary Number Theory and Its Applications*, pg. 30; *The Book of Numbers*, pp. 111.

[ccxc] "Fibonacci numbers in popular culture," *Wikipedia*, https://en.wikipedia.org/wiki/Fibonacci_numbers_in_popular_culture, downloaded Dec. 13, 2018.

[ccxci] Chandra, Pravin; Weisstein, Eric W. "Fibonacci Number," *MathWorld--A Wolfram Web Resource*, http://mathworld.wolfram.com/FibonacciNumber.html, downloaded Dec. 14, 2018; TV Show *NUMB3RS,* Season 1, Episode 6, "Sabotage," Aired 10:00 PM, Feb. 25, 2005 on CBS, as per http://www.tv.com/shows/numb3rs/sabotage-398596/; Brown, Dan. (2003) *The Da Vinci Code* (Doubleday, New York) pp. 43, 60-61, and 189-192.

[ccxcii] *The Book of Numbers*, pp. 111ff.

[ccxciii] *Discrete Mathematics and Its Applications*, pp. 148-151.

[ccxciv] Mark; Rota, Gian-Carlo; Schwartz, Jacob T. (1992) *Discrete Thoughts* (Springer Science & Business Media, New York) pg. 263.

[ccxcv] "FLOPs," *Wikipedia*, https://en.wikipedia.org/wiki/FLOPS, downloaded Jan. 22, 2019.

[ccxcvi] Wilf, Herbert S. "Algorithms and complexity," *University of Pennsylvania*, https://www.math.upenn.edu/~wilf/AlgoComp.pdf, downloaded Dec. 17, 2018; Green, Jessica L.; Hastings, Alan; Arzberger, Peter; Ayala, Francisco J.; Cottingham, Kathryn L.; Cuddington, Kim; … Neubert, Michael (2005), "Complexity in ecology and conservation: Mathematical, statistical, and computational challenges," *BioScience*, 55, 501-510; Carja, Oana; Sellis, Diamantis; Thompson, Joel; Longo, Mark D. "The mathematics of complexity (BIO 131, Fall 2012) (videos)," *Stanford University, Stanford Complexity Group*, http://complexity.stanford.edu/classes, downloaded Dec. 17, 2018.

[ccxcvii] Mastin, Luke. *The Human Memory*, http://www.human-memory.net/brain_neurons.html, downloaded Dec. 17, 2018.

[ccxcviii] Strogatz, Steven H. (2001), "Exploring complex networks," Nature, 410, 268-276.

[ccxcix] West, Geoffrey. The surprising math of cities and corporations, *TEDGlobal 2011*, https://www.ted.com/talks/geoffrey_west_the_surprising_math_of_cities_and_corporations?language=en, downloaded Dec. 17, 2018.

[ccc] Adi Shamir, in: B. Pullman, editor (1996) *The Emergence of Complexity in Mathematics, Physics, Chemistry and Biology: Proceedings, Plenary Session of the*

Pontifical Academy of Sciences, 27-31 October 1992 (Pontificia Academia of Scientiarum, Vatican City; Princeton University Press, Princeton, N.J.).

[ccci] Fortnow, Lance (2009) "The status of the P versus NP problem" (PDF). *Communications of the ACM*, 52, 78-86; Cook, Stephen (1971) "The complexity of theorem proving procedures," *Proceedings of the Third Annual ACM Symposium on Theory of Computing*, pp. 151-158; "P versus NP problem," *Wikipedia*, https://en.wikipedia.org/wiki/P_versus_NP_problem, downloaded Dec. 13, 2018.

[cccii] Cook, William J. (2012) *In Pursuit of the Traveling Salesman: Mathematics at the Limits of Computation* (Princeton University Press, Princeton).

[ccciii] Babai, László "Graph isomorphism in quasipolynomial time" https://arxiv.org/abs/1512.03547; Babai, László, "László Babai's Home Page," http://people.cs.uchicago.edu/~laci/, downloaded March 6, 2019; Cho, Adrian "Mathematician claims breakthrough in complexity theory" *Science*, Nov. 10, 2015, 5:45 PM, http://www.sciencemag.org/news/2015/11/mathematician-claims-breakthrough-complexity-theory.

[ccciv] Wierman, Mark J. "An introduction for the mathematics of uncertainty," *Creighton University*, August 20, 2010, https://www.creighton.edu/fileadmin/user/CCAS/programs/fuzzy_math/docs/MOU.pdf, downloaded Dec. 17, 2018; "Bayesian probability", *Wikipedia*, https://en.wikipedia.org/wiki/Bayesian_probability, downloaded Dec. 13, 2018.

[cccv] "statistics (n.)," *Online Etymology Dictionary*, http://www.etymonline.com/index.php?term=statistics, downloaded Dec. 17, 2018.

[cccvi] *2015 CUPM Curriculum Guide to Majors in the Mathematical Sciences*, pg. 31.

[cccvii] "Regression toward the mean," *Wikipedia*, https://en.wikipedia.org/wiki/Regression_toward_the_mean, downloaded Dec. 13, 2018.

[cccviii] *A Modern Course in Statistical Physics*, Ch.5.

[cccix] *2015 CUPM Curriculum Guide to Majors in the Mathematical Sciences*, pg. 31.

[cccx] The Phrase Finder, https://www.phrases.org.uk/meanings/lies-damned-lies-and-statistics.html, downloaded June 10, 2019.

[cccxi] Gian Carlo Rota, as quoted in "Doron Zeilberger's Collection of Quotes," updated Oct. 19, 2018, http://www.math.rutgers.edu/~zeilberg/quotes.html.

[cccxii] Gelman, Andrew (2012) "Statistics for sellers of cigarettes," *Chance*, 25, 43-46.

[cccxiii] Richard P. Feynman, as quoted in *Goodreads*, http://www.goodreads.com/work/quotes/736922-six-easy-pieces-essentials-of-physics-explained-by-its-most-brilliant-t, downloaded June. 10, 2019; from Feynman, Richard P. (2011) *The Six Easy Pieces: Essentials of Physics by Its Most Brilliant Teacher* (Basic Books, New York).

[cccxiv] "Vehicle registration plates of California," *Wikipedia*, https://en.wikipedia.org/wiki/Vehicle_registration_plates_of_California, downloaded Dec. 13, 2018.

[cccxv] Slotnik, Daniel E. Richard Jarecki, "Doctor who conquered foulette, dies at 86," *New York Times*, Aug. 8, 2018, https://www.nytimes.com/2018/08/08/obituaries/richard-jarecki-doctor-who-conquered-roulette-dies-at-86.html.

[cccxvi] *A Practical Guide to Quantitative Finance Interviews*, pp. 61-62.

[cccxvii] "Winning at slots - What are my odds?," *VegasSlotsOnline*, http://www.vegasslots online.com/odds/, downloaded Dec. 17, 2018.

[cccxviii] "The Vegas tripping guide to casino slots," *Vegas Tripping*, http://www.vegastripp ing.com/playersclub/slots.php, downloaded Dec. 17, 2018.

[cccxix] Weisstein, Eric W. "Bernoulli Distribution,: *MathWorld--A Wolfram Web Resource*, http://mathworld.wolfram.com/BernoulliDistribution.html, downloaded Dec. 14, 2018.

[cccxx] Weisstein, Eric W. "Geometric Distribution," *MathWorld--A Wolfram Web Resource*, http://mathworld.wolfram.com/GeometricDistribution.html, downloaded Dec. 14, 2018.

[cccxxi] Weisstein, Eric W. "Poisson Distribution," *MathWorld--A Wolfram Web Resource*, http://mathworld.wolfram.com/PoissonDistribution.html, downloaded Dec. 14, 2018.

[cccxxii] Cohen, Ben "The 'Hot Hand' debate gets flipped on its head," *Wall Street Journal*, updated Sept. 30, 2015, 8:43 PM ET, https://www.wsj.com/articles/the-hot-hand-debate-gets-flipped-on-its-head-144346571;Miller, Joshua B.; Sanjurjo, Sanjurjo (2018), "Surprised by the hot hand fallacy? A truth in the law of small numbers," *Econometrica*, 86, 2019–2047.

[cccxxiii] Gusfield, Dan "Hot hands, streaks and coin-flips: How the New York Times got it wrong," *SIAM News*, March 01, 2016, https://sinews.siam.org/Details-Page/hot-hands-streaks-and-coin-flips-how-the-new-york-times-got-it-wrong.

[cccxxiv] *Elementary Number Theory and Its Applications*, pp. 371-371; *A Practical Guide to Quantitative Finance Interviews,* pg. 71.

[cccxxv] *In Pursuit of the Unknown: 17 Equations That Changed the World*, Equation 7: Normal Distribution.

[cccxxvi] Weisstein, Eric W. "Normal distribution," *MathWorld--A Wolfram Web Resource*, http://mathworld.wolfram.com/NormalDistribution.html, downloaded Dec. 20, 2018.

[cccxxvii] Gan, K. K. "Lecture 1: Probability and statistics," *Ohio State*, https://www.physics .ohio-state.edu/~gan/teaching/spring04/Chapter1.pdf, downloaded Jan.7, 2019.

[cccxxviii] Tiouririne, Nedjl "NORMAL distribution: Origin of the name," *DePaul University*, http://condor.depaul.edu/ntiourir/NormalOrigin.htm, which cites Pearson, Karl "Contributions to the Mathematical Theory of Evolution," *Philosophical Transactions of the Royal Society of London. A*, **185**, (1894) p. 72, downloaded Dec. 17, 2018.

[cccxxix] Cohen, Ben "Shawn Bradley Is really, really tall. But why?," *Wall Street Journal*, Sept. 18, 2018. 10:01 am, https://www.wsj.com/articles/shawn-bradley-genetic-test-height-1537278144; Sexton, Corinne E.; Ebbert, Mark T. W.; Miller, Ryan H.; Ferrel, Meganne; Tschanz, Jo Ann T.; Corcoran, Christopher D.; … Kauwe, John S. K. (2018),

"Common DNA variants accurately rank an individual of extreme height," *International Journal of Genomics*, Article ID 5121540.

[cccxxx] "Intelligence quotient," *Wikipedia*, https://en.wikipedia.org/wiki/Intelligence_quotient, downloaded Dec. 13, 2018.

[cccxxxi] McGinty, Jo Craven "Variety of IQ tests means measuring gray natter is a gray area," *Wall Street Journal*, Aug. 5, 2016, 3:58 PM ET, https://www.wsj.com/articles/variety-of-iq-tests-means-measuring-gray-matter-is-a-gray-area-1470421472.

[cccxxxii] Kepner, Tyler "Give baseball scouts a (perfect) 80 for tradition," *New York Times*, June 12, 2017, https://www.nytimes.com/2017/06/12/sports/baseball/give-baseball-scouts-a-perfect-80-for-tradition.html; McDaniel, Kiley "Scouting explained: The 20-80 scouting scale," *FanGraphs*, Sept. 4, 2014, https://www.fangraphs.com/blogs/scouting-explained-the-20-80-scouting-scale/.

[cccxxxiii] Feller, William, *An Introduction to Probability Theory and Its Applications*, Volumes 1 (3rd edition, 1967) and 2 (1st edition, 1966), (Wiley, New York).

[cccxxxiv] *Computational and Inferential Thinking*, pg. 403; Chebyshev's inequality, *Wikipedia*, https://en.wikipedia.org/wiki/Chebyshev%27s_inequality, downloaded Dec. 18, 2018.

[cccxxxv] Chernick, Michael R. (2011) *The Essentials of Biostatistics for Physicians, Nurses, and Clinicians* (John Wiley & Sons, New York), pg. 50.

[cccxxxvi] "Log-normal distribution," *Wikipedia*, https://en.wikipedia.org/wiki/Log-normal_distribution., downloaded Dec. 13, 2018.

[cccxxxvii] "Blood test results - Normal ranges bloodbook.com," *docshare.tips*, http://docshare.tips/blood-test-results-normal-range-reference-chart-blood-book-bloodinformati_587d5882b6d87fe25c8b55a7.html, downloaded Dec. 18, 2018; "Reference ranges for blood tests," *Wikipedia*, https://en.wikipedia.org/wiki/Reference_ranges_for_blood_tests, downloaded Dec. 13, 2018; Marshall; William J. (2008) in *Clinical biochemistry: Metabolic and clinical aspects*, Bangert, Stephen K.; Marshall; William J., editors (Churchill Livingstone/Elsevier, Edinburgh), Ch. 3, pg. 19.

[cccxxxviii] Limpert, Eckhard; Stahel, Werner A.; Abbt, Markus (2001), "Log-normal distributions across the sciences: Keys and clues,"*BioScience*, 51, 341-351.

[cccxxxix] Gottstein, Gunter; Shvindlerman, Lasar S. (1999) *Grain Boundary Migration in Metals: Thermodynamics, Kinetics, Applications*, 2nd edition (CRC Press, Boca Raton).

[cccxl] "Arcsine distribution," *Wikipedia*, https://en.wikipedia.org/wiki/Arcsine_distribution, downloaded Dec. 13, 2018.

[cccxli] "Gamma distribution," *Wikipedia*, https://en.wikipedia.org/wiki/Gamma_distribution, downloaded Dec. 13, 2018.

[cccxlii] "Chi-squared distribution," *Wikipedia*, https://en.wikipedia.org/wiki/Chi-squared_distribution, downloaded Dec. 13, 2018.

[cccxliii] "Cauchy distribution," *Wikipedia*, https://en.wikipedia.org/wiki/Cauchy_distribution, downloaded Dec. 13, 2018.

[cccxliv] "An Introduction for the Mathematics of Uncertainty;" "Bayesian probability," *Wikipedia*.

[cccxlv] Navarro, Mireya "Long lines, and odds, for New York's subsidized housing lotteries," *New York Times*, Jan. 29, 2015, https://www.nytimes.com/2015/01/30/nyregion/long-lines-and-low-odds-for-new-yorks-subsidized-housing-lotteries.html.

[cccxlvi] Spears, Marc J. "Expect a happy ending for OKC in Durant's free-agent drama," *The Undefeated*, June 30, 2016, http://theundefeated.com/features/expect-a-happy-ending-for-okc-in-durants-free-agent-drama/.

[cccxlvii] McCleland, Jacob "Kevin Durant to sign with Golden State Warriors," *KGOU*, July 4, 2016, http://kgou.org/post/kevin-durant-sign-golden-state-warriors#stream.

[cccxlviii] Coelho, Caio "Terciles and probabilities in the context ofseasonal forecasts," *CPTEC/INPE*, MedCOF Training Workshop on Verification of Operational Seasonal Forecasts in the Mediterranean region. Rome, Nov. 15-18, 2016, http://medcof.aemet.es/images/doc_events/training2/docTraining2/presentaciones/MedCOF_Training2_Coelho_Tercile-probabilities.pdf, downloaded Dec. 19, 2018.

[cccxlix] Quote 1,511 attributed to H. L. Mencken, as quoted in Byrne, Robert (2012) *The 2,548 Wittiest Things Anybody Ever Said* (Touchstone, New York).

[cccl] Quote 2,319 attributed to Henny Youngman, as quoted by Milton Berle in *B.S. I Love You*, 1988, as quoted in Byrne, Robert (2012) *The 2,548 Wittiest Things Anybody Ever Said* (Touchstone, New York).

[cccli] Quote 183 attributed to Damon Runyon, as quoted in Byrne, Robert (2003) *The 2,548 Best Things Anybody Ever Said* (Touchstone, New York).

[ccclii] TV Show *Inspector Morse,* Season 1, Episode 2, "The Silent World of Nicholas Quinn," 1987.

[cccliii] "Estimated probability of competing in professional athletics," *NCAA*, http://www.ncaa.org/about/resources/research/estimated-probability-competing-professional-athletics, last updated April, 2019.

[cccliv] "American League batting year-by-year averages: League year-by-year batting--averages," *Baseball Reference*, https://www.baseball-reference.com/leagues/AL/bat.shtml, downloaded Dec. 19, 2018; "National League batting year-by-year averages: League year-by-year batting--averages," *Baseball Reference*, https://www.baseball-reference.com/leagues/NL/bat.shtml, downloaded Dec. 19, 2018.

[ccclv] https://twitter.com/ESPNStatsInfo, downloaded Oct. 9, 2016.

[ccclvi] Frushour, Casey; Frushour, Andy. "2014 Super Bowl squares odds," *Casey's Head*, Jan. 26, 2014, http://caseyshead.com/2014-super-bowl-squares-odds/; Frushour, Casey. "The odds of every Super Bowl box pool pair (AKA: How screwed are you with 6 and 2?*), Jan. 27, 2014, https://www.si.com/extra-mustard/2014/01/27/odds-super-bowl-box-pool-probabilities.

[ccclvii] Chairusmi, Jim "In a Super Bowl box pool? Here are your chances," *World Street Journal*, Feb. 5, 2016, 11:55 a.m. ET, https://www.wsj.com/articles/in-a-super-bowl-box-

pool-here-are-your-chances-1454691348. Also see "Your 2019 Super Bowl squares odds: How historically weird scores and Patriots-Rams win probabilities impact your chances," *ELDORADO*, Jan. 24. 2019, https://www.eldo.co/super-bowl-squares-odds-probabilities-best-worst-numbers.html.

[ccclviii] "U.S. Tornado Climatology," *NOAA*, https://www.ncdc.noaa.gov/climate-informatio
n/extreme-events/us-tornado-climatology, downloaded Jan. 2, 2019.

[ccclix] "Historical Records and Trends," *NOAA: National Centers for Environmental Information, National Oceanic and Atmospheric Administration*, https://www.ncdc.noaa.gov/climate-information/extreme-events/us-tornado-climatology/trends, downloaded Dec. 19, 2018.

[ccclx] Asimov, Isaac (2008, reprint) *Foundation* (Del Rey, New York).

[ccclxi] TV Show *Scorpion,* Season 3, Episode 10, "This is the Pits," Aired 10:00 PM, Dec. 12, 2016 on CBS.

[ccclxii] Dexter, Colin (1975) *Last Bus to Woodstock* (Random House, New York), pp. 112-115.

[ccclxiii] "Drake equation," *Wikipedia*, https://en.wikipedia.org/wiki/Drake_equation, downloaded Dec. 14, 2018; Tu, Chau "Frank Drake is still searching for E.T.," *Science Friday*, July 28, 2016, http://www.sciencefriday.com/articles/frank-drake-is-still-searching-for-e-t/.

[ccclxiv] Bialik, Carl "Even with millions typing away, aping Shakespeare ain't easy," *Wall Street Journal*, Oct. 8, 2011, https://www.wsj.com/articles/SB10001424052970203388804576615530390710482; Bialik, Carl "No monkey business: Typing Shakespeare," *Wall Street Journal*, Oct 7, 2011, 8:03 PM, https://blogs.wsj.com/numbers/no-monkey-business-typing-shakespeare-1092.

[ccclxv] Mazur, Joseph (2016) *Fluke: The Math and Myth of Coincidence* (Basic Books, New York); Bialik, Carl "The crash calculations," *Wall Street Journal*, March 3, 2009, 7:50 PM ET, https://blogs.wsj.com/numbers/the-crash-calculations-621; Bialik, Carl "The odds when birds, subs and satellites collide," Wall Street Journal, March 4, 2009, 11:59 PM ET, https://www.wsj.com/articles/SB123612666324824087; Alemi, Farrokh "HAP 525: Risk analysis in healthcare: Probability of rare events," *George Mason University Open Online Courses*, http://openonlinecourses.com/RiskAnalysis/ProbabilityRareEvent.asp, downloaded Dec. 19, 2018.

[ccclxvi] Gelman, Andrew; Silver, Nate; Edlin, Aaron (2012), "What is the probability your vote will make a difference?," Economic Inquiry, 50, 321-326.

[ccclxvii] Bialik, Carl "DNA and Bin Laden's positive ID," *Wall Street Journal*, May 6, 2011, 8:28 PM ET, https://blogs.wsj.com/numbers/dna-and-bin-ladens-positive-id-1057; Bialik, Carl "Matching science of DNA with art of identification," *Wall Street Journal*, May 7, 2011, 12:01 AM ET, https://www.wsj.com/articles/SB1000142405274870399270

4576305062901952434; Barrett, Devlin "FBI's new DNA process produces more matches in suspect database," *Wall Street Journal*, Aug. 25, 2016, 5:30 AM ET, https://www.wsj.com/articles/fbis-new-dna-process-produces-more-matches-1472117404.

[ccclxviii] Bialik, Carl "Are disinfectant germ-killing claims clean?," *Wall Street Journal*, Dec 15, 2009, 8:10 PM ET, https://blogs.wsj.com/numbers/are-disinfectant-germ-killing-claims-clean-861; Bialik, Carl "Kills 99.9% of germs -- Under some lab conditions," *Wall Street Journal*, Dec. 16, 2009, 12:01 AM ET, https://www.wsj.com/articles/SB1260 92257189692937.

[ccclxix] From the 1950 stage and 1955 film version of *Guys and Dolls* and, with slightly different wording, in the Damon Runyon short story *The Idyll of Miss Sarah Brown*, *Collier's Weekly*, Jan. 28, 1933. Damon Runyon, *Wikipedia*, https://en.wikiquote.org/ wiki/Damon_Runyon, downloaded Dec. 18 , 2018.

[ccclxx] Mlodinow, Leonard. (2008) *The Drunkard's Walk: How Randomness Rules Our Lives* (Vintage, U.S.A.).

[ccclxxi] "John Edmund Kerrich" *Wikipedia*, https://en.wikipedia.org/wiki/John_Edmund_ Kerrich, downloaded Jan. 8, 2019.

[ccclxxii] Stoppard, Tom (2017, 50th Anniversary Edition) *Rosencrantz and Guildenstern Are Dead* (Grove Press, New York).

[ccclxxiii] |Andrews, Edmund L.; Zwiebel, Jeffrey "Why the "Hot Hand" may be real after all," *Stanford Business*, March 25, 2014, https://www.gsb.stanford.edu/insights/jeffrey-zwiebel-why-hot-hand-may-be-real-after-all; Green, Brett S.; Zwiebel, Jeffrey, "The Hot-Hand Fallacy: Cognitive mistakes or equilibrium adjustments? Evidence from Major League Baseball," *SSRN*, last revised May 17, 2017, http://papers.ssrn.com/sol3/papers. cfm?abstract_id=2358747.

[ccclxxiv] *A Practical Guide to Quantitative Finance Interviews*, pg. 73.

[ccclxxv] *Naked Statistics*, pg. 90; *Computational and Inferential Thinking*, pp. 242-248; *A Practical Guide to Quantitative Finance Interviews*, pg. 78.

[ccclxxvi] McKean, Kevin "Decisions, Decisions," *Discover Magazine*, as from http://d1m3qhodv9fjlf.cloudfront.net/wp-content/uploads/2013/01/Decisions_Decisions .pdf, downloaded Dec. 27, 2018.

[ccclxxvii] Beyth-Marom, Ruth; Dekel, Shlomith; Gombo, Ruth; Shaked, Moshe (1985) *Elementary Approach to Thinking Under Uncertainty* (L. Erlbaum Associates, Hillsdale, N.J.).

[ccclxxviii] "Syphilis testing algorithms using Treponemal Tests for initial screening --- Four Laboratories, New York City, 2005--2006," *CDC/Centers for Disease Control and Prevention/MMWR Weekly*, Aug. 15, 2008, http://www.cdc.gov/mmwr/preview/mmwrht ml/mm5732a2.htm.

[ccclxxix] *A Practical Guide to Quantitative Finance Interviews*, pp. 72-73.

[ccclxxx] "Likelihood ratios in diagnostic testing," *Wikipedia*, https://en.wikipedia.org/wiki/

Likelihood_ratios_in_diagnostic_testing, downloaded Dec. 14, 2018.
[ccclxxxi] Sanders, Lisa "Why did this man lose his memory, words and even his ability to walk?," *New York Times*, April 13, 2017, https://www.nytimes.com/2017/04/13/magazine/why-did-this-man-lose-his-memory-words-and-even-his-ability-to-walk.html.
[ccclxxxii] Croswell, Jennifer M.; Kramer, Barnett S.; Kreimer, Aimee R.; Prorok, Phil C.; Xu, Jina-Lu; Baker, Stuart G.; … Schoen, Robert E. (2009), "Cumulative incidence of false-positive results in repeated, multimodal cancer screening," *Ann. Fam. Med.*, 7, 212-222.
[ccclxxxiii] Schröder, Fritz H.; Hugosson, Jonas; Roobol, Monique J.; Tammela, Teuvo L. J.; Ciatto, Stefano; Nelen, Vera; … Recker, Franz; *et al.*, for the ERSPC Investigators* (2009), "Screening and prostate-cancer mortality in a randomized European study," *N. Engl. J. Med.*, 360, 1320-1328.
[ccclxxxiv] "Cumulative incidence of false-positive results in repeated, multimodal cancer screening."
[ccclxxxv] Welch H. G., Schwartz L. M., Woloshin, S. (2005), "Prostate-specific antigen levels in the United States: implications of various definitions for abnormal," *J. Natl. Cancer Inst.*, 97, 1132-1137.
[ccclxxxvi] "Archived final recommendation statement: Prostate cancer: Screening," *U.S. Preventive Services: Task Force*, current as of May 2012, http://www.uspreventiveservicestaskforce.org/Page/Document/RecommendationStatementFinal/prostate-cancer-screening.
[ccclxxxvii] Lagnado, Lucette "Prostate-cancer gene test helps patients decide on treatment," *Wall Street Journal*, March 31, 2018, 7:05 a.m. ET, https://www.wsj.com/articles/prostate-cancer-gene-test-helps-patients-decide-on-treatment-1522494300.
[ccclxxxviii] Murphy, Heather "Most white Americans' DNA can be identified through genealogy databases," *New York Times*, Oct. 11, 2018, https://www.nytimes.com/2018/10/11/science/science-genetic-genealogy-study.html, downloaded June 13, 2019; Joh, Elizabeth "Want to see my genes? Get a warrant," *New York Times*, June 11, 2019, https://www.nytimes.com/2019/06/11/opinion/police-dna-warrant.html, downloaded June 13, 2019.
[ccclxxxix] Finley, Allysia "The making of a DNA detective," *Wall Street Journal*, Feb. 15, 2019, 6:14 PM ET, https://www.wsj.com/articles/the-making-of-a-dna-detective-11550272449.
[cccxc] Brenner, Charles H. "Forensic mathematics," http://dna-view.com/, downloaded Dec. 19, 2018; Helm, H. J. van der; Hische, E. A. H. (1979), "Application of Bayes's Theorem to results of quantitative clinical chemical determination," *Clin. Chem.*, 25, 985-988; "National Research Council. 1996. The evaluation of forensic DNA evidence," Washington, DC: The National Academies Press. https://doi.org/10.17226/5141.
[cccxci] "DNA and Bin Laden's Positive ID."
[cccxcii] "Forensic mathematics."

[cccxciii] Gabrielson, Ryan and Sanders, Topher "How a $2 roadside drug test sends innocent people to jail," *New York Times*, July 7, 2016, https://www.nytimes.com/2016/07/10/magazine/how-a-2-roadside-drug-test-sends-innocent-people-to-jail.html; Gabrielson, Ryan; Sanders, Topher "Busted," *ProPublica*, July 7, 2016, https://www.propublica.org/article/common-roadside-drug-test-routinely-produces-false-positives; Gabrielson, Ryan " 'No Field Test is Fail Safe': Meet the Chemist behind Houston's police drug kits," *ProPublica*, July 11, 2016, 8 AM EDT, https://www.propublica.org/article/no-field-test-is-fail-safe-meet-the-chemist-behind-houston-police-drug-kits, downloaded Dec. 19, 2018; "Cobalt(II) thiocyanate" *Wikipedia*, https://en.wikipedia.org/wiki/Cobalt(II)_thiocyanate, downloaded Dec. 14, 2018.

[cccxciv] Durrett, Richard (2012) *Essentials of Stochastic Processes* (Springer-Verlag, New York).

[cccxcv] Metropolis, Nicholas; Rosenbluth, Arianna W.; Rosenbluth, Marshall N.; Teller, Augusta H.; Teller, Edward (1953), "Equation of state calculations by fast computing machines," *J. Chem. Phys.*, 21, 1087-1092.

[cccxcvi] Redner, Sidney (2001) *A Guide to First-Passage Processes* (Cambridge University Press, Cambridge).

[cccxcvii] *A Guide to First-Passage Processes*

[cccxcviii] *An Introduction to Probability Theory and Its Applications.*

[cccxcix] *An Introduction to Probability Theory and Its Applications*, Vol. 1, pg. 88.

[cd] Weisstein, Eric W. "Pólya's Random Walk Constants," *MathWorld--A Wolfram Web Resource*, http://mathworld.wolfram.com/PolyasRandomWalkConstants.html, downloaded Dec. 14, 2018.

[cdi] Ben-Naim, E.; Vazquez, F.; Redner, S. (2007), "What is the most competitive sport?," *Journal of the Korean Physical Society*, 50, 124-126.

[cdii] Vergin, Roger C. (2000), "Winning streaks in sports and the misperception of momentum," *Journal of Sport Behavior*, 23, 181-197.

[cdiii] Bradley, Ralph Allan; Terry, Milton E. (1952), "Rank analysis of incomplete block designs: I. The method of paired comparisons," *Biometrika*, 39, 324-345.

[cdiv] Sire, Clément; Redner, Sidney. (2009), "Understanding baseball team standings and streaks," *Eur. Phys. J. B*, 67, 473-481.

[cdv] Gabel, Alan; Redner, Sidney (2012), "Random walk picture of basketball scoring," *J. Quantitative Analysis in Sports*, 8, Manuscript 1416.

[cdvi] Clauset, A.; Kogan, M.; Redner, S. (2015), "Safe leads and lead changes in competitive team sports," *Phys. Rev. E*, 91, 062815.

[cdvii] Palmer, T. N. (2002), "Predicting uncertainty in numerical weather forecasts," *Int. Geophysics*, 83, 3-13.

[cdviii] McGinty, Jo Craven "As forecasts go, you can bet on Monte Carlo simulations," *Wall Street Journal*, Aug. 12, 2016, 5:30 AM ET, https://www.wsj.com/articles/as-forecasts-go-you-can-bet-on-monte-carlo-1470994203.

[cdix] "IBM SPSS Statistics," *IBM,* https://www.ibm.com/us-en/marketplace/spss-statistics, downloaded Dec. 19, 2018; Why use StatsDirect for your research?, *StatsDirect*, https://www.statsdirect.com/, downloaded Dec. 19, 2018.
[cdx] "Labor force statistics from the current population survey: How the government measures unemployment," *United States Department of Labor, Bureau of Labor Statistics*, updated Oct. 8, 2015, http://www.bls.gov/cps/cps_htgm.htm.
[cdxi] Winslow, Ron "The small warnings before cardiac arrest," *Wall Street Journal*, Feb. 1, 2016 1:04 PM ET, https://www.wsj.com/articles/the-small-warnings-before-cardiac-arrest-1454349863; Marijon, Eloi; Uy-Evanado, Audrey; Dumas, Florence; Karam, Nicole; Reinier, Kyndaron; Teodorescu, Carmen; … Chugh, Sumeet S. (2016), "Warning symptoms are associated with survival from sudden cardiac arrest," *Ann. Intern. Med.*, 164, 23-29.
[cdxii] Wheelan, Charles. (2013) *Naked Statistics* (Norton, New York).
[cdxiii] Valliant, Richard; Dever, Jill A.; Kreuter, Frauke (2013) *An Overview of Sample Design and Weighting*, Volume 51 of the series *Statistics for Social and Behavioral Sciences* (Springer, Germany).
[cdxiv] *Computational and Inferential Thinking*, pg. 348.
[cdxv] *Computational and Inferential Thinking.*
[cdxvi] "Framingham Heart Study," *Wikipedia*, https://en.wikipedia.org/wiki/Framingham_Heart_Study, downloaded Dec. 14, 2018.
[cdxvii] *Naked Statistics*, pg. 115.
[cdxviii] Wilson, Richard; Crouch, Edmund A.C. (2001) *Risk–Benefit Analysis* (Harvard University Press, Cambridge, Mass), pg. 231.
[cdxix] *Naked Statistics*, pg. 112.
[cdxx] Brody, Leslie "NYC school crimes drop, while confiscated weapons increase," *Wall Street Journal*, Aug. 1, 2017, 6:20 PM ET, https://www.wsj.com/articles/nyc-school-crimes-drop-while-confiscated-weapons-increase-1501626024; Harris, Elizabeth A. "Crime in New York City schools is at a record low, City says, *New York Times*, Aug. 1, 2017, https://www.nytimes.com/2017/08/01/nyregion/crime-in-new-york-city-schools-is-at-a-record-low-city-says.html.
[cdxxi] McCartney, Scott "Your lost bags may not count as lost" *Wall Street Journal*, Nov. 15, 2017, 7:40 PM ET, https://www.wsj.com/articles/your-lost-bags-may-not-count-as-lost-1510761097.
[cdxxii] Holmes, Elizabeth "When shopping online, can you trust the reviews?," *Wall Street Journal*, Nov. 29, 2016, 11:47 AM ET, https://www.wsj.com/articles/when-shopping-online-can-you-trust-the-reviews-1480438071.
[cdxxiii] Freeman, James "The incredible campaign against plastic straws," *Wall Street Journal*, July 25, 2018, 2:59 PM ET, https://www.wsj.com/articles/the-incredible-campaign-against-plastic-straws-1532545186.
[cdxxiv] Central Park, NY Historical Data, *National Weather Service*, http://www.weather.

gov/okx/CentralParkHistorical, downloaded Dec. 19, 2018.; "National Weather Service Forecast Office New York, NY, Central Park, New York, NY November Records: 1869 - Present," *National Weather Service*, http://www.weather.gov/media/okx/Climate/Almanacs/nyc/nycnov.pdf, downloaded Dec. 19, 2018.

[cdxxv] O'Connor, Anahad "For coffee drinkers, the buzz may be in your genes," *New York Times*, July 12, 2016, 8:59 AM, https://well.blogs.nytimes.com/2016/07/12/for-coffee-drinkers-the-buzz-may-be-in-your-genes; Cornelis, Marilyn C.; El-Sohemy, Ahmed; Kabagambe, Edmond K.; Campos, Hannia (2006), "Coffee, CYP1A2 genotype, and risk of myocardial infarction," *JAMA*, 295, 1135-1141; Palatini, P.; Ceolotto, G.; Ragazzo, F.; Dorigatti, F.; Saladini, F.; Papparella, I.; Mos, L.; Zanata, G.; Santonastaso, M. (2009), "CYP1A2 genotype modifies the association between coffee intake and the risk of hypertension," *J Hypertens.*, 27, 1594-1601.

[cdxxvi] Mrazek, David A. (2010) *Psychiatric Pharmacogenomics* (Oxford University Press, Oxford), pg. 88, Table 7.2.

[cdxxvii] "Statistical*H*elp: Variance, standard deviation and spread," *StatsDirect*, http://www.statsdirect.com/help/default.htm#basic_descriptive_statistics/standard_deviation.htm, downloaded Dec. 19, 2018.

[cdxxviii] "Statistical*H*elp: Regression and correlation," *StatsDirect*, http://www.statsdirect.com/help/default.htm#regression_and_correlation/regression_and_correlation.htm, downloaded Dec. 19, 2018.

[cdxxix] *Computational and Inferential Thinking*, pp. 396-431.

[cdxxx] "STAT 501 Regression methods: Lesson 1: Simple linear regression," *PennState Eberly College of Science*, https://onlinecourses.science.psu.edu/stat501/node/250, downloaded Dec. 20, 2018; "STAT 501 Regression methods: 1.1 - What is simple linear regression?," *PennState Eberly College of Science*, https://onlinecourses.science.psu.edu/stat501/node/251, downloaded Dec. 20, 2018; "Linear regression," *Wikipedia*, https://en.wikipedia.org/wiki/Linear_regression, downloaded Dec. 14, 2018; "Linear regression," *Yale*, http://www.stat.yale.edu/Courses/1997-98/101/linreg.htm, downloaded Dec. 20, 2018; Nau, Robert "Linear regression models," *Duke*, http://people.duke.edu/~rnau/regintro.htm, downloaded Dec. 20, 2018; Weisstein, Eric W. "Linear regression," *MathWorld--A Wolfram Web Resource*, http://mathworld.wolfram.com/LinearRegression.html, downloaded Dec. 14, 2018; Weisstein, Eric W. "Least squares fitting," *MathWorld--A Wolfram Web Resource*, http://mathworld.wolfram.com/LeastSquaresFitting.html, downloaded Dec. 14, 2018. "Regression analysis," *Wikipedia*, https://en.wikipedia.org/wiki/Regression_analysis, downloaded Dec. 14, 2018.

[cdxxxi] Statistical*H*elp: Variance, Regression and Correlation; Computational and Inferential Thinking, pp. 185, 446-453, 459, 479, and 581.

[cdxxxii] Weisstein, Eric W. "Correlation coefficient," *MathWorld--A Wolfram Web Resource*, http://mathworld.wolfram.com/CorrelationCoefficient.html, downloaded Dec. 14, 2018; Pearson correlation coefficient, *Wikipedia*, https://en.wikipedia.org/wiki/Pears

on_product-moment_correlation_coefficient, downloaded Dec. 14, 2018.

[cdxxxiii] "Standard error," *Wikipedia*, https://en.wikipedia.org/wiki/Standard_error, downloaded Dec. 14, 2018.

[cdxxxiv] *Computational and Inferential Thinking*, pg. 432.

[cdxxxv] "Standard error," *Wikipedia*.

[cdxxxvi] Dallal, Gerlald E. "The standard error of a proportion," http://www.jerrydallal.com/lhsp/psd.htm, last modified May 22, 2012.

[cdxxxvii] "Margin of error," *Wikipedia*, https://en.wikipedia.org/wiki/Margin_of_error, downloaded Dec. 14, 2018.

[cdxxxviii] Bialik, Carl "What's a statistical tie, anyway?," *Wall Street Journal*, Dec. 6, 2007, 1:50 pm ET, https://blogs.wsj.com/numbers/whats-a-statistical-tie-anyway-234.

[cdxxxix] Mercer, Andrew "5 key things to know about the margin of error in election polls," *Pew Research Center: FactTank*, Sept. 8, 2016, http://www.pewresearch.org/fact-tank/2016/09/08/understanding-the-margin-of-error-in-election-polls/.

[cdxl] Jamrisko, Michelle; Dopp, Terrence "Failed polls in 2016 call into question a profession's precepts," *Bloomberg*, Nov. 9, 2016, 2:45 AM EST, https://www.bloomberg.com/politics/articles/2016-11-09/failed-polls-in-2016-call-into-question-a-profession-s-precepts, downloaded Dec. 20, 2018.

[cdxli] "Failed polls in 2016 call into question a profession's precepts."

[cdxlii] Rothschild, David; Goel, Sharad "When you hear the margin of error is plus or minus 3 percent, think 7 instead," New York Times, Oct. 5, 2016, https://www.nytimes.com/2016/10/06/upshot/when-you-hear-the-margin-of-error-is-plus-or-minus-3-percent-think-7-instead.html; Shirani-Mehr, Houshmand; Rothschild, David; Goel, Sharad; Gelman, Andrew (2016), "Disentangling bias and variance in election polls," *J. Am. Stat. Assoc.*, 113, 607-614.

[cdxliii] Bialik, Carl "When Polls Turn Up the Wrong Number."

[cdxliv] McDonald, Michael P. "2016 November general election turnout rates," *United States Elections Project*, http://www.electproject.org/2016g, downloaded Dec. 20, 2018.

[cdxlv] Assessing the Representativeness of Public Opinion Surveys, *Pew Research Center*, May 15, 2012, http://www.people-press.org/2012/05/15/assessing-the-representativeness-of-public-opinion-surveys/.

[cdxlvi] Bialik, Carl "In polling, sometimes it's not all in the question," *Wall Street Journal*, April 10, 2007, 4:00 PM ET, https://blogs.wsj.com/numbers/in-polling-sometimes-its-not-all-in-the-question-81.

[cdxlvii] Shirani-Mehr, Houshmand; Rothschild, David; Goel, Sharad; Gelman, Andrew. (2016), "Disentangling bias and variance in election polls," *J. Am. Stat. Assoc.*, 113, 607-614.

[cdxlviii] "Failed polls in 2016 call into question a profession's precepts"; Berley, Max "Perils of polling," *Bloomberg*, updated Oct. 29, 2018, 3:09 PM EDT, https://www.bloomberg.com/quicktake/perils-of-polling, downloaded Dec. 23, 2018.

[cdxlix] "Chi-squared test," *Wikipedia*, https://en.wikipedia.org/wiki/Chi-squared_test, downloaded Dec. 14, 2018; Gan, K. K. :Lecture 6: Chi square distribution (χ^2) and least squares fitting," *Ohio State*, https://www.physics.ohio-state.edu/~gan/teaching/spring04/Chapter6.pdf, downloaded Dec. 26, 2018; "Pearson's chi-squared test," *Wikipedia*, https://en.wikipedia.org/wiki/Pearson%27s_chi-squared_test, downloaded Dec. 14, 2018.

[cdl] "Chi-squared distribution," *Wikipedia.*

[cdli] "Pearson's chi-squared test," *Wikipedia.*

[cdlii] "U.S. Department of Health and Human Services study *Health, United States, 2017," National Center for Health Statistics. Health, United States, 2017: With special feature on mortality. Hyattsville, MD. 2018*, https://www.cdc.gov/nchs/data/hus/hus17.pdf, Table 21, downloaded June 7, 2019.

[cdliii] "Life table," *Wikipedia*, https://en.wikipedia.org/wiki/Life_table, downloaded Dec. 14, 2018.

[cdliv] "Actuarial life table," *Social Security*, https://www.ssa.gov/oact/STATS/table4c6.html, downloaded Dec. 26, 2018.

[cdlv] Maier, Mark; Imazeki, Jennifer (2014) *The Data Game*, 4th edition (Routledge, London), pg. 11.

[cdlvi] Callaway, Ewen (2016), "The visualizations transforming biology," *Nature*, 535, 187-188.

[cdlvii] Huff, Darrell (1993) *How to Lie with Statistics* (W. W. Norton & Company, New York City).

[cdlviii] "Curve-fitting methods and the messages they send," *xkcd: A Webcomic of Romance, Sarcasm, Math, and Language*, https://www.xkcd.com/2048/, downloaded Jan. 21, 2019.

[cdlix] Finley, Allysia "*'How to lie with Statistics': Teachers union edition*," *Wall Street Journal*, May 2, 2018, 7:17 p.m. ET, https://www.wsj.com/articles/how-to-lie-with-statistics-teachers-union-edition-1525303025.

[cdlx] Berg, Robbie "Hurricane Joaquin," *National Hurricane Center Tropical Cyclone Report*, Jan. 12, 2016, https://www.nhc.noaa.gov/data/tcr/AL112015_Joaquin.pdf, Fig. 9.

[cdlxi] From Michael Tippett, private comunication; "Statistical humor," http://ceadserv1.nku.edu/longa//public_html/htmls/statjokes.html, downloaded Jan. 2, 2019.

[cdlxii] "Price Changes 1996 to 2016: Selected consumer goods and services, Do you hear that? It might be the growing sounds of pocketbooks snapping shut and the chickens coming home.....," *AEI*, http://www.aei.org/publication/do-you-hear-that-it-might-be-the-growing-sounds-of-pocketbooks-snapping-shut-and-the-chickens-coming-home/, downloaded Dec. 26, 2018.

[cdlxiii] Belkin, Douglas; Mitchell, Josh; Korn, Melissa "House GOP to propose sweeping changes to higher education," *Wall Street Journal*, Nov. 29, 2017, 5:39 PM ET,

https://www.wsj.com/articles/house-gop-to-propose-sweeping-changes-to-higher-education-1511956800.

[cdlxiv] "How does my 5k pace compare to others?," *Pace Calculator*, http://www.pace-calculator.com/5k-pace-comparison.php, downloaded Dec. 26, 2018; "How does my 10k pace compare to others?," *Pace Calculator*, http://www.pace-calculator.com/10k-pace-comparison.php, downloaded Dec. 26, 2018; "How does my half marathon pace compare to others?," *Pace Calculator*, http://www.pace-calculator.com/half-marathon-pace-comparison.php, downloaded Dec. 26, 2018; "How does my marathon pace compare to others?," *Pace Calculator*, http://www.pace-calculator.com/marathon-pace-comparison.php, downloaded Dec. 26, 2018.

[cdlxv] Bialik, Carl "Seeking a poverty measure for the next 50 years," *Wall Street Journal*, Sept. 20, 2013, 10:56 pm ET, https://blogs.wsj.com/numbers/seeking-a-poverty-measure-for-the-next-50-years-1275; Bialik, Carl "New way of calculating poverty rate faces hurdles,? *Wall Street Journal*, Sept. 20, 2013, 11:01 PM ET, https://www.wsj.com/articles/SB10001424127887324807704579085860737840606.

[cdlxvi] "Actuarial life table," *Social Security*.

[cdlxvii] Gould, Stephen Jay "The median isn't the message," *CancerGuide*, updated May 31, 2002, https://www.cancerguide.org/median_not_msg.html; Gould, Stephen Jay "The median isn't the message," *YouTube*, https://www.youtube.com/watch?v=cH6XuiOBbkc, downloaded Dec. 26, 2018; Gawande, Atul (2014) *Being Mortal: Medicine and What Matters* (Deckle Edge, South Carolina), pp. 170-171; Gould, Stephen Jay (2011) *Full House* (Harvard University Press. Cambridge), p. 233.

[cdlxviii] Eugenia, Cheng "Calculating a faster checkout line," *Wall Street Journal* May 19, 2017, 11:35 AM ET, https://www.wsj.com/articles/calculating-a-faster-checkout-line-1495208140.

[cdlxix] "Pearson's chi-squared test," *Wikipedia*.

[cdlxx] *Computational and Inferential Thinking*, pp. 305, 318, and 334.

[cdlxxi] "Engineering Statistics Handbook:1.3.6.7.4. Critical values of the chi-square distribution," *NIST*, https://www.itl.nist.gov/div898/handbook/eda/section3/eda3674.htm, downloaded Dec. 26, 2018; Lane, David M. "Online statistics education: An interactive multimedia course of study," *Rice University, University of Houston Clear Lake, and Tufts University*, http://onlinestatbook.com/2/logic_of_hypothesis_testing/significance.html, downloaded Dec. 26, 2018; "Statistical significance," *Wikipedia*, https://en.wikipedia.org/wiki/Statistical_significance, downloaded Dec. 14, 2018; "Variance: Distribution of the sample variance," *Wikipedia*, https://en.wikipedia.org/wiki/Variance#Distribution_of_the_sample_variance, downloaded Dec. 14, 2018; "Chi-squared test," *Wikipedia*; Pearson's chi-squared test, *Wikipedia*.

[cdlxxii] Halsey, Lewis G.; Curran-Everett, Douglas; Vowler, Sarah L.; Drummond, Gordon B. (2015), "The fickle P value generates irreproducible results," *Nature Methods*, 12, 179-185.

[cdlxxiii] Ziliak, Stephen T.; McCloskey, Deirdre N. (2008) *The Cult of Statistical Significance: How the Standard Error Costs Us Jobs, Justice, and Lives (Economics, Cognition, and Society)* (University of Michigan Press, Ann Arbor).

[cdlxxiv] *Computational and Inferential Thinking*, pg. 311.

[cdlxxv] *The Cult of Statistical Significance*; Box, Joan F. (1987), "Guinness, Gosset, Fisher, and small samples," *Statistical Science*, 2, 45-52, accessed from https://www.jstor.org/stable/2245613.

[cdlxxvi] "Student's t-test," *Wikipedia*, https://en.wikipedia.org/wiki/Student%27s_t-test, downloaded Dec. 18 , 2018.

[cdlxxvii] *The Cult of Statistical Significance*, pg. 23.

[cdlxxviii] Muller, Jerry Z. "A cure for our fixation on metrics," *Wall Street Journal*, Jan. 12, 2018, 10:50 AM ET, https://www.wsj.com/articles/a-cure-for-our-metric-fixation-1515772238; Jerry Z. Muller (2018) *The Tyranny of Metrics*, (Princeton University Press, Princeton, NJ).

[cdlxxix] McGinty, Jo Craven "How do energy companies measure the temperature? Not in Fahrenheit or Celsius," *Wall Street Journal*, Sept. 14, 2018, 7:00 AM ET, https://www.wsj.com/articles/how-do-energy-companies-measure-the-temperature-not-in-fahrenheit-or-celsius-1536922800.

[cdlxxx] "Dow Jones Industrial Average," *Wikipedia*, https://en.wikipedia.org/wiki/Dow_Jones_Industrial_Average, downloaded June 28, 2019/ Santilli, Peter; DeStefano, Tom "The Dow's lightweight title," *Wall Street Journal*, Nov. 26, 2017 12:08 pm ET, https://www.wsj.com/articles/the-dows-lightweight-title-1511716109.

[cdlxxxi] Mackintosh, James "We're already at Dow 30000, you just don't know it," *Wall Street Journal*, Jan. 25, 2017, 2:06 PM ET, https://www.wsj.com/articles/were-already-at-dow-30000-you-just-dont-know-it-1485362316.

[cdlxxxii] Wursthorn, Michael; Otani, Akane "What about Amazon? Dow Industrials dumping GE for Walgreens reflects index's dilemma," *Wall Street Journal*, June 20, 2018, 6:24 PM ET, https://www.wsj.com/articles/what-about-amazon-dow-industrials-dumping-ge-for-walgreens-reflects-indexs-dilemma-1529533470.

[cdlxxxiii] Elmerraji, Jonas "5 must-have metrics for value investors," *INVESTOPEDIA*, updated May 25, 2018, http://www.investopedia.com/articles/fundamental-analysis/09/five-must-have-metrics-value-investors.asp.

[cdlxxxiv] "S&P 500 PE ratio," *Multp*, http://www.multpl.com/, downloaded Dec. 26, 2018, from Shiller, Robert J. (2016) *Irrational Exuberance*, 3rd edition (Princeton University Press, Princeton).

[cdlxxxv] Francis, Theo; Fuhrmans, Vanessa "Are you underpaid? In a first, U.S. firms reveal how much they pay workers," *Wall Street Journal*, March 12, 2018, https://www.wsj.com/articles/are-you-underpaid-in-a-first-u-s-firms-reveal-how-much-they-pay-workers-1520766000.

[cdlxxxvi] Irwin, Neil "Is capital or labor winning at your favorite company? Introducing the Marx ratio," *New York Times*, May 21, 2018, https://www.nytimes.com/interactive/2018/05/21/upshot/marx-ratio-median-pay.html.

[cdlxxxvii] Gramm, Phil; Early, John F. "The myth of American inequality," *Wall Street Journal* Aug. 9, 2018, 6:51 PM ET, https://www.wsj.com/articles/the-myth-of-american-inequality-1533855113.

[cdlxxxviii] Glassman, Brian; "Income inequality metrics and economic well-being in U.S. metropolitan statistical areas," *U.S. Census Bureau*, https://www.census.gov/content/dam/Census/library/working-papers/2016/demo/SEHSD-WP2016-19.pdf, downloaded June 28, 2019.

[cdlxxxix] "Advertising metrics - Management by the numbers," www.management-by-the-numbers.com/concepts/advmetrics.pptx.

[cdxc] Holliday, Simon "Uncovering video performance metrics," *COULL*ET, http://coull.com/our-blog/uncovering-video-performance-metrics/, downloaded Dec. 26, 2018; Batra, Anil "21 metrics for measuring online display advertising," *Digital Marketing and Analytics by Anil Batra*, June 10, 2018, http://webanalysis.blogspot.com/2014/05/21-metrics-for-measureing-online.html#axzz4QIE8X3xd.

[cdxci] Shields. Mike "Facebook says it found more miscalculated metrics," *Wall Street Journal*, Nov. 16, 2016, 10:03 AM ET, https://www.wsj.com/articles/facebook-says-it-found-more-miscalculated-metrics-1479303984.

[cdxcii] "Poverty Measure, The New York City Government poverty measure, 2005-2017," *The Mayor's Office for Economic Opportunity, NYC Opportunity*, https://www1.nyc.gov/site/opportunity/poverty-in-nyc/poverty-measure.page, downloaded June 10, 2019.

[cdxciii] Mele, Christopher " 'Five-Second Rule' for food on floor is untrue, study finds," *New York Times*, Sept. 19, 2016, https://www.nytimes.com/2016/09/20/science/five-second-rule.html; Robyn C. Miranda; Donald W. Schaffner (2016), "Longer contact times increase cross-, contamination of Enterobacter aerogenes from surfaces to food," *Appl. Enviro., Microbiology*, 82, 6491-6496.

[cdxciv] Hirsch, J. E. (2005), "An index to quantify an individual's scientific research output," *Proc. Nat. Acad. Sci.*, 46, 16569-16572.

[cdxcv] Redner, S. (2010), "On the meaning of the h-index," *J. Stat. Mech: Theory and Exp.*, 2010, L03005.

[cdxcvi]Isaac, Mike "The ratio establishes itself on Twitter," *New York Times*, Feb. 14, 2018, https://www.nytimes.com/interactive/2018/02/09/technology/the-ratio-trends-on-twitter.html.

[cdxcvii] Passy, Charles "Zagat overhauls restaurant review ratings." *Wall Street Journal*, July 26, 2016, 12:00 AM ET, https://www.wsj.com/articles/zagat-overhauls-restaurant-review-ratings-1469505614.

[cdxcviii] Fritz, Ben "Hollywood now worries about viewer scores, not reviews," Wall Street Journal, July 20, 2016, 2:16 PM, https://www.wsj.com/articles/hollywood-turns-spotlight-on-websites-that-aggregate-movie-reviews-1469038587.

[cdxcix] "How we create the Metascore magic," *Metacritic*, https://www.metacritic.com/about-metascores, downloaded Dec. 26, 2018.

[d] Roberts, Sam "Joseph B. Keller, Mathematician with whimsical curiosity, dies at 93," *New York Times*, Sept. 16, 2016, https://www.nytimes.com/2016/09/17/us/joseph-b-keller-mathematician-with-whimsical-curiosity-dies-at-93.html; Keller, Joseph B. (1982), "Time-dependent Queues," *SIAM Rev.*, 24, 401-412.

[di] McCartney, Scott "Which airlines pad their schedules the most?," *Wall Street Journal*, June 28, 2017, 11:15 am ET, https://www.wsj.com/articles/which-airlines-pad-their-schedules-the-most-1498662950.

[dii] Birnbaum, Phil "A guide to Sabermetric research," *SABR: Society for American Baseball Research*, https://sabr.org/sabermetrics, downloaded Dec. 26, 2018.

[diii] Smith, Sean "Projection Blog," *BaseballProjection*, http://www.baseballprojection.com/, downloaded Dec. 26, 2018.

[div] "Win shares," *Wikipedia*, https://en.wikipedia.org/wiki/Win_Shares, downloaded Dec. 14, 2018; James, Bill; Henzler, Jim (2002) *Win Shares* (Stats Inc., U.S.A.).

[dv] Studeman, Dan "WAR vs. Win shares," *The Hardball Times*, Sept. 14, 2009, https://www.fangraphs.com/tht/war-vs-win-shares/.

[dvi] James, Bill (1985) *The Bill James Historical Baseball Abstract* (Villard, New York City).

[dvii] *Moneyball.*

[dviii] Diamond, Jared "The chemistry experiment behind the Astros winning the World Series," *Wall Street Journal*, Nov. 2, 2017, 5:05 PM ET, https://www.wsj.com/articles/how-the-astros-hit-on-a-winning-formula-1509648139.

[dix] Costa, Brian; Diamond, Jared "The downside of baseball's data revolution," *Wall Street Journal*, Oct. 3, 2017, 11:18 AM ET, https://www.wsj.com/articles/the-downside-of-baseballs-data-revolutionlong-games-less-action-1507043924.

[dx] "Player efficiency rating," *Wikipedia*, https://en.wikipedia.org/wiki/Player_efficiency_rating#Career_PER_leaders, downloaded June 28, 2019; "Calculating PER," *Basketball Reference*, https://www.basketball-reference.com/about/per.html, downloaded June 28, 2019; "2018-19 Hollinger NBA player statistics - All players," ESPN, http://insider.espn.go.com/nba/hollinger/statistics, downloaded June 28, 2019.

[dxi] McGinty, Jo Craven "Math madness: College hoops fans hope to day 'RIP' to the RPI," *Wall Street Journal*, March 9, 2018 9:00 AM ET, https://www.wsj.com/articles/math-madness-college-hoops-fans-hope-to-say-rip-to-the-rpi-1520604000.

[dxii] Higgins, Laine "College basketball says goodbye to the RPI, Hello NET," *Wall Street Journal*, Feb. 14, 2019 9:57 AM ET, https://www.wsj.com/articles/college-basketball-says-goodbye-to-the-rpi-hello-net-11550156261.

[dxiii] "NFL Quarterback rating formula," *NFL*, http://www.nfl.com/help/quarterbackrating formula, downloaded Dec. 26, 2018.

[dxiv] Oliver, Dean "Guide to the Total Quarterback Rating," *ESPN*, Aug. 4, 2011, http://espn.go.com/nfl/story/_/id/6833215/explaining-statistics-total-quarterback-rating.

[dxv] "What is age-graded scoring?," *RaceCentral*, https://www.runraceresults.com/AgeGrade.htm, downloaded Dec. 26, 2018; Mateo, Ashley "What is age grading and why does it matter?," *Runner's World*, Aug. 31, 2018, https://www.runnersworld.com/tools/age-grade-calculator, downloaded Dec. 26, 2018; Jones, Alan "Age grading running races," https://www.runscore.com/Alan/AgeGrade.html. downloaded Dec. 26, 2018.

[dxvi] Sowell, Thomas (1996) *The Vision of the Anointed: Self-Congratulation as a Basis for Social Policy* (Basic Books, New York).

[dxvii] *Computational and Inferential Thinking*, pg. 28.

[dxviii] Vigen, Tyler "Spurious correlations," http://tylervigen.com/spurious-correlations, downloaded Dec. 26, 2018; Vigen, Tyle (2015) *Spurious Correlations* (Hachette, New York City); Wilson, Mark "Hilarious graphs prove that correlation isn't causation," *Fast Company*, May 13, 2014, http://www.fastcodesign.com/3030529/infographic-of-the-day/hilarious-graphs-prove-that-correlation-isnt-causation.

[dxix] Messerli, Franz H. (2012), "Chocolate consumption, cognitive function, and Nobel laureates," *N. Engl. J. Med.*, 367, 1562-1564.

[dxx] Pepe, Margaret Sullivan; Janes, Holly; Longton, Gary; Leisenring, Wendy; Newcomb, Polly (2004), "Limitations of the odds ratio in gauging the performance of a diagnostic, prognostic, or screening marker," *Am. J. Epidemiology*, 159, 882-890; Liu, C.C.; Manzi, S.; Ahearn, J. M. (2005), "Biomarkers for systemic lupus erythematosus: a review and perspective," *Curr. Opin. Rheumatol.*, 17, 543-549; Illei, G.G.; Tackey, E.; Lapteva, L.; Lipsky, P.E (2004), "Biomarkers in systemic lupus erythematosus. I. General overview of biomarkers and their applicability," *Arthritis Rheum.*, 50, 1709-1720; Surrogate endpoint, *Wikipedia*, https://en.wikipedia.org/wiki/Surrogate_endpoint, downloaded Dec. 14, 2018.

[dxxi] Angwin, Julia; Scheiber, Noam; Tobin, Ariana "Facebook job ads raise concerns about age discrimination," *New York Times*, Dec. 20, 2017, https://www.nytimes.com/2017/12/20/business/facebook-job-ads.html.

[dxxii] "The Age Discrimination in Employment Act of 1967," *U.S. Equal Employment Opportunity Commission*, https://www.eeoc.gov/laws/statutes/adea.cfm, downloaded Dec. 26, 2018.

[dxxiii] Hill, Theodore P. (1995), "A statistical derivation of the Significant-digit Law," *Statistical Science*, 10, 354-363; Berger, Arno; Hill, Theodore P. (2011), "Benford's Law strikes back: No simple explanation in sight for mathematical gem," *The Mathematical Intelligencer*, 33, 85-91; Berger, Arno; Hill, Theodore P. (2015) *An Introduction to Benford's Law* (Princeton University Press, Princeton); Amiram, Dan; Bozanic, Zahn; Rouen, Ethan (2015), "Financial statement errors: Evidence from the distributional

properties of financial statement numbers," *Rev. Account. Stud.*, 20, 1540-1593. Erratum (2015), 20, 1594-1595.

[dxxiv] McGinty, Jo Craven "When using math to catch crooks, you can't jump to conclusions," *Wall Street Journal*, Dec 5, 2014, 2:15 PM ET, https://blogs.wsj.com/numbers/when-using-math-to-catch-crooks-you-cant-jump-to-conclusions-1870; McGinty, Jo Craven "Accountants increasingly use data analysis to catch fraud," *Wall Street Journal*, Dec. 5, 2014, 6:48 p.m. ET, https://www.wsj.com/articles/accountants-increasingly-use-data-analysis-to-catch-fraud-1417804886; Nigrini, Mark (2012) *Benford's Law: Applications for Forensic Accounting, Auditing, and Fraud Detection* (Wiley, New York); "Financial statement errors."

[dxxv] *Computational and Inferential Thinking.*

[dxxvi] Pankanti, Sharath; Prabhakar, Salil; Jain, Anil K. (2002), "On the individuality of fingerprints," *IEEE Transactions on Pattern Analysis and Machine Intelligence*, 24, 1010-1025; Boyd, Dakota; Short, Dustin; Lee, Elizabeth; Huppenthal, John; Proft, Shelby; Teller, Wacey "Statistics of fingerprints," *Arizona State, Tempe campus*, https://math.la.asu.edu/~dieter/courses/Math_Modeling_2013/StatsPresentation8.pptx, downloaded Dec. 26, 2018; Russell, Sue "Why fingerprints aren't the proof we thought they were," *Pacific Standard*, Sept. 20, 2012, https://psmag.com/news/why-fingerprints-arent-proof-47079.

[dxxvii] Weintraub, Karen "Steven Pinker thinks the future is bright," *New York Times*, Nov. 19, 2018, https://www.nytimes.com/2018/11/19/science/steven-pinker-future-science.html; Pinker, Steven (2018) *Enlightenment Now: The Case for Reason, Science, Humanism, and Progress* (Viking, New York).

[dxxviii] *Computational and Inferential Thinking*, pp. 25-26.

[dxxix] Hempel, Sandra (2007) *The Strange Case of the Broad Street Pump: John Snow and the Mystery of Cholera* (University of California Press, Oakland).

[dxxx] Nesterak, Max "How Delta masters the game of overbooking flights," *PBS News Hour*, April 11, 2017, 10:30 AM EST, https://www.pbs.org/newshour/making-sense/how-delta-masters-the-game-of-overbooking-flights/.

[dxxxi] Russell, Karl "Why we feel so squeezed when we fly," *New York Times*, April 17, 2017, updated May 2, 2017, https://www.nytimes.com/interactive/2017/04/17/business/how-flying-changed.html.

[dxxxii] Victor, Daniel; Stevens, Matt "United Airlines passenger is dragged from an overbooked flight," *New York Times*, April 10, 2017, https://www.nytimes.com/2017/04/10/business/united-flight-passenger-dragged.html;Creswell, Julie; Maheshwari, Sapna "United grapples with PR crisis over videos of man being dragged off plane," *New York Times*, April 11, 2017, https://www.nytimes.com/2017/04/11/business/united-airline-passenger-overbooked-flights.html.

[dxxxiii] McCartney, Scott, "The push to end bumping passengers from flights," *Wall Street Journal*, Dec. 5, 2018, 9:11 AM ET, https://www.wsj.com/articles/the-push-to-end-bumping-passengers-from-flights-1544019076.

[dxxxiv] "Census," *Wikipedia*, https://en.wikipedia.org/wiki/Census, downloaded Dec. 14, 2018.

[dxxxv] Wines, Michael "With 2020 census looming, worries about fairness and accuracy," *New York Times*, Dec. 9, 2017, https://www.nytimes.com/2017/12/09/us/census-2020-redistricting.html.

[dxxxvi] *Being Mortal: Medicine and What Matters*, pg. 18.

[dxxxvii] Dwyer-Lindgren, Laura; Bertozzi-Villa, Amelia; Stubbs, Rebecca W.; Morozoff, Chloe; Kutz, Michael J.; Huynh, Chantal; … Murray, Christopher J. L. (2016), "US county-level trends in mortality rates for major causes of death, 1980-2014," *J. Am. Med. Assoc.*, 316, 2385-2401; Adamy, Janet "What kills Americans varies widely by region," *Wall Street Journal*, Dec. 13, 2016, 11:00 AM ET, https://www.wsj.com/articles/what-kills-americans-varies-widely-by-region-1481644837.

[dxxxviii] Reed, Stanley "A Danish wind turbine maker harnesses data in a push to stay ahead," *New York Times*, Aug. 18, 2016, https://www.nytimes.com/2016/08/19/business/energy-environment/denmark-vestas-wind-renewable-energy.html.

[dxxxix] Malone, Michael S. "The Big-Data future has arrived" *Wall Street Journal*, Feb. 22, 2016, 6:47 PM ET, https://www.wsj.com/articles/the-big-data-future-has-arrived 1456184869.

[dxl] Lane, Julia; Stodden, Victoria; Bender, Stefan; Nissenbaum, Helen, editors (2014) *Privacy, Big Data, and the Public Good: Frameworks for Engagement* (Cambridge University Press, Cambridge).

[dxli] *IMDB*, https://www.imdb.com, downloaded Dec. 26, 2018.

[dxlii] *Box Office Mojo*, https://www.boxofficemojo.com/, downloaded Dec. 26, 2018.

[dxliii] *The Numbers®: Where Data and the Movie Business Meet*, http://www.the-numbers.com, downloaded Dec. 26, 2018.

[dxliv] *Computational and Inferential Thinking.*

[dxlv] *Baseball Reference*, https://www.baseball-reference.com/, downloaded Dec. 26, 2018.

[dxlvi] Berkon, Ben "Baseball's data revolution is elevating defensive dynamos," *New York Times*, April 17, 2017, https://www.nytimes.com/2017/04/17/sports/baseball/baseballs-data-revolution-is-elevating-defensive-dynamos.html.

[dxlvii] Bowers, Jeremy; Pearce, Adam; Ward, Joe "Baseball's upward trend is leaving some players grounded," *New York Times*, July 9, 2017, https://www.nytimes.com/inter active/2017/07/09/sports/baseball/BASEBALL-LAUNCH-ANGLE.html.

[dxlviii] Lohr, Steve "A 10-digit key code to your private life: Your cellphone number," *New York Times*, Nov. 12, 2016, https://www.nytimes.com/2016/11/13/business/cellphone-number-social-security-number-10-digit-key-code-to-private-life.html; Lohr, Steve "When phones are lifelines, start-ups spy opportunities," *New York Times*, Nov. 12, 2016,

https://www.nytimes.com/2016/11/13/business/cellphones-lifelines-start-ups-spy-opportunities.html.

[dxlix] "The unreasonable effectiveness of mathematics in the natural sciences."

[dl] Halevy, Alon; Norvig, Peter; Pereira, Fernando (2009), "The unreasonable effectiveness of data," *IEEE Intelligent Systems*, March/April 2009, 8-12.

[dli] Wheelan, Charles (2013) *Naked Statistics* (Norton, New York).

[dlii] Tukey, John W. (1962), "The future of data analysis," *Ann. Math. Statist.*, 33, 1-67; boyd, danah; Crawford, Kate (2012), "Critical questions for big data," *Information, Communication & Society*, 15, 662-679; Freedman, David A. (1991), "Statistical models and shoe leather," *Sociological Methodology*, 21, 291-313; Chambers, J.M. (1993), "Greater or lesser statistics: a choice for future research," *Stat. Comput.*, 3, 182-184.

[dliii] Irizarry, Rafa; Peng, Roger; Leek, Jeff "Tukey talks turkey #futureofstats," Simply Statistics, https://simplystatistics.org/2013/10/29/tukey-talks-turkey-futureofstats/, downloaded Dec. 26, 2018.

[dliv] Breiman, Leo (2001), "Statistical modeling: The two cultures," *Statist. Sci.*, 16, pp. 199-231.

[dlv] Ezell, Barry Charles; Bennett, Steven P.; von Winterfeldt, Detlof; Sokolowski, John; Collins, Andrew J. (2010), "Probabilistic risk analysis and terrorism risk," *Risk Analysis*, 30, 575-589.

[dlvi] "Machine learning," *Wikipedia*, https://en.wikipedia.org/wiki/Machine_learning, downloaded Dec. 14, 2018; *Computational and Inferential Thinking*, pg. 534.

[dlvii] Marr, Bernard "What is the difference between artificial intelligence and machine learning?," *Forbes*, Dec 6, 2016, 02:24am, https://www.forbes.com/sites/bernardmarr/2016/12/06/what-is-the-difference-between-artificial-intelligence-and-machinelearning/#1dfdd7712742.

[dlviii] Lecun, Yann; Candela, Joaquin Quiñonero "Artificial intelligence, revealed," *ML APPLICATIONS*, posted on Dec. 1, 2016, https://code.facebook.com/posts/384869298519962/artificial-intelligence-revealed; Paid for and posted by Facebook, "Artificial intelligence: how we help machines learn," *New York Times*, https://paidpost.nytimes.com/facebook/artificial-intelligence-how-we-help-machines-learn.html. Accessed Dec. 12, 2018. 3:35 PM ET.

[dlix] Knuth, Donald (1990), "Arthur Lee Samuel, 1901-1990," *TUGboat*, 11, 497–498.

[dlx] "Statistical Modeling: The Two Cultures"; "Machine learning," *Wikipedia*, https://en.wikipedia.org/wiki/Machine_learning, downloaded Dec. 14, 2018.

[dlxi] Kuang, Cliff "Can A.I. be taught to explain itself?," *New York Times*, Nov. 21, 2017, https://www.nytimes.com/2017/11/21/magazine/can-ai-be-taught-to-explain-itself.html.

[dlxii] Schechner, Sam "Why do gas station prices constantly change? Blame the algorithm" *Wall Street Journal*, May 8, 2017, 6:41 PM ET, https://www.wsj.com/articles/why-do-gas-station-prices-constantly-change-blame-the-algorithm-1494262674.

[dlxiii] Hope, Bradley; Chung, Juliet "The future is bumpy: High-tech hedge fund hits limits of robot stock picking," *Wall Street Journal*, Dec. 11, 2017, 7:23 PM ET, https://www.wsj.com/articles/the-future-is-bumpy-high-tech-hedge-fund-hits-limits-of-robot-stock-picking-1513007557.

[dlxiv] Ip, Greg "How robots may make radiologists' jobs easier, not redundant," *Wall Street Journal*, Nov. 22, 2017, 12:56 PM ET, https://www.wsj.com/articles/how-robots-may-make-radiologists-jobs-easier-not-redundant-1511368729; "Deep learning," *Wikipedia*, https://en.wikipedia.org/wiki/Deep_learning, downloaded Dec. 14, 2018.

[dlxv] Topol, Eric J. (2019), "High-performance medicine: The convergence of human and artificial intelligence," *Nature Medicine*, 25, 44-56.

[dlxvi] Metz, Cade "A.I. shows promise assisting physicians," *New York Times*, Feb. 11, 2019, https://www.nytimes.com/2019/02/11/health/artificial-intelligence-medical-diagnosis.html.

[dlxvii] Mukherjee, Siddhartha "This cat sensed death. What if computers could, too?," *New York Times*, Jan. 3, 2018, https://www.nytimes.com/2018/01/03/magazine/the-dying-algorithm.html; Avati, Anand; Jung, Kenneth; Harman, Stephanie; Downing, Lance; Ng, Andrew; Shah, Nigam H. (2017), "Improving palliative care with deep learning," *arXiv*, arXiv:1711.06402.

[dlxviii] Lohr, Steve "Facial recognition is accurate, if you're a white guy," *New York Times*, Feb. 9, 2018, https://www.nytimes.com/2018/02/09/technology/facial-recognition-race-artificial-intelligence.html.

[dlxix] Kim, Pauline T. (2017), "Data-driven discrimination at work," *William & Mary Law Review*, 48, 857-936.

[dlxx] O'Neil, Cathy (2016) *Weapons of Math Destruction: How Big Data Increases Inequality and Threatens Democracy* (Crown, New York).

[dlxxi] Wing, Jeannette M. "Data for good: FATES, elaborated," *Columbia University Data Science Institute*, Jan. 23, 2018, https://datascience.columbia.edu/FATES-Elaborated.

[dlxxii] Hardt, Moritz; Price, Eric; Srebro, Nathan (2016), "Equality of opportunity in supervised learning," *arXiv*, arXiv:1610.02413 [cs.LG].

[dlxxiii] Wines, Michael "With 2020 census looming, worries about fairness and accuracy" *New York Times*, Dec. 9, 2017, https://www.nytimes.com/2017/12/09/us/census-2020-redistricting.html.

[dlxxiv] Moritz, Mark "Big data's 'streetlight effect': Where and how we look affects what we see," *The Conversation*, May. 17, 2016, 6:11 PM EDT, https://theconversation.com/big-datas-streetlight-effect-where-and-how-we-look-affects-what-we-see-58122.

[dlxxv] *Weapons of Math Destruction.*

[dlxxvi] "Data for good: FATES, elaborated."

[dlxxvii] Bui, Quoctrung; Cohn, Nate "Adventures in extreme gerrymandering: See the fair and wildly unfair maps we made for Pennsylvania," *New York Times*, Jan. 17, 2018, https://www.nytimes.com/interactive/2018/01/17/upshot/pennsylvania-gerrymandering.ht

ml; Gabriel, Trip "In a comically drawn Pennsylvania district, the voters are not amused," *New York Times*, Jan. 26, 2018, https://www.nytimes.com/2018/01/26/us/pennsylvania-gerrymander-goofy-district.html; Cohn, Nate "How big a deal is a new Congressional map for Pennsylvania?," *New York Times*, Jan. 22, 2018, https://www.nytimes.com/2018/01/22/upshot/pennsylvania-congressional-map-ruling.html; Kendall, Brent; Epstein, Reid J. "Pennsylvania gets a new district map," *Wall Street Journal*, Feb. 20, 2018, 11:21 AM ET, https://www.wsj.com/articles/pennsylvania-gets-a-new-district-map-1519089329; Gabriel, Trip; Bidgood, Jess "Court-drawn map in Pennsylvania may lift Democrats' House chances," *New York Times*, Feb. 19, 2018, https://www.nytimes.com/2018/02/19/us/pennsylvania-map.html.

[dlxxviii] "Court-drawn map in Pennsylvania may lift Democrats' House chances."

[dlxxix] Ellenberg, Jordan "How computers turned Gerrymandering into a science," *New York Times*, Oct. 6, 2017, https://www.nytimes.com/2017/10/06/opinion/sunday/computers-gerrymandering-wisconsin.html.

[dlxxx] Knuth, Donald E. (quoted in 'The Unix Programming Environment' by Kernighan and Pine, p. 91), as quoted in http://www.math.rutgers.edu/~zeilberg/quotes.html.

[dlxxxi] Wald, Matthew L. "Wind energy bumps into power grid's limits," *New York Times*, Aug. 26, 2008, http://www.nytimes.com/2008/08/27/business/27grid.html.

[dlxxxii] Robinson, Joshua "The secret of Dutch speedskating—It's not what you think," *Wall Street Journal*, Feb. 23, 2018, 4:45 am ET, https://www.wsj.com/articles/the-secret-of-dutch-speedskatingits-not-what-you-think-1519379130.

[dlxxxiii] Nocedal, Jorge; Wright, S. (1999) *Numerical Optimization* (Springer, New York).

[dlxxxiv] "Cost curve," *Wikipedia*, https://en.wikipedia.org/wiki/Cost_curve, downloaded Dec. 26, 2018.

[dlxxxv] "Cobb–Douglas production function," *Wikipedia*, https://en.wikipedia.org/wiki/Cobb%E2%80%93Douglas_production_function, downloaded Dec. 14, 2018.

[dlxxxvi] Cheng, Eugenia "I am not a perfectionist—really," *Wall Street Journal*, March 16, 2017, 11:42 AM, https://www.wsj.com/articles/i-am-not-a-perfectionistreally-1489678930.

[dlxxxvii] "Travelling salesman problem," *Wikipedia*, https://en.wikipedia.org/wiki/Travelling_salesman_problem, downloaded Dec. 14, 2018;_Johnson, David S.; McGeoch, Lyle A. (2003) "The Traveling Salesman Problem: A case study in local optimization," in *Local Search in Combinatorial Optimization*, Aarts, Emile; Lenstra, Jan Karel, editors (Princeton University Press, Princeton), Chapter 8, pp. 215ff.

[dlxxxviii] McGinty, Jo Craven "How do you fix a school-bus problem? Call MIT," *Wall Street Journal*, Aug. 11, 2017, 9:00 AM ET, https://www.wsj.com/articles/how-do-you-fix-a-school-bus-problem-call-mit-1502456400.

[dlxxxix] Cook, William "The problem of the traveling politician," *New York Times*, Dec. 21, 2011, 9:00 PM, https://campaignstops.blogs.nytimes.com/2011/12/21/the-problem-of-the-traveling-politician.

[dxc] Cook, William J. (2012) *In Pursuit of the Traveling Salesman: Mathematics at the Limits of Computation* (Princeton University Press, Princeton).

[dxci] Puget, Jean Francois "No, the TSP isn't NP complete," *IBM Community*, Dec. 29, 2013, https://www.ibm.com/developerworks/community/blogs/jfp/entry/no_the_tsp_isn_t_np_complete?lang=en; Hilton, Rod "Traveling salesperson: The most misunderstood problem," *Absolutely No Machete Juggling*, Sept. 14, 2012, https://www.nomachetejuggling.com/2012/09/14/traveling-salesman-the-most-misunderstood-problem/.

[dxcii] *In Pursuit of the Traveling Salesman.*

[dxciii] "How do you fix a school-bus problem? Call MIT."

[dxciv] *In Pursuit of the Traveling Salesman.*

[dxcv] *In Pursuit of the Traveling Salesman.*

[dxcvi] Swanson, Ana "What really drives you crazy about waiting in line (it actually isn't the wait at all)," *The Washington Post*, Nov. 27, 2015, https://www.washingtonpost.com/news/wonk/wp/2015/11/27/what-you-hate-about-waiting-in-line-isnt-the-wait-at-all/?utm_term=.aad5bdefef89.

[dxcvii] Stone, Alex "Why waiting is torture," *New York Times*, Aug. 18, 2012, http://www.nytimes.com/2012/08/19/opinion/sunday/why-waiting-in-line-is-torture.html.

[dxcviii] Badger, Emily "What's the right number of taxis (or Uber or Lyft cars) in a city?," *New York Times*, Aug. 10, 2018, https://www.nytimes.com/2018/08/10/upshot/ub er-lyft-taxi-ideal-number-per-city.html.

[dxcix] Newell, G. F. (1982) *Applications of Queueing Theory* (Dordrecht: Springer, Netherlands), 2nd edition.

[dc] Swanson, Ana "Researchers have discovered a better way to wait in line, and you're going to hate it," *The Washington Post*, Sept. 9, 2015, https://www.washingtonpost.com/news/wonk/wp/2015/09/09/researchers-have-discovered-a-better-way-to-wait-in-line-and-youre-going-to-hate-it/.

[dci] Cheng, Eugenia "Calculating a faster checkout line," *Wall Street Journal*, May 19, 2017, 11:35 AM ET, https://www.wsj.com/articles/calculating-a-faster-checkout-line-1495208140.

[dcii] McGinty, Jo Craven. "The science of standing in line," *Wall Street Journal*, Oct. 7, 2016, 10:30 a.m. ET, http://www.wsj.com/articles/the-science-of-standing-in-line-1475850601; Bialik, Carl "The waiting game," *Wall Street Journal*, Aug 18, 2009, 8:07 pm ET, https://blogs.wsj.com/numbers/the-waiting-game-782; Bialik, Carl "Justice -- Wait for it -- on the checkout line," *Wall Street Journal*, Aug. 19, 2009, 12:01 a.m. ET, https://www.wsj.com/articles/SB125063608198641491; Bialik, Carl "The science of lines, and other reading," *Wall Street Journal*, Mar 26, 2007 1:19 pm ET, https://blogs.wsj.com/numbers/the-science-of-lines-and-other-reading-67.

[dciii] Allen, Theodore; Bernshteyn, Mikhail (2006), "Mitigating voter waiting times," *CHANCE*, 19, 25-34.

[dciv] "Agner Krarup Erlang (1878 - 1929)," *+ plus magazine*, https://plus.maths.org/cont

ent/agner-krarup-erlang-1878-1929, downloaded Dec. 26, 2018; Erlang C Calculator, *Westbay Engineers*, https://www.erlang.com/calculator/erlc/, downloaded Dec. 26, 2018; "The Erlang C Traffic Model," *KoolToolz*, https://www.kooltoolz.com/erlang-c.htm, downloaded Dec. 26, 2018; Erlang C Calculator, *Agenses*, http://www.agenses.com/489/free_erlangc_calclator/, downloaded Dec. 26, 2018.

[dcv] Tanner, Mike "The Erlang-C formula: The widely-used Erlang-C formula is given here for people who want to see the actual mathematical definition," *MITAN*, http://www.mitan.co.uk/erlang/elgcmath.htm, downloaded Dec. 26, 2018.

[dcvi] Whitt, Ward (1992), "Understanding the efficiency of multi-server service systems," *Management Science*, 38, 708-723; Whitt, Ward (2013), "Offered load analysis for staffing," *Management & Service Operations Management*, 15, 166-169.

[dcvii] "Mitigating voter waiting times."

[dcviii] McGinty, Jo Craven "A long wait to vote? Odds are, you were in the A-to-M line," *Wall Street Journal*, Nov. 16, 2018, 1:01 PM ET, https://www.wsj.com/articles/a-long-wait-to-vote-odds-are-you-were-in-the-a-to-m-line-1542369601.

[dcix] "Goldilocks principle," *Wikipedia*, https://en.wikipedia.org/wiki/Goldilocks_principle, downloaded Dec. 14, 2018.

[dcx] Steffen, Jason "There's a better way to board planes," *The Atlantic*, Nov. 26, 2014, http://www.theatlantic.com/business/archive/2014/11/theres-a-better-way-to-board-planes/383181/; Steffen, Jason (2008), "Optimal boarding method for airline passengers," *J. Air Transport Management*, 14, 146-150.

[dcxi] Fitzsimmons, Emma G; Fessenden, Ford; Lai, K.K. Rebecca "Every New York City subway line is getting worse. Here's why," *New York Times*, June 28, 2017, https://www.nytimes.com/interactive/2017/06/28/nyregion/subway-delays-overcrowding.html.

[dcxii] Pearce, Adam "How 2 M.T.A. decisions pushed the subway into crisis," *New York Times*, May 9, 2018, https://www.nytimes.com/interactive/2018/05/09/nyregion/subway-crisis-mta-decisions-signals-rules.html.

[dcxiii] McGinty, Jo Craven "One rail-station design may be just the ticket to ease congestion," *Wall Street Journal*, May 19, 2017, 5:30 AM ET, https://www.wsj.com/articles/one-rail-station-design-may-be-just-the-ticket-to-ease-congestion-1495186204.

[dcxiv] Bernard J. Arseneau in McGinty, Jo Craven "Slowing down will get you through a traffic jam faster," *Wall Street Journal*, Nov. 7, 2014, 1:50 PM ET, https://www.wsj.com/articles/traffic-engineers-say-slowing-down-will-get-you-through-a-jam-faster-1415386073; Shellenbarger, Sue "One driver can prevent a traffic jam," *Wall Street Journal*, Oct. 12, 2016 9:28 AM ET, https://www.wsj.com/articles/one-driver-can-prevent-a-traffic-jam-1476204858.

[dcxv] Maher, Kris "States to drivers: Avoid the urge to merge," *Wall Street Journal*, Jan. 10, 2017, 2:51 PM ET, https://www.wsj.com/articles/states-to-drivers-avoid-the-urge-to-merge-1484077888.

[dcxvi] "Signalized intersections," *University of Washington*, http//www.courses.washington.edu/cee320ag/Lecture/Signalized%20 Intersections.ppt, downloaded Dec. 26, 2018.
[dcxvii] "New York's award-winning traffic control system," *ITS International*, https://http://www.itsinternational.com/sections/nafta/features/new-yorks-award-winning-traffic-control-system/, downloaded Dec. 26, 2018.
[dcxviii] McGinty, Jo Craven "In love, probability calculus suggests only fools rush in," *Wall Street Journal*, updated Feb. 10, 2017, 11:24 AM ET, https://www.wsj.com/articles/in-love-probability-calculus-suggests-only-fools-rush-in-1486722600; Ferguson, Thomas S. (1989), "Who solved the secretary problem?," *Statistical Science*, 4, 282-289; "Secretary problem," *Wikipedia*, https://en.wikipedia.org/wiki/Secretary_problem, downloaded March 18, 2019.
[dcxix] Bialik, Carl "Clemson controversy calls into question US News College rankings," *Wall Street Journal*, Jun 4, 2009, 4:51 PM ET, https://blogs.wsj.com/numbers/clemson-controversy-calls-into-question-us-news-college-rankings-717.
[dcxx] Sanger-Katz, Margot. "You'd be surprised at how many foods contain added sugar," *New York Times*, May 21, 2016, http://www.nytimes.com/2016/05/22/upshot/it-isnt-easy-to-figure-out-which-foods-contain-sugar.html; Popkin, Barry M.; Hawkes, Corinna (2016), "Sweetening of the global diet, particularly," The Lancet Diabetes & Endocrinology, 4, 174–186.
[dcxxi] *Grant*, pp. 23 and 27.
[dcxxii] Gellman, Lindsay; Baer, Justin "Goldman Sachs to stop rating employees with numbers," *Wall Street Journal*, May 26, 2016, 6:53 PM, http://www.wsj.com/articles/goldman-sachs-dumps-employee-ranking-system-1464272443.
[dcxxiii] *Weapons of Math Destruction*, Chapter 3, pg. 60.
[dcxxiv] "New York City mayoral elections," *Wikipedia*, https://en.wikipedia.org/wiki/New_York_City_mayoral_elections, downloaded Dec. 14, 2018.
[dcxxv] "Selection process FAQ," *Pro Football Hall of Fame*, https://www.profootballhof.com/heroes-of-the-game/selection-process-faq/, downloaded Feb. 8, 2017.
[dcxxvi] Martin, Jeremy "Math, fairness and social choice," *University of Kansas*, March 25, 2014, http://jlmartin.faculty.ku.edu/~jlmartin/talks/science-on-tap.pdf; Brams, Steven J. (2008) *Mathematics and Democracy: Designing better voting and fair-division procedures* (Princeton University Press, Princeton); Bialik, Carl "Numbers Guy interview: Steven Brams," *Wall Street Journal*, May 16, 2008, 12:08 PM ET, https://blogs.wsj.com/numbers/numbers-guy-interview-steven-brams-340.
[dcxxvii] "Cy Young Award," *Wikipedia*, https://en.wikipedia.org/wiki/Cy_Young_Award, downloaded Dec. 14, 2018.
[dcxxviii] Lauber, Scott "Rick Porcello wins AL Cy Young, despite fewer first-place votes than Justin Verlander," *ESPN*, Nov. 17, 2016, http://www.espn.com/mlb/story/_/id/18067034/rick-porcello-boston-red-sox-wins-american-league-cy-young-award.

dcxxix "Bucklin voting," *Wikipedia*, https://en.wikipedia.org/wiki/Bucklin_voting, downloaded Dec. 14, 2018.

dcxxx McGinty, Jo Craven "Third-party candidates don't have to be spoilers," *Wall Street Journal*, Sept. 9, 2016, 1:24 PM ET, https://www.wsj.com/articles/third-party-candidates-dont-have-to-be-spoilers-1473436855.

dcxxxi Maskin, Eric; Sen, Amartya "How majority rule might have stopped Donald Trump," *New York Times*, April 28, 2016, http://www.nytimes.com/2016/05/01/opinion/sunday/how-majority-rule-might-have-stopped-donald-trump.html.

dcxxxii *Mathematics and Democracy*; "Numbers Guy Interview: Steven Brams."

dcxxxiii Laslier, Jean-François; Sanver, M. Remzi, editors (2010) *Handbook on Approval Voting* (Springer, Heidelberg).

dcxxxiv "Proportional approval voting," *Wikipedia*, https://en.wikipedia.org/wiki/Proportional_approval_voting, downloaded Dec. 14, 2018; Kilgour, D. Marc "Approval balloting for multi-winner elections," in Laslier, Jean-François; Sanver, M. Remzi, editors (2010) *Handbook on Approval Voting* (Springer, Heidelberg), pgs. 105-124; Brams, Steven J.; Sanver, M. Remzi "Voting systems that combine approval and preference" in (2009) *The Mathematics of Preference, Choice and Order: Essays in Honor of Peter C. Fishburn*, Brams, Steven, Gehrlein, William V., Roberts, Fred S. editors (Springer, Heidelberg), pp. 215-237.

dcxxxv "It's time for ranked choice voting (RCV) in Maine," *Fairvote*, http://www.lwvme.org/files/RCV_Whos_Using_Fact_Sheet.pdf, downloaded Dec. 26, 2018.

dcxxxvi Sharp, David "New way of voting faces test in Maine congressional district," *AP*, Nov. 12, 2018, https://www.apnews.com/c2b64b9efe5f403d9401ca9d59fe060d; "Computer algorithm will decide whether a Democrat unseats Republican in Maine where 'ranked-choice voting' is being used for first ever - but could end up in court," *Daily Mail from the Associated Press*, Nov.12, 2008, https://dailym.ai/2Didq5C.

dcxxxvii Shepherd, Michael "Golden defeats Poliquin in contested 2nd District ranked-choice count," *Bangor Daily News*, updated Nov. 15, 2018, 5:36 PM, https://bangordailynews.com/2018/11/15/politics/golden-defeats-poliquin-in-contested-2nd-district-ranked-choice-count; "Ranked choice voting," *League of Women Voters® of Maine*, http://www.lwvme.org/RCV.html, downloaded Dec. 26, 2018.

dcxxxviii Tullis, Tracy "How game theory helped improve New York City's high school application process," *New York Times*, Dec. 5, 2014, http://www.nytimes.com/2014/12/07/nyregion/how-game-theory-helped-improve-new-york-city-high-school-application-process.html; Roth, A. F. (1985), "The college admissions problem is not equivalent to the marriage problem," *J. Economic Theory*, 36, 277–288; "Alvin E. Roth Nobel Prize Lecture: The theory and practice of market design," *The Nobel Prize*, https://www.nobelprize.org/nobel_prizes/economic-sciences/laureates/2012/roth-lecture.html, downloaded Jan. 2, 2019.

[dcxxxix] "How game theory helped improve New York City's high school application process"; "Alvin E. Roth Nobel Prize Lecture: The theory and practice of market design."
[dcxl] Gale, D.; Shapley, L. (1962), "College admissions and the stability of marriage," *Amer. Math. Monthly*, 69, 9–15; McGinty, Jo Craven "You nay now kiss the algorithm," *Wall Street Journal*, Feb. 9, 2018, 9:00 AM ET, https://www.wsj.com/articles/you-may-now-kiss-the-algorithm-1518184800.
[dcxli] "Alvin E. Roth Nobel Prize Lecture: The theory and practice of market design"; "Lloyd S. Shapley Nobel Prize Lecture: Allocation games - the deferred acceptance algorithm," *The Nobel Prize*, https://www.nobelprize.org/prizes/economic-sciences/2012/shapley/lecture/, downloaded Jan. 2, 2019.
[dcxlii] Brams, Steven J. (1993) *Theory of Moves* (Cambridge University, Cambridge); *Mathematics and Democracy.*
[dcxliii] Patel; Yogita "I gave my kidney to a stranger to save my brother's life," *Wall Street Journal*, March 9, 2019, 9:00 AM ET, https://www.wsj.com/articles/i-gave-my-kidney-to-a-stranger-to-save-my-brothers-life-11552140000.
[dcxliv] Engineering Systems Division "Allocating flu vaccines to maximize number of people remaining healthy," *MIT News*, June 14, 2013, http://news.mit.edu/2013/allocating-flu-vaccines-to-maximize-the-number-of-people-remaining-healthy.
[dcxlv] "Pareto efficiency," *Wikipedia*, https://en.wikipedia.org/wiki/Pareto_efficiency, downloaded Dec. 14, 2018; "Kenneth Arrow," *Wikipedia*, https://en.wikipedia.org/wiki/Kenneth_Arrow, downloaded Dec. 14, 2018; "Amartya Sen," *Wikipedia*, https://en.wikipedia.org/wiki/Amartya_Sen, downloaded Dec. 14, 2018; *The Mathematics Lover's Companion*, Chapter 22.
[dcxlvi] Barbanel, Julius, B.; Brams, Steven J. (2014), "Two-person cake cutting: The optimal number of cuts," *Springer Science+Business Media New York*, 36 (3), 23-35.
[dcxlvii] Colman, Andrew M. (1998) *Game Theory and its Applications: In the Social and Biological Sciences* (Psychology Press, London), 2nd edition.
[dcxlviii] Yildiz, Muhamet "14.12 Game theory lecture notes: Introduction," *MIT*, web.mit.edu/14.12/www/02F_lecture102.pdf, downloaded March 7, 2019.
[dcxlix] Poundstone, William (1993) *Prisoner's Dilemma* (Anchor, New York City).
[dcl] *Game Theory and its Applications.*
[dcli] *Theory of Moves.*
[dclii] "Game theory," *Wikipedia*, https://en.wikipedia.org/wiki/Game_theory, downloaded Dec. 14, 2018.
[dcliii] *Game Theory and its Applications.*
[dcliv] *Game Theory and its Applications.*
[dclv] *Theory of Moves.*
[dclvi] Brams, Steven J. "Game theory and the Cuban missile crisis," *+plus Magazine*, https://plus.maths.org/content/game-theory-and-cuban-missile-crisi, downloaded Dec. 27, 2018.

[dclvii] Quote from the film *Little Big Man* (1970).

[dclviii] Walker, Mark "Mixed strategies: Minimax/Maximin and Nash Equilibrium," *University of Arizona*, http://www.u.arizona.edu/~mwalker/MixedStrategy3.pdf, downloaded Dec. 27, 2018.

[dclix] von Neumann, John; Morgenstern, Oskar (1944) *Theory of Games and Economic Behavior* (Princeton University Press, Princeton); Copeland, Arthur H. (1945), "Review: John von Neumann and Oskar Morgenstern, Theory of games and economic behavior," *Bull. Amer. Math. Soc.*, 51, 498-504.

[dclx] "Mixed strategies: Minimax/Maximin and Nash Equilibrium."

[dclxi] *A Beautiful Mind.* The film *A Beautiful Mind* (2001).

[dclxii] "John F. Nash Jr. Nobel Prize Seminar: The work of John Nash in Game Theory," *The Nobel Prize*, https://www.nobelprize.org/prizes/economic-sciences/1994/nash/lecture /, downloaded Jan. 2, 2019.

[dclxiii] *Mathematics and Democracy*, pg. 23.

[dclxiv] *Game Theory and its Applications*.

[dclxv] *Prisoner's Dilemma*.

[dclxvi] *Theory of Moves*; *Prisoner's Dilemma*.

[dclxvii] TV Show *Law and Order: Special Victim's Unit,* Season 1, Episode 3, "… Or Just Look Like One," Aired 10:00 PM, Oct. 4, 1999 on NBC, as per http://www.tv.com/ shows/law-order-special-victims-unit/or-just-look-like-one-12328/recap/.

[dclxviii] *Theory of Moves*.

[dclxix] *Prisoner's Dilemma*; Rapoport, Anatol; Guyer, Melvin (1966) *A Taxonomy of 2 x 2 Games* (Bobbs-Merrill, Indianapolis); *Game Theory and its Applications*.

[dclxx] *Game Theory and its Applications*, pp. 62-67.

[dclxxi] *Game Theory and its Applications*, Matrix 4.6, pg. 66.

[dclxxii] *Game Theory and its Applications*.

[dclxxiii] *Theory of Moves*, "14.12 Game theory lecture notes: Introduction."

[dclxxiv] Quealy, Kevin "Lessons from game theory: What keeps Kasich in the race?," *New York Times*, Feb. 24, 2016, http://www.nytimes.com/2016/02/25/upshot/john-kasich-republican-nomination.html; Schelling, Thomas C. (1956), "An essay on bargaining," *The American Economic Review*, 46, 281-306; Bliss, Christopher; Nalebuff, Barry (1984), "Dragon-slaying and ballroom dancing: The private supply of a public good," *Journal of Public Economics*, 25, l-12; *Prisoner's Dilemma*.

[dclxxv] Synolakis, Costas; Karagiannis, George "The global lessons of Italy's earthquake," *Wall Street Journal*, Aug. 28, 2016, 5:54 PM ET, https://www.wsj.com/articles/the-global-lessons-of-italys-earthquake-1472421264.

[dclxxvi] Bernstein, Peter L. (1998) *Against the Gods: The Remarkable Story of Risk* (Wiley, New York), pg. 1.

[dclxxvii] *Risk–Benefit Analysis*.

[dclxxviii] *Risk–Benefit Analysis*; Curtin, François; Schulz, Pierre (2011), "Assessing the benefit:risk ratio of a drug - randomized and naturalistic evidence," *Dialogues Clin. Neurosci.*, 13, 183–190.

[dclxxix] "HAP 525: Risk analysis in healthcare."

[dclxxx] *Risk–Benefit Analysis*, pg. 85.

[dclxxxi] "HAP 525: Risk analysis in healthcare."

[dclxxxii] Harris, Jr., Charles E.; Pritchard, Michael S.; Rabins, Michael J.; James, Ray; Englehardt, Elaine (2013) *Engineering Ethics: Concepts and Cases*, 5th edition (Cengage Learning, New York).

[dclxxxiii] Carroll, Aaron E. "The power of simple life changes to prevent heart disease," *New York Times*, Dec. 12, 2016, http://www.nytimes.com/2016/12/12/upshot/the-power-of-simple-life-changes-to-prevent-heart-disease.html; Khera, Amit V.; Emdin, Connor A.; Drake, Isabel; Natarajan, Pradeep; Bick, Alexander G.; Cook, Nancy R.; … Kathiresan, Sekar (2016), "Genetic risk, adherence to a healthy lifestyle, and coronary disease," *N. Engl. J. Med.*, 375, 2349-2358.

[dclxxxiv] Zeltser, Marina V. (2009), "Influenza vaccination: Financial burden or public health solution?," *Biotechnol Healthc.*, 6, 29-31.

[dclxxxv] "Seasonal influenza vaccine effectiveness, 2004-2018," *CDC/Centers for Disease Control and Prevention/Influenza (Flu)*, https://www.cdc.gov/flu/professionals/vaccination/effectiveness-studies.htm; downloaded Dec. 17, 2018.

[dclxxxvi] Carroll, Aaron E. "Why it's still worth getting a flu shot," *New York Times*, Jan. 11, 2018, https://www.nytimes.com/2018/01/11/upshot/flu-shot-risks-benefits-strain.html; "Vaccine effectiveness - How well does the flu vaccine work?," *CDC/Centers for Disease Control and Prevention/ Influenza (Flu)*, https://www.cdc.gov/flu/about/qa/vaccineeffect.htm, downloaded Dec. 17, 2018.

[dclxxxvii] Mukherjee, Siddhartha "Can doctors choose between saving lives and saving a fortune?," *New York Times*, April 3, 2018, https://www.nytimes.com/2018/04/03/magazine/can-doctors-choose-between-saving-lives-and-saving-a-fortune.html.

[dclxxxviii] Sarah Knapton "Blood test shows chance of suffering heart attack within five years," *The Telegraph*, June 20, 2016, 12:01 AM, https://www.telegraph.co.uk/science/2016/06/19/blood-test-shows-chance-of-suffering-heart-attack-within-five-ye/; Khamis, Ramzi Y.; Hughes, Alun D.; Caga-Anan, Mikhail; Chang, Choon L.; Boyle, Joseph J.; Kojima, Chiari; … Haskard, Dorian O. (2016), "High serum Immunoglobulin G and M levels predict freedom from adverse cardiovascular events in hypertension: A nested case-control study of the Anglo-Scandinavian Cardiac Outcomes Trial," *EBioMedicine*, 9, 372-380.

[dclxxxix] Reddy, Sumathi "Test your genes to find your best diet," *Wall Street Journal*, Aug. 22, 2016, 1:36 PM ET, https://www.wsj.com/articles/test-your-genes-to-find-your-best-diet-1471887390.

[dcxc] Celis-Morales, Carlos; Livingstone, Katherine M; Marsaux, Cyril F. M.; Macready, Anna L; Fallaize, Rosalind; O'Donovan, Clare B.; … Mathers, John C. (2017), "Effect of personalized nutrition on health-related behaviour change: evidence from the Food4me European randomized controlled trial," *Int. J. Epidemiol.*, 46, 578-588.

[dcxci] Linderman, Michael D.; Nielsen Daiva E.; Green, Robert C. (2016), "Review: Personal genome sequencing in ostensibly healthy individuals and the PeopleSeq Consortium," *J. Pers. Med.*, 6, 14.

[dcxcii] Palazzolo, Joe "Wisconsin Supreme Court to rule on predictive algorithms used in sentencing," *Wall Street Journal*, June 5, 2016, 5:30 AM ET, https://www.wsj.com/articles/wisconsin-supreme-court-to-rule-on-predictive-algorithms-used-in-sentencing-1465119008; Brennan, Tim; Dieterich, William "Correctional offender management profiling for alternative sanctions (COMPAS)," in (2018) *Handbook of Recidivism Risk/Needs Assessment Tools* (Wiley-Blackwell, Hoboken), Singh, Jay P.; Kroner, Daryl G.; Wormith, J. Stephen; Desmarais, Sarah L.; Hamilton, Zachary, editors, pp. 49-75; Larson, Jeff; Mattu, Surya; Kirchne, Lauren; Angwin, Julia "How we analyzed the COMPAS recidivism algorithm," *Pro Publica*, May 23, 2016, https://www.propublica.org/article/how-we-analyzed-the-compas-recidivism-algorithm; Blomberg, Thomas; Bales, William; Mann, Karen; Meldrum, Ryan; Nedelec, Joe "Validation of the Compas Risk Assessment Classification Instrument, Prepared for the Broward Sheriff's Office Department of Community Control," Sept. 2010, downloaded on Dec. 27, 2018 from http://criminology.fsu.edu/wp-content/uploads/Validation-of-the-COMPAS-Risk-Assessment-Classification-Instrument.pdf.

[dcxciii] Diakopoulos, Nicholas "We need to know the algorithms the government uses to make important decisions about us," *The Conversation*, May 23, 2016, 8.48 PM EDT, https://theconversation.com/we-need-to-know-the-algorithms-the-government-uses-to-make-important-decisions-about-us-57869; Israni, Ellora Thadaney "When an algorithm helps send you to prison," *New York Times*, By Oct. 26, 2017, https://www.nytimes.com/2017/10/26/opinion/algorithm-compas-sentencing-bias.html; Liptak, Adam "Sent to prison by a software program's secret algorithms," *New York Times*, May 1, 2017, https://www.nytimes.com/2017/05/01/us/politics/sent-to-prison-by-a-software-programs-secret-algorithms.html; Angwin, Julia "Make algorithms accountable," *New York Times*, Aug. 1, 2016, https://www.nytimes.com/2016/08/01/opinion/make-algorithms-accountable.html.

[dcxciv] Leopold, George (2016) *Calculated Risk: The Supersonic Life and Times of Gus Grissom* (Purdue University Press, West Lafayette, IN).

[dcxcv] "Risk assessment basics," *Airsafe.com*, http://www.airsafe.com/risk/basics.htm, downloaded Dec. 27, 2018.

[dcxcvi] *Risk–Benefit Analysis.*

[dcxcvii] Kaplan, Stanley; Garrick, B. John (1981), "On the quantitative definition of risk," *Risk Analysis*, 1, 11-27.

[dcxcviii] *Risk–Benefit Analysis*, Ch. 7.

[dcxcix] *Risk–Benefit Analysis*, pg. 65, Eq. 2-6.

[dcc] *Risk–Benefit Analysis*, pg. 115.

[dcci] *Risk–Benefit Analysis*, pp. 231-232.

[dccii] *Risk–Benefit Analysis*, pg. 48.

[dcciii] Tubiana, Maurice; Feinendegen, Ludwig E.; Yang, Chichuan ; Kaminski, Joseph M. (2009), "The linear no-threshold relationship is inconsistent with radiation biologic and experimental data," *Radiology*, 251, 13–22.

[dcciv] Little, Mark P.; Wakeford, Richard; Tawn, E. Janet; Bouffler, Simon D.; de Gonzalez, Amy Berrington. (2009), "Risks associated with low doses and low dose rates of ionizing radiation: Why linearity may be (almost) the best we can do," *Radiology*, 251, 6–12.

[dccv] Johnson, George "When radiation isn't the real risk," *New York Times*, Sept. 21, 2015, http://www.nytimes.com/2015/09/22/science/when-radiation-isnt-the-real-risk.html.

[dccvi] *Risk–Benefit Analysis*, pg. 60.

[dccvii] *Risk–Benefit Analysis*, Appendix 5, pp. 341ff.

[dccviii] "Cancer slope factor," *Wikipedia*, https://en.wikipedia.org/wiki/Cancer_slope_factor, downloaded Dec. 14, 2018.

[dccix] "When radiation isn't the real risk."

[dccx] Bialik, Carl "Radiation numbers in damaged Japan plant," *Wall Street Journal*, Mar 22, 2011, 9:22 PM ET, https://blogs.wsj.com/numbers/radiation-numbers-in-damaged-japan-plant-1047; Bialik, Carl "Radiation math: How do we count the rays?," *Wall Street Journal*, March 23, 2011, 12:01 AM ET, https://www.wsj.com/articles/SB1000142405274870446130457621682030949 4008.

[dccxi] "Background radiation," *Wikipedia*, https://en.wikipedia.org/wiki/Background_radiation, downloaded Dec. 14, 2018.

[dccxii] *Risk–Benefit Analysis*; "Radiation exposure during commercial airline flights," *Health Physics Society*, https://hps.org/publicinformation/ate/faqs/commercialflights.html, downloaded Dec. 27, 2018.

[dccxiii] Gilberta, Ethel S. (2009), "Ionising radiation and cancer risks: What have we learned from epidemiology?," *Int. J. Radiat. Bio.*, 85, 467-482.

[dccxiv] Friedman, Lisa "E.P.A. plans to get thousands of pollution deaths off the books by changing its math," *New York Times*, May 20, 2019, https://www.nytimes.com/2019/05/20/climate/epa-air-pollution-deaths.html.

[dccxv] Cooke, Roger M. (1991) *Experts in Uncertainty: Opinion and subjective probability in science* (Oxford University Press, New York), as from "HAP 525: Risk analysis in healthcare."

[dccxvi] Colglazier, E. W.; Weatherwax, R. K. "Failure estimates for the Space Shuttle," Abstracts for Society for Risk Analysis Annual Meeting 1986, Boston MA, Nov. 9-12, 1986, pg. 80, as from "HAP 525: Risk analysis in healthcare."

[dccxvii] Feynman, R.P. "What do you care what other people think?," Appendix F, http://science.ksc.nasa.gov/shuttle/missions/51-l/docs/rogers-commission/Appendix-F.txt, 1988.

[dccxviii] Fragola, Joseph R.; Maggio, Gaspare; Frank, Michael V.; Gerez, Luis; Mcfadden, Richard H.; Collins, Erin P.; Ballesio, Jorge; Appignani, Peter L.; Karns, James J. "Probabilistic risk assessment of the Space Shuttle. Phase 3: A study of the potential of losing the vehicle during nominal operation," Volume 5: Auxiliary shuttle risk analyses, *NASA Technical Reports Server*, Feb. 28, 1995, https://ntrs.nasa.gov/search.jsp?R=19950019983.

[dccxix] "Early Space Shuttle flights riskier than estimated," *NPR Talk of the Nation*, March 4, 2011, 1:00 PM ET, https://www.npr.org/2011/03/04/134265291/early-space-shuttle-flights-riskier-than-estimated.

[dccxx] "Space flight related mishaps," *Airsafe.com*, http://www.airsafe.com/events/space.htm, downloaded Dec. 27, 2018.

[dccxxi] TV Show *Through the Decades,* Aired July 20, 2016 on the DECADES TV Channel.

[dccxxii] "List of Space Shuttle missions," *Wikipedia*, https://en.wikipedia.org/wiki/List_of_Space_Shuttle_missions, downloaded May 28, 2019.

[dccxxiii] "Space Shuttle fatal event risk," *Airsafe.com*, http://www.airsafe.com/risk/shuttle.html. Revised June 15, 2015, downloaded July 19, 2016; "HAP 525: Risk analysis in healthcare."

[dccxxiv] Rausand, Marvin (2011) *Risk Assessment: Theory, Methods, and Applications* (John Wiley & Sons, New York).

[dccxxv] *Risk Assessment*, Chapter 4.

[dccxxvi] *Risk Assessment*, adapted from Fig. 4.6, pg. 97.

[dccxxvii] *Risk Assessment*, pg. 88.

[dccxxviii] "On the quantitative definition of risk."

[dccxxix] *Risk Assessment*, Chapter 4.

[dccxxx] Cooke, R.; Jager, E. (1998), "A probabilistic model for the failure frequency of underground gas pipelines," *Risk Anal.*, 18, 511-527.

[dccxxxi] Willis, Henry H.; Morral, Andrew R.; Kelly, Terrence K.; Medby, Jamison Jo "Estimating terrorism risk," *RAND Center for Terrorism Risk Management Policy*, 2005, https://www.rand.org/content/dam/rand/pubs/monographs/2005/RAND_MG388.pdf; "Probabilistic risk analysis and terrorism risk."

[dccxxxii] *Risk Assessment*, Chapter 7.

[dccxxxiii] "Weibull distribution," *Wikipedia*, https://en.wikipedia.org/wiki/Weibull_distribution, downloaded Dec. 13, 2018.

[dccxxxiv] "HAP 525: Risk analysis in healthcare;" *Risk–Benefit Analysis; Risk Assessment.*

[dccxxxv] *Risk–Benefit Analysis*, pp. 41-46; *Risk Assessment, Chapter 10.*

[dccxxxvi] "Wilhelm Röntgene," *Wikipedia*, https://en.wikipedia.org/wiki/Wilhelm_R%C3% B6ntgene, downloaded Dec. 14, 2018.

[dccxxxvii] "Radium Girls," *Wikipedia*, https://en.wikipedia.org/wiki/Radium_Girls, downloaded Dec. 14, 2018.

[dccxxxviii] *Now It Can Be Told*, pg. 57.

[dccxxxix] *Risk Assessment*, Chapter 4.

[dccxl] Ip, Greg "The unstable economics in Obama's health law," *Wall Street Journal*, Aug. 17, 2016, 12:55 PM ET, https://www.wsj.com/articles/the-unstable-economics-in-obamas-health-law-1471452938.

[dccxli] Scism, Leslie "Millions bought insurance to cover retirement health costs. Now they face an awful choice," *Wall Street Journal*, Jan. 17, 2018, 11:31 AM ET, https://www.wsj.com/articles/millions-bought-insurance-to-cover-retirement-health-costs-now-they-face-an-awful-choice-1516206708.

[dccxlii] Benartzi, Shlomo; Bhargava, Saurabh "Common errors when buying insurance," *Wall Street Journal*, Feb. 11, 2018, 10:11 PM ET, https://www.wsj.com/articles/common -errors-when-buying-insurance-1518405060.

[dccxliii] Johnson, Eric J.; Hershey, John; Meszaros, Jacqueline; Kunreuther, Howard (1993), "Framing, probability distortions, and insurance decisions," *J. Risk Uncertainty*, 7, 35–51.

[dccxliv] "Framing, probability distortions, and insurance decisions;" Bialik, Carl. When Polls Turn Up the Wrong Number.

[dccxlv] "Framing, probability distortions, and insurance decisions."

[dccxlvi] Vinod, Hrishikesh (Rick) D.; Reagle, Derrick, (2004) *Preparing for the Worst: Incorporating Downside Risk in Stock Market Investments* (John Wiley & Sons, New York).

[dccxlvii] Joshi, Mark (2003) *The Concepts and Practice of Mathematical Finance* (Wiley, New York); "Robert C. Merton Nobel Prize Lecture: Applications of option-pricing theory: Twenty-five years later," *The Nobel Prize*, https://www.nobelprize.org/prizes/ economic-sciences/1997/merton/lecture/, downloaded Jan. 2, 2019.

[dccxlviii] Statman, Meir "Your tolerance for investment risk is probably not what you think," *Wall Street Journal*, Sept. 10, 2017, 10:16 PM ET, https://www.wsj.com/articles/ your-tolerance-for-investment-risk-is-probably-not-what-you-think-1505096160.

[dccxlix] Derman, Emanuel (2007) *My Life as a Quant* (Wiley, New York); Patterson, Scott (2011) *The Quants: How a New Breed of Math Whizzes Conquered Wall Street and Nearly Destroyed It* (Crown Business, New York); Patterson, Scott "The minds behind the meltdown," *Wall Street Journal*, updated Jan. 22, 201, 12:01 AM ET, https://www.wsj.com/articles/SB10001424052748704509704575019032416477138.

[dccl] Zuckerman, Gregory; Hope, Bradley "The Quants run Wall Street now" *Wall Street Journal*, May 21, 2017, 4:30 PM ET, https://www.wsj.com/articles/the-quants-run-wall-street-now-1495389108.

[dccli] *Preparing for the Worst*, pp. 9ff.

[dcclii] *Preparing for the Worst*, pg. 41.

[dccliii] *Preparing for the Worst*, pp. 38-39.

[dccliv] Statman, Meir "How financially literate are you really? Let's find out," *Wall Street Journal*, Oct. 19, 2017, 10:01 AM ET, https://www.wsj.com/articles/how-financially-literate-are-you-really-lets-find-out-1508421702.

[dcclv] *The Concepts and Practice of Mathematical Finance.*

[dcclvi] *The Concepts and Practice of Mathematical Finance.*

[dcclvii] *The Concepts and Practice of Mathematical Finance.*

[dcclviii] *In Pursuit of the Unknown: 17 Equations That Changed the World*, Equation 17: Black-Scholes Equation.

[dcclix] *The Concepts and Practice of Mathematical Finance*; "Robert C. Merton Nobel Prize Lecture: Applications of option-pricing theory: Twenty-five years later."

[dcclx] Lewis, Michael (2010) *The Big Short: Inside the Doomsday Machine* (W. W. Norton & Company, New York).

[dcclxi] *Weapons of Math Destruction*, Chapter 2, pg. 32.

[dcclxii] Taleb, Nassim Nicholas (2007) *The Black Swan: The Impact of the Highly Improbable* (Random House, New York).

[dcclxiii] Stewart, Ian "The mathematical equation that caused the banks to crash," *The Guardian*, Feb.11, 2012, 19:05 EST, https://www.theguardian.com/science/2012/feb/12/black-scholes-equation-credit-crunch; Worstall, Tim "Black-Scholes didn't cause the financial crash," Forbes, Feb. 13, 2012, 08:56 AM, https://www.forbes.com/sites/timworstall/2012/02/13/black-scholes-didnt-cause-the-financial-crash/#4d1f01741360; Adapted from Udemy "The history of the Black-Scholes formula," Priceonomics, Dec. 15, 2015, https://priceonomics.com/the-history-of-the-black-scholes-formula; Salmon, Felix. "Recipe for disaster: The formula that killed Wall Street," Wired, Feb. 23. 2009, 12:00 PM, https://www.wired.com/2009/02/wp-quant/; "Copula (probability theory)," *Wikipedia*, https://en.wikipedia.org/wiki/Copula_(probability_theory)#Quantitative_finance), downloaded Dec. 14, 2018; "David X. Li," *Wikipedia*, https://en.wikipedia.org/wiki/David_X._Li, downloaded Dec. 14, 2018; Li, David X. (2000), "On default correlation: A Copula Function approach," *Journal of Fixed Income*, 9, 43–54; Berenson, Alex "A year later, little change on Wall St.," *New York Times*, Sept. 11, 2009, http://www.nytimes.com/2009/09/12/business/12change.html; Patterson, Scott (2011) *The Quants: How a New Breed of Math Whizzes Conquered Wall Street and Nearly Destroyed It* (Crown Business, New York); Patterson, Scott "The minds behind the meltdown," *Wall Street Journal*, updated Jan. 22, 2010, 12:01 AM ET, https://www.wsj.com/articles/SB10001424052748704509704575019032416477138;

Patterson, Scott "How math whizzes helped sink the economy [Book Excerpt]," *Scientific American*, Sept. 22, 2011.

[dcclxiv] *The Black Swan: The Impact of the Highly Improbable.*

[dcclxv] McGinty, Jo Craven "Rough guides to the golden years: Retirement calculators," *Wall Street Journal*, July 22, 2016, 8:39 AM ET, https://www.wsj.com/articles/rough-guides-to-the-golden-years-retirement-calculators-1469191146; McGinty, Jo Craven "Retirement calculators, assumptions and statistical methods," *Wall Street Journal*, Jul 22, 2016, 2:16 PM ET, https://blogs.wsj.com/numbers/retirement-calculators-assumptions-and-statistical-methods-2214.

[dcclxvi] Tergesen, Anne "How to monitor your retirement readiness," *Wall Street Journal*, June 28, 2017, 9:35 AM ET, https://www.wsj.com/articles/how-to-monitor-your-retirement-readiness-1498656937; Ruffenach, Glenn "Beyond the '70/80' rule: How much do you really need in retirement?, "*Wall Street Journal*, March 22, 2018, 11:22 a.m. ET, https://www.wsj.com/articles/beyond-the-70-80-rule-how-much-do-you-really-need-in-retirement-1521732162.

[dcclxvii] Ariely, Dan; Holzwarth, Aline "How much money will you really spend in retirement? Probably a lot more you think," *Wall Street Journal*, Sept. 3, 2018, 11:28 PM ET, https://www.wsj.com/articles/how-much-money-will-you-really-spend-in-retirement-probably-a-lot-more-than-you-think-1536026820.

[dcclxviii] Siegel, Jeremy J. (2014) *Stocks for the Long Run: The Definitive Guide to Financial Market Returns & Long-Term Investment Strategies* (McGraw-Hill Education, New York), 5th edition, pp. 77, 82 and 96.

[dcclxix] Malanga, Steve "Covering up the pension crisis," *Wall Street Journal*, Aug. 25, 2016, 6:39 PM ET, http://www.wsj.com/articles/covering-up-the-pension-crisis-147216 4758; Dow Jones Industrial Average, Wikipedia, https://en.wikipedia.org/wiki/Dow_Jon es_Industrial_Average, downloaded Dec. 14, 2018; *Stocks for the Long Run;* Roth, Allan S. "Stocks for the long run? Not so far this century," *Wall Street Journal*, Feb. 12, 2016, 8:03 AM ET, http://www.wsj.com/articles/stocks-for-the-long-run-not-so-far-this-century-1455282180.

[dcclxx] Actuarial life table, *Social Security.*

[dcclxxi] Bengen, William P. (1994), "Determining withdrawal rates using historical data," *J. Financial Planning*, 7, 171-180; Bengen, William P. (August 1996) "Asset allocation for a lifetime," *J. Financial Planning*, pp. 58-67; "How to monitor your retirement readiness."

[dcclxxii] Tergesen, Anne "Forget the 4% Rule: Rethinking common retirement beliefs," *Wall Street Journal*, Feb. 9, 2018, 5:30 am ET, https://www.wsj.com/articles/forget-the-4-rule-rethinking-common-retirement-beliefs-1518172201; Tergesen, Anne "Retiring soon? Plan for market downturns," *Wall Street Journal*, Sept. 21, 2018, 5:30 AM ET, https://www.wsj.com/articles/retiring-soon-plan-for-market-downturns-1537522201.

[dcclxxiii] Tergesen, Anne "A better way to tap your retirement savings," *Wall Street Journal*, May 31, 2015, https://www.wsj.com/articles/a-better-way-to-tap-your-retirement-savings-1432836119.

[dcclxxiv] "Forget the 4% Rule: Rethinking common retirement beliefs."

[dcclxxv] "IRA required minimum distribution worksheet, *IRS*, https://www.irs.gov/pub/irs-tege/uniform_rmd_wksht.pdf, downloaded Dec. 27, 2018. These life expectancy factors were in place in 2018, when the initial withdrawing age was "70½." Look for updated rules.

[dcclxxvi] Smetters, Kent "Pay off student loans or save for retirement? The answer may surprise you," *Wall Street Journal*, Sept. 11, 2017, 12:27 PM ET, https://blogs.wsj.com/experts/2017/09/11/pay-off-student-loans-or-save-for-retirement-the-answer-may-surprise-you.

[dcclxxvii] *Risk–Benefit Analysis*, pg. 111; Slovic, P. (1987), "Perception of risk," *Science*, 236, 280-285.

[dcclxxviii] Wamsley, Vanessa "The psychology of anti-Vaxers: How story trumps science," *The Atlantic*, Oct. 19, 2014, https://www.theatlantic.com/health/archive/2014/10/how-anti-vaccine-fear-takes-hold/381355, downloaded Dec. 27, 2018; "Vaccine myths debunked," *Public Health*, https://www.publichealth.org/public-awareness/understanding-vaccines/vaccine-myths-debunked/, downloaded Dec. 27, 2018.

[dcclxxix] *Risk–Benefit Analysis*, pg. 111; Slovic, Paul (1998), "The risk game," *Reliability Engineering & System Safety*, 59, 73-77.

[dcclxxx] Slovic, Paul (2001), "The risk game," *J. Hazardous Materials*, 86, 17-24.

[dcclxxxi] Kahneman, Daniel; Tversky, Amos (1979), "Prospect Theory: An analysis of decision under risk," *Econometrica*, 47, 263-292; Lewis, Michael (2016) *The Undoing Project: A Friendship That Changed Our Minds* (W.W. Norton, New York).

[dcclxxxii] "Prospect Theory: An analysis of decision under risk."

[dcclxxxiii] Kahneman, Daniel; Tversky, Amos (1984), "Choices, values, and frames," *Am. Psychologist*, 39, 341-350.

[dcclxxxiv] "Choices, values, and frames."

[dcclxxxv] Fredrickson, Barbara L., Kahneman, Daniel (1993) "Duration neglect in retrospective evaluations of affective episodes," *Journal of Personality and Social Psychology*, 65, 45–55; *Being Mortal: Medicine and What Matters*, pg. 237.

[dcclxxxvi] Tversky, Amos; Koehler, Derek J. (1994), "Support Theory: A nonextensional representation of subjective probability," *Psychological Rev.*, 101, 547-567.

[dcclxxxvii] "Prospect Theory: An analysis of decision under risk"; Tversky, Amos; Kahneman, Daniel (1992), "Advances in Prospect Theory: Cumulative representation of uncertainty," *J. Risk and Uncertainty*, 5, 297-323; "Decisions, Decisions"; Kahneman, Daniel; Tversky, Amos (2000) *Choices, Values, and Frames* (Cambridge University Press, Cambridge); "Daniel Kahneman Nobel Prize Lecture: Maps of bounded rationality," *The Nobel Prize*, https://www.nobelprize.org/prizes/economic-sciences/2002/kahneman/lecture/, downloaded Jan. 2, 2019.

[dcclxxxviii] "Richard H. Thaler Nobel Prize Lecture: From cashews to fudges: The evolution of behavioral economics," *The Nobel Prize*, https://www.nobelprize.org/prizes/economic-sciences/2017/thaler/lecture/, downloaded Jan. 2, 2019.

[dcclxxxix] Benartzi, Shlomo; Hershfield, Hal E. "Would you rather have $1 Million or $5,000 monthly in retirement?," *Wall Street Journal*, March 27, 2017, 10:33 AM ET, https://www.wsj.com/articles/would-you-rather-have-1-million-or-5-000-monthly-in-retirement-1490582208; Goldstein, Daniel G.; Hershfield, Hal E.; Benartzi, Shlomo (2016), "The illusion of wealth and its reversal," *J. Marketing Research*, 53, 804-813.

[dccxc] Benartzi, Shlomo; Shu, Suzanne; "Why retirees are wary of annuities," *Wall Street Journal*, Feb. 10, 2019, 10:09 PM ET, https://www.wsj.com/articles/why-retirees-are-wary-of-annuities-1154985454.

[dccxci] Sommer, Jeff "The billion-dollar jackpot: Engineered to drain your wallet," *New York Times*, Aug. 12, 2016, http://www.nytimes.com/2016/08/14/your-money/the-billion-dollar-lottery-jackpot-engineered-to-drain-your-wallet.html.

[dccxcii] "Lottery madness," *Statistical Ideas*, July 3, 2016, http://statisticalideas.blogspot.com/search?q=loser%27s+lottery, downloaded Dec. 27, 2018; "A loser's lottery," *Statistical Ideas*, April 24, 2016, http://statisticalideas.blogspot.com/2016/04/a-losers-lottery.html, downloaded Dec. 27, 2018.

[dccxciii] "The billion-dollar jackpot: Engineered to drain your wallet."

Index

A

B

C

D

E

F

G

H

I

J

K

L

M

N

Q

R

S

T

U

V

W